The IMA Volumes in Mathematics and its Applications

Volume 42

Series Editors
Avner Friedman Willard Miller, Jr.

Institute for Mathematics and its Applications

IMA

The **Institute for Mathematics and its Applications** was established by a grant from the National Science Foundation to the University of Minnesota in 1982. The IMA seeks to encourage the development and study of fresh mathematical concepts and questions of concern to the other sciences by bringing together mathematicians and scientists from diverse fields in an atmosphere that will stimulate discussion and collaboration.

The IMA Volumes are intended to involve the broader scientific community in this process.

Avner Friedman, Director
Willard Miller, Jr., Associate Director

* * * * * * * * * *

IMA ANNUAL PROGRAMS

1982–1983	**Statistical and Continuum Approaches to Phase Transition**
1983–1984	**Mathematical Models for the Economics of Decentralized Resource Allocation**
1984–1985	**Continuum Physics and Partial Differential Equations**
1985–1986	**Stochastic Differential Equations and Their Applications**
1986–1987	**Scientific Computation**
1987–1988	**Applied Combinatorics**
1988–1989	**Nonlinear Waves**
1989–1990	**Dynamical Systems and Their Applications**
1990–1991	**Phase Transitions and Free Boundaries**
1991–1992	**Applied Linear Algebra**
1992–1993	**Control Theory and its Applications**
1993–1994	**Emerging Applications of Probability**

IMA SUMMER PROGRAMS

1987	**Robotics**
1988	**Signal Processing**
1989	**Robustness, Diagnostics, Computing and Graphics in Statistics**
1990	**Radar and Sonar**
1990	**Time Series**
1991	**Semiconductors**
1992	**Environmental Studies: Mathematical, Computational, and Statistical Analysis**

* * * * * * * * * *

SPRINGER LECTURE NOTES FROM THE IMA:

The Mathematics and Physics of Disordered Media
Editors: Barry Hughes and Barry Ninham
(Lecture Notes in Math., Volume 1035, 1983)

Orienting Polymers
Editor: J.L. Ericksen
(Lecture Notes in Math., Volume 1063, 1984)

New Perspectives in Thermodynamics
Editor: James Serrin
(Springer-Verlag, 1986)

Models of Economic Dynamics
Editor: Hugo Sonnenschein
(Lecture Notes in Econ., Volume 264, 1986)

B. Dahlberg E. Fabes R. Fefferman
D. Jerison C. Kenig J. Pipher
Editors

Partial Differential Equations with Minimal Smoothness and Applications

Springer-Verlag
New York Berlin Heidelberg London Paris
Tokyo Hong Kong Barcelona Budapest

B. Dahlberg
Department of Mathematics
University of South Carolina
Columbia, SC 29303
USA

Eugene Fabes
Department of Mathematics
University of Minnesota
Minneapolis, MN 55455
USA

R. Fefferman
Department of Mathematics
University of Chicago
Chicago, IL 60637
USA

David Jerison
Department of Mathematics
Massachusetts Institute of Technology
Cambridge, MA 02139
USA

Carlos Kenig
Department of Mathematics
University of Chicago
Chicago, IL 60637
USA

J. Pipher
Department of Mathematics
University of Chicago
Chicago, IL 60637
USA

Mathematics Subject Classifications: 35-02, 42-02

Library of Congress Cataloging-in-Publication Data
Partial differential equations with minimal smoothness and applications / B. Dahlberg . . . [et al.].
p. cm. — (The IMA volumes in mathematics and its applications ; v. 42)
"Based on the proceedings of an IMA Participating Institutions (PI) Conference held at the University of Chicago in March 1990" —Fwd.
Includes bibliographical references.
ISBN 0-387-97774-0. — ISBN 3-540-97774-0
1. Differential equations, Partial—Congresses I. Dahlberg, B. II. Series.
QA370.P37 1992
515'.353—dc20 91-45326

Printed on acid-free paper.

Production managed by Hal Henglein; manufacturing supervised by Robert Paella.
Camera-ready copy prepared by the IMA.
Printed and bound by Edwards Brothers, Inc., Ann Arbor, MI.
Printed in the United States of America.

9 8 7 6 5 4 3 2 1

ISBN 0-387-97774-0 Springer-Verlag New York Berlin Heidelberg
ISBN 3-540-97774-0 Springer-Verlag Berlin Heidelberg New York

The IMA Volumes
in Mathematics and its Applications

Current Volumes:

Volume 1: Homogenization and Effective Moduli of Materials and Media
Editors: Jerry Ericksen, David Kinderlehrer, Robert Kohn, J.-L. Lions

Volume 2: Oscillation Theory, Computation, and Methods of Compensated Compactness
Editors: Constantine Dafermos, Jerry Ericksen,
David Kinderlehrer, Marshall Slemrod

Volume 3: Metastability and Incompletely Posed Problems
Editors: Stuart Antman, Jerry Ericksen, David Kinderlehrer, Ingo Muller

Volume 4: Dynamical Problems in Continuum Physics
Editors: Jerry Bona, Constantine Dafermos, Jerry Ericksen,
David Kinderlehrer

Volume 5: Theory and Applications of Liquid Crystals
Editors: Jerry Ericksen and David Kinderlehrer

Volume 6: Amorphous Polymers and Non-Newtonian Fluids
Editors: Constantine Dafermos, Jerry Ericksen, David Kinderlehrer

Volume 7: Random Media
Editor: George Papanicolaou

Volume 8: Percolation Theory and Ergodic Theory of Infinite Particle Systems
Editor: Harry Kesten

Volume 9: Hydrodynamic Behavior and Interacting Particle Systems
Editor: George Papanicolaou

Volume 10: Stochastic Differential Systems, Stochastic Control Theory and Applications
Editors: Wendell Fleming and Pierre-Louis Lions

Volume 11: Numerical Simulation in Oil Recovery
Editor: Mary Fanett Wheeler

Volume 12: Computational Fluid Dynamics and Reacting Gas Flows
Editors: Bjorn Engquist, M. Luskin, Andrew Majda

Volume 13: Numerical Algorithms for Parallel Computer Architectures
Editor: Martin H. Schultz

Volume 14: Mathematical Aspects of Scientific Software
Editor: J.R. Rice

Volume 15: Mathematical Frontiers in Computational Chemical Physics
Editor: D. Truhlar

Volume 16: Mathematics in Industrial Problems
by Avner Friedman

Volume 17: Applications of Combinatorics and Graph Theory to the Biological and Social Sciences
Editor: Fred Roberts

Volume 18: q-Series and Partitions
Editor: Dennis Stanton

Volume 19: Invariant Theory and Tableaux
Editor: Dennis Stanton

Volume 20: Coding Theory and Design Theory Part I: Coding Theory
Editor: Dijen Ray-Chaudhuri

Volume 21: Coding Theory and Design Theory Part II: Design Theory
Editor: Dijen Ray-Chaudhuri

Volume 22: Signal Processing: Part I - Signal Processing Theory
Editors: L. Auslander, F.A. Grünbaum, J.W. Helton, T. Kailath, P. Khargonekar and S. Mitter

Volume 23: Signal Processing: Part II - Control Theory and Applications of Signal Processing
Editors: L. Auslander, F.A. Grünbaum, J.W. Helton, T. Kailath, P. Khargonekar and S. Mitter

Volume 24: Mathematics in Industrial Problems, Part 2
by Avner Friedman

Volume 25: Solitons in Physics, Mathematics, and Nonlinear Optics
Editors: Peter J. Olver and David H. Sattinger

Volume 26: Two Phase Flows and Waves
Editors: Daniel D. Joseph and David G. Schaeffer

Volume 27: Nonlinear Evolution Equations that Change Type
Editors: Barbara Lee Keyfitz and Michael Shearer

Volume 28: Computer Aided Proofs in Analysis
Editors: Kenneth Meyer and Dieter Schmidt

Volume 29: Multidimensional hyperbolic problems and computations
Editors Andrew Majda and Jim Glimm

Volume 30: Microlocal Analysis and Nonlinear Waves
Editors: Michael Beals, R. Melrose and J. Rauch

Volume 31: Mathematics in Industrial Problems, Part 3
by Avner Friedman

Volume 32: Radar and Sonar, Part 1
by Richard Blahut, Willard Miller, Jr. and Calvin Wilcox

Volume 33: Directions in Robust Statistics and Diagnostics: Part 1
Editors: Werner A. Stahel and Sanford Weisberg

Volume 34: Directions in Robust Statistics and Diagnostics: Part 2
Editors: Werner A. Stahel and Sanford Weisberg

Volume 35: Dynamical Issues in Combustion Theory
Editors: P. Fife, A. Liñán and F.A. Williams

Volume 36: Computing and Graphics in Statistics
Editors: Andreas Buja and Paul Tukey

Volume 37: Patterns and Dynamics in Reactive Media
Editors: Harry Swinney, Gus Aris and Don Aronson

Volume 38: Mathematics in Industrial Problems, Part 4
by Avner Friedman

Volume 39: Radar and Sonar, Part 2
Editors: F. Alberto Grünbaum, Marvin Bernfeld and Richard E. Blahut

Volume 40: Nonlinear Phenomena in Atmospheric and Oceanic Sciences
Editors: F. Carnevale and R.T. Pierrehumbert

Volume 41: Chaotic Processes in the Geological Sciences
Editor: David Yuen

Volume 42: Partial Differential Equations with Minimal Smoothness and Applications
Editors: B. Dahlberg, E. Fabes, R. Fefferman, D. Jerison, C. Kenig and J. Pipher

Forthcoming Volumes:

1989-1990: *Dynamical Systems and Their Applications*

Twist Mappings and Their Applications

Dynamical Theories of Turbulence in Fluid Flows

Summer Program 1990: *Time Series in Time Series Analysis*

Time Series (2 volumes)

1990-1991: *Phase Transitions and Free Boundaries*

On the Evolution of Phase Boundaries

Shock Induced Transitions and Phase Structures

Microstructure and Phase Transitions

Statistical Thermodynamics and Differential Geometry of Microstructured Material

Free Boundaries in Viscous Flows

Summer Program 1991: *Semiconductors*

Semiconductors (2 volumes)

1991-1992: *Phase Transitions and Free Boundaries*

Sparse Matrix Computations: Graph Theory Issues and Algorithms

FOREWORD

This IMA Volume in Mathematics and its Applications

PARTIAL DIFFERENTIAL EQUATIONS
WITH MINIMAL SMOOTHNESS AND APPLICATIONS

is based on the proceedings of an IMA Participating Institutions (PI) Conference held at the University of Chicago in March 1990. Each year the 24 Participating Institutions select, through a competitive process, several conference proposals from the PIs, for partial funding. This conference brought together leading researchers interested in lack of smoothness results for PDEs. We thank B. Dahlberg, E. Fabes, R. Fefferman, D. Jerison, C. Kenig and J. Pipher for organizing the meeting and editing the proceedings.

Avner Friedman

Willard Miller, Jr.

PREFACE

In recent years there has been a great deal of activity in both the theoretical and applied aspects of partial differential equations, with emphasis on realistic engineering applications, which usually involve lack of smoothness.

On March 21–25, 1990, the University of Chicago hosted a workshop that brought together approximately forty five experts in theoretical and applied aspects of these subjects.

This workshop was a vehicle for summarizing the current status of research in these areas, and for defining new directions for future progress.

Many of the participants in the workshop have contributed articles to these Proceedings. We hope that this volume will stimulate further interest in these exciting areas.

Acknowledgements. This workshop was made possible by generous funding from the Institute for Mathematics and its Applications, the National Science Foundation, the Department of Mathematics at the University of Chicago and Volvo Data Corporation. We are very grateful for their support.

In addition, we would like to thank Ms. Dorothy Frazier and Ms. Lorraine Kubiak, in the staff of the Department of Mathematics at the University of Chicago, for their invaluable help in organizing the workshop.

B. Dahlberg
E. Fabes
R. Fefferman
D. Jerison
C. Kenig
J. Pipher

Organizing Committee for the Workshop, and Editors of the Proceedings

CONTENTS

WEAKLY ELLIPTIC SYSTEMS WITH OBSTACLE CONSTRAINTS
PART I - A 2 × 2 MODEL PROBLEM*

DAVID R. ADAMS**

1. Introduction. This note is intended to be the first in a series of papers treating linear elliptic systems of partial differential operators subject to obstacle type constraints. There is a large literature concerning solutions to linear and nonlinear elliptic systems of partial differential equations, but there seems to be much less work devoted exclusively to the understanding of solutions to such systems when they are subject to constraints. These constrained systems often take the form of a system of variational inequalities. Such problems have been treated, for example in [F1] and [HW] at least for strongly elliptic, i.e. Legendre-Hadamard elliptic, systems of variational inequalities. In this note we want to begin a study of a broader class of such systems, what we shall refer to as weakly elliptic systems - elliptic in the standard sense that the characteristic form of the principal part has no real zeros. One very important feature of this larger class is that the solution vector(s) can not in general, have the same degree of regularity in each component direction, as is generally the case for strongly elliptic systems. And as is generally well understood, solutions to variational inequalities only inherit a very limited amount of regularity from the data. For weakly elliptic linear systems, this inherited regularity is intimately tied up with certain algebraic structure considerations, considerations that do not appear when the obstacle constraints are removed.

In this note we will concentrate on a special 2×2 system that will reflect the type of behavior of interest. We will investigate existence, uniqueness, and especially the regularity under rather weak assumptions on the data, the obstacle ψ. Our prototypical 2×2 situation arises from the special linear operator

$$\mathcal{L} = A\Delta - B \tag{1}$$

where A and B are 2×2 matrices of real numbers, Δ is the usual Laplace operator defined on smooth functions of $x = (x_1, ..., x_n) \in \Omega$, a smooth bounded domain in Euclidean n-space $\boldsymbol{R}^n$, $n \geq 1$. Further, we set

$$\boldsymbol{K} = \{v \in H_0^1(\Omega) \times H_0^1(\Omega) : \quad v^1(x) > \psi(x), \text{ a.e. } \Omega\}$$

where $v = (v^1(x), v^2(x))$, and $H_0^1(\Omega)$ is the usual Sobolev class of square integrable functions in Ω, vanishing on the boundary of Ω, $\partial\Omega$, and whose first order derivatives are also square integrable in Ω. Here ψ is a given function on Ω with some smoothness properties to be specified later. Our main concern is with investigating the solutions of:

$$\begin{cases} \text{find } u \in \boldsymbol{K} \text{ such that} \\ \\ \langle \mathcal{L}u, V - u\rangle \geq 0, \text{ for all } V \in \boldsymbol{K}. \end{cases} \tag{2}$$

---

* Partially supported by NSF-grant DMS-8702755 and by the Commonwealth of Kentucky through the Kentucky EPSCoR program

** University of Kentucky, Lexington, KY 40506

Here the brackets $\langle\cdot,\cdot\rangle$ denote the H^{-1}, H_0^1 duality pairing.

Now if the matrix A is positive definite, then $\mathcal{L}$ is strongly elliptic and the solution to (2) exists and belongs to the class $W^{2,\infty}(\Omega) \times W^{2,\infty}(\Omega)$, provided the eigenvalues of B are sufficiently large i.e. both components have bounded second partials in Ω. However, if we merely assume that det $(A) \neq 0$, then $\mathcal{L}$ is just weakly elliptic. If we set $A = \begin{bmatrix} 0 & 1 \\ 1 & 0 \end{bmatrix}$ and $B = \begin{bmatrix} 0 & 0 \\ 0 & 1 \end{bmatrix}$, then (2) is equivalent to the biharmonic obstacle problem for the "hinged plate", see especially [CF]. In this case the component regularity is $H^3(\Omega) \cap W^{2,\infty}(\Omega) \times H^1(\Omega) \cap L^\infty(\Omega)$, and in fact, one generally can do no better than this. Other examples that can be rewritten so as to fit very nicely into our framework are: the two membrane problem of G. Vegara-Cafferelli and a sixth-order obstacle problem with an obstacle constraint on the Laplacian; see [KS] pages 79 and 80.

This work is a natural extension of that begun in [A1] and [A2] for single equations of order 2 and 4, and the basic role that capacity plays in obtaining a priori estimates on solutions. We will continue with this idea here, i.e. obtaining estimates in terms of Choquet integrals (see below). Here we treat the 2×2 case. In Part II, we will discuss the general $N \times N$ case, and in subsequent papers, the variable coefficient operator case and limiting relationships between the various solution classes as well as various cases where the matrix A is singular.

2. Notation. Here we want the notion of the conductor capacity of a subset $E \subset \Omega$. First for K compactly contained in Ω, we set

$$C_1(K;\Omega) = \inf \int_\Omega |\nabla\phi|^2 dx$$

where the infimum is over all $\phi \in C_0^\infty(\Omega)$ such that $\phi = 1$ on K; $C_0^\infty(\Omega)$ = infinitely differential functions with compact support in Ω. For $E \subset \Omega$, we define

$$C_1(E;\Omega) = \sup_{K \subset E} C_1(K;\Omega).$$

Similarly, for $C_2(\cdot;\Omega)$ we first set $C_2(K;\Omega) = \inf \int_\Omega |\nabla^2\phi|^2 dx$ where $\nabla^2\phi$ represents the $n(n+1)/2$-vector of all second derivatives of ϕ. And $C_2(\cdot;\Omega)$ is extended to all $E \subset \Omega$ as before.

A function f defined C_k-a.e. on Ω, is called C_k-quasi continuous (C_k-q.c.) on Ω if for every $\epsilon > 0$ there is an open set $G \subset \Omega$ such that $C_k(G;\Omega) < \epsilon$ and f restricted to $\Omega \setminus G$ is continuous.

With this notation we can define the Choquet integral of a non-negative function f on Ω by

$$\int_\Omega f dC_k(\cdot;\Omega) \equiv \int_0^\infty C_k(\{x \in \Omega : \ f(x) > \lambda\};\Omega) d\lambda.$$

From [A2] Theorems 1 and 2 (for case $k = 1$), or more generally from [H] (covering both cases $k = 1, 2$), we have

THEOREM 2.1. *There is a constant Q independent of f such that*

$$\frac{1}{Q} \cdot P_k(f)^2 \leq \int_\Omega f^2 dC_k(\cdot;\Omega) \leq Q \cdot P_k(f)^2 \tag{3}$$

for all C_k-q.c. f. Here

$$P_k(f) = \inf \int |\nabla^k \phi|^2 dx$$

where the infimum is over all $\phi \in H_0^2(\Omega)$ such that $\phi \geq f$ C_k-a.e. on Ω. $P_k(f) = +\infty$ if there are no such ϕ's, $k = 1, 2$. $H_0^k(\Omega) = W_0^{k,2}(\Omega)$, i.e. all k-th order derivatives belong to $L^2(\Omega)$, and up to $k-1$ of them vanish on $\partial\Omega$.

The interested reader might want to note that in establishing Theorem 2.1 from the work [H], it is convenient to consider an equivalent form of C_2 replacing $|\nabla^2\phi|$ by $|\Delta\phi|$. The potentials of Hansson, in this case, can be taken to be $GG\mu$, where G is the usual Green's function for $-\Delta$ in Ω.

3. The penalized system. For a matrix A we will denote the element of the i-th row and j-th column by $A_{i,j}$ and since det $(A) \neq 0$, we set $M = A^{-1}B$. Next let $\eta(t) \in C^\infty(\boldsymbol{R})$ such that $\eta(t) \leq 0$ for all t, $\eta(t) = 0$ for all $t \geq 0$, and $\eta'(t) \geq 0$ for all t. We now consider the system

$$(4) \qquad \begin{cases} A\Delta u_\epsilon - Bu_\epsilon = -e_1 \cdot \frac{1}{\epsilon}\eta(u_\epsilon^1 - \psi), \text{ in } \Omega, \\ u_\epsilon = 0, \quad \text{on } \partial\Omega. \end{cases}$$

Here $\epsilon > 0$ is the penalty parameter and $e_1 = (1,0)^T$, T = transpose. The usual procedure is to solve (4) for each $\epsilon > 0$, obtain estimates on u_ϵ independent of ϵ, and then pass to the limit as $\epsilon \to 0$. The limiting vector u will then, hopefully, solve (2).

Equations (4) are easily seen to have a C^2-solution in Ω by applying the Schauder fixed point theorem. The main difficulty in carrying out the penalization program is, of course, obtaining appropriate estimates. These we get by first reducing (4) via the change of dependent variable $v_\epsilon = Cu_\epsilon$, where C is a 2×2 invertible matrix from the group $\mathcal{G}_2$; $C \in \mathcal{G}_2$ iff $C_{1,1} = 1$, $C_{1,2} = 0$, and $C_{2,2} \neq 0$. System (4) now becomes

$$(5) \qquad \begin{cases} \Delta v_\epsilon - Jv_\epsilon = -(A^1, A^2)^T \cdot \frac{1}{\epsilon}\eta(v_\epsilon^1 - \psi), \quad \text{in } \Omega, \\ v_\epsilon = 0, \quad \text{on } \partial\Omega, \end{cases}$$

where $J = \begin{bmatrix} \lambda_1 & 1 \\ 0 & \lambda_2 \end{bmatrix}$, λ_k = eigenvalues of $M = A^{-1}B$, provided $M_{1,2} \neq 0$. This is accomplished by choosing $C_{2,1} = \lambda_2 - M_{2,2}$ and $C_{2,2} = M_{1,2}$. It then follows that $A^1 = A_{2,2}/\det(A)$ and $A^2 = (\lambda_2 A_{2,2} - B_{2,2})/\det(A)$. Notice that $M_{1,2} = 0$ iff $A_{2,2}B_{1,2} - A_{1,2}B_{2,2} = 0$ in which case system (4) uncouples. The first equation there is then

$$(6) \qquad \Delta u_\epsilon^1 - M_{1,1}u_\epsilon^1 = -\frac{A_{2,2}}{\det(A)} \cdot \frac{1}{\epsilon}\eta(u_\epsilon^1 - \psi).$$

Below, we will always assume $M_{1,2} \neq 0$ since estimates for (6) follow directly from the literature (cf. [C]) and with no recourse to the group $\mathcal{G}_2$.

4. A priori estimates. We first prove

THEOREM 4.1. *There exists a constant Q_1, independent of ϵ, such that if*

(i) $A_{2,2}/\det(A) < 0$,

(ii) $\mu_1 < B_{2,2}/A_{2,2}$,

(iii) $Re(\lambda_k) > \mu_1, \quad k = 1,2,$

(iv) ψ *is* C_1*-q.c. and* $\int_\Omega \psi_+^2 dC_1(\cdot;\Omega) < \infty$,

then any solution u_ϵ *to (4) satisfies*

$$||u_\epsilon||_{H^1(\Omega)\times H^1(\Omega)} \leq Q_1 \int_\Omega \psi_+^2 dC_1(\cdot;\Omega). \tag{7}$$

In the above, μ_1 is the first eigenvalue of Δ on Ω; $0 > \mu_1 > \mu_2 \geq \cdots$. Also $\psi_+ = \max(\psi, 0)$.

THEOREM 4.2. *There is a constant Q_2, independent of ϵ, such that if*

(i) $A_{2,2} = 0$

(ii) $B_{2,2}/\det(A) < 0$

(iii) $Re(\lambda_k) > \mu_1, \quad k = 1,2,$

(iv) ψ *is* C_2*-q.c. and* $\int_\Omega \psi_+^2 dC_2(\cdot;\Omega) < \infty$,

then any solution u_ϵ *to (4) satisfies*

$$||u_\epsilon||_{H^2(\Omega)\times H^0(\Omega)} \leq Q_2 \int_\Omega \psi_+^2 dC_2(\cdot;\Omega). \tag{8}$$

Remarks 4.3. (a) Notice that conditions (ii) and (iii) of (4.1) are sharp in the sense that if $A = \begin{bmatrix} -1 & 0 \\ 0 & A_{2,2} \end{bmatrix}$ and $B = \begin{bmatrix} 0 & 1 \\ 0 & B_{2,2} \end{bmatrix}$, with $B_{22}/A_{22} = \mu_1$, then there are infinitely many solutions to (4) all of which can not satisfy (7).

(b) Notice that conditions (i) and (ii) of (4.2) imply $B_{22}/\det(A) \leq 0$ when $A_{22} \to 0$, and the case $A_{22} = B_{22} = 0$ leads to $M_{1,2} = 0$.

(c) Using the theory contained in [A2] and/or [H], it is easy to see that (iv) of (4.1) is finite when $\int_\Omega |\nabla\psi_+|^2 dx < \infty$ and (iv) of (4.2) is finite when $\int_\Omega (-\Delta\psi)_+^2 dx < \infty$; $\psi \in H^1(\Omega)$ or $H^2(\Omega)$, respectively (cf. [A1], theorems 6.1 and 7.1).

Proof of Theorem 4.1. We first assume that λ_1, λ_2 are real and that $A^2 > 0$, i.e. that there is an eigenvalue, call it λ_2, such that $\lambda_2 < B_{22}/A_{22}$. In fact there is no loss in generality in assuming that $\lambda_2 = \min(\lambda_1, \lambda_2)$. System (5) is just

$$\text{(9)} \qquad \begin{cases} -\Delta v_\epsilon^1 + \lambda_1 v_\epsilon^1 + v_\epsilon^2 & = \mu_\epsilon \\ & \qquad\qquad \text{in } \Omega, \\ -\Delta v_\epsilon^2 + \lambda_2 v_\epsilon^2 & = -\xi \cdot \mu_\epsilon \\ v_\epsilon^1 = v_\epsilon^2 = 0, & \qquad\qquad \text{on } \partial\Omega. \end{cases}$$

where $\xi = -A^2/A^1 > 0$, and $d\mu_\epsilon = A^1 \cdot \frac{1}{\epsilon}\eta(V_\epsilon^1 - \psi)dx$. Since $\mu_1 < \lambda_2 \leq \lambda_1$, we can set G_k to be the Green's function for the operator $-\Delta + \lambda_k$ on Ω, $k = 1, 2$. System (9) is equivalent to

$$\text{(10)} \qquad \begin{cases} v_\epsilon^1 = -G_1 v_\epsilon^2 + G_1 \mu_\epsilon \\ v_\epsilon^2 = -\xi \cdot G_2 \mu_\epsilon \end{cases}$$

The maximum principle applied to (9) implies that $v_\epsilon^2 \leq 0$ and $v_\epsilon^1 \geq 0$ on Ω. Thus (10) gives $v_\epsilon^1 \geq G_1 \mu_\epsilon$. Hence

$$\text{(11)} \qquad \int_S |G_1 \mu_\epsilon|^2 dC_1 \leq \int_S |v_\epsilon^1|^2 \leq \int_\Omega \psi_+^2 dC_1$$

where $S = \text{supp } \mu_\epsilon = \{v_\epsilon^1 \leq \psi\}$.

Next, following [A2], we solve the harmonic obstacle problem in Ω for the obstacle $G_1 \mu_\epsilon \cdot \chi_S$, where χ_S is the characteristic function of S. If G denotes the usual Green's function for $-\Delta$ on Ω, then the solution to this problem can be represented as a potential $G\nu$, where ν is a positive Radon measure supported on S. Thus we can write

$$\begin{aligned} \int_\Omega |G_1 \mu_\epsilon|^2 dC_1 &\leq Q \int_\Omega |\nabla G_1 \mu_\epsilon|^2 dx \\ &\leq Q \left[\int_\Omega |\nabla G_1 \mu_\epsilon|^2 dx + \lambda_1 \int_\Omega |G_1 \mu_\epsilon|^2 dx \right] \\ &= Q \int_S G_1 \mu_\epsilon d\mu_\epsilon = Q \int_S G\nu d\mu_\epsilon. \end{aligned} \qquad \text{(12)}$$

And

$$\begin{aligned} \int_S G\nu d\mu_\epsilon &= \int \nabla G\nu \cdot \nabla G_1 \mu_\epsilon dx + \lambda_1 \int G\nu G_1 \mu_\epsilon dx \\ &\leq Q ||\nabla G\nu||_2 \cdot ||\nabla G_1 \mu||_2 \leq \delta ||\nabla G_1 \mu_\epsilon||_2^2 + Q_\delta ||\nabla G\nu||_2^2 \\ &\leq \delta \cdot Q \int G\nu d\mu_\epsilon + Q_\delta ||\nabla G\nu||_2^2, \end{aligned}$$

first by Poincaré's inequality, then Young's inequality where we introduced a positive number δ to be specified later, and then finally by the argument used to get (12). Thus for sufficiently small δ, we have

$$\int_S G\nu d\mu_\epsilon \leq Q \int_\Omega |\nabla G\nu|^2 dx$$

and consequently,

$$\int_S G\nu d\mu_\epsilon \leq Q \int_S |G_1 \mu_\epsilon|^2 dC_1.$$

Hence from (12) and (11)

$$\text{(13)} \qquad \int_\Omega |G_1 \mu_\epsilon|^2 dC_1 \leq Q \int \psi_+^2 dC_1.$$

Next, we look at $\int |G_1G_2\mu_\epsilon|^2 dC_1$. Set $z = G_1G_2\mu_\epsilon$ and note that $(-\Delta + \lambda_2)z = G_1\mu_\epsilon \leq w$, for some $w \in H_0^1(\Omega)^+$ since we have already seen that $\int_\Omega |G_1\mu|^2 dC_1 < \infty$; cf. [A2], theorem 4. Thus

$$(-\mu_1 + \lambda_2)\int |z|^2 dx \leq \int (|\nabla z|^2 + \lambda_2|z|^2)dx \leq \int wz dx \leq ||w||_2 \cdot ||z||_2.$$

Therefore $||z||_2 \leq Q||w||_2$, for some constant Q. A similar argument then gives $||\nabla z||_2 \leq Q||w||_2$. Hence we get

$$\int_\Omega |G_1G_2\mu_\epsilon|^2 dC_1 \leq Q\int \psi_+^2 dC_1. \tag{14}$$

Thus (10), (13), and (14) imply

$$\int_\Omega |v_\epsilon^1|^2 dx \leq Q\int_\Omega \psi_+^2 dC_1. \tag{15}$$

Next, we use (15), first to obtain L^2-estimates, and then H^1-estimates for v_ϵ^2. From (9), we have

$$\int |v_\epsilon^2|^2 dx = \int v_\epsilon^2(\mu_\epsilon + \Delta v_\epsilon^1 - \lambda_1 v_\epsilon^1)dx$$
$$= \frac{1}{\xi}\int v_\epsilon^2(\Delta v_\epsilon^2 - \lambda_2 v_\epsilon^2)dx - \lambda_1\int v_\epsilon^1 v_\epsilon^2 dx + \int v_\epsilon^1 \Delta v_\epsilon^2 dx.$$

Thus

$$\frac{1}{\xi}\int |\nabla v_\epsilon^2|^2 dx + \left(1 + \frac{\lambda_2}{\xi}\right)\int |v_\epsilon^2|^2 dx = (\lambda_2 - \lambda_1)\int v_\epsilon^1 v_\epsilon^2 dx + \xi\int v_\epsilon^1 d\mu_\epsilon. \tag{16}$$

Now we need the

LEMMA 4.4. *Let v_ϵ^1 and μ_ϵ be as in (9), then there is a constant Q (independent of ϵ) such that*

$$\xi\int v_\epsilon^1 d\mu_\epsilon \leq Q\left(\int \psi_+^2 dC_1\right)^{1/2} \cdot ||v_\epsilon^2||_{H^1} \tag{17}$$

Proof of lemma. Choose $w \in H_0^1(\Omega)$ such that $w \geq \psi_+$ on Ω, then

$$\xi\int v_\epsilon^1 d\mu_\epsilon \leq \xi\int \psi_+ d\mu_\epsilon \leq \xi\int w d\mu_\epsilon = \int w(\Delta v_\epsilon^2 - \lambda_2 v_\epsilon^2)dx$$

and the result follows.

Thus continuing with (16), we apply Lemma 4.4, (15), and Young's inequality to get: if

$$-\frac{\mu_1}{\xi} + 1 + \frac{\lambda_2}{\xi} > 0, \tag{18}$$

then there is a constant Q such that

$$||v_\epsilon^2||_{H^1}^2 \leq Q\int \psi_+^2 dC_1. \tag{19}$$

Notice that condition (18) is just condition (ii) of the theorem.

Finally, we get an estimate on $||\nabla v_\epsilon^1||_2$ from the first equation of (9), (15), (19) and Lemma 4.4.

So now we assume $A^2 < 0$ with λ_1, λ_2 still real. Of course $\xi < 0$, so we multiply thru (16) by -1. Since we are assuming $\lambda_2 \leq \lambda_1$ we need only notice that

$$\int v_\epsilon^1 v_\epsilon^2 dx \leq \int (|\nabla v_\epsilon^1|^2 + \lambda_1 |v_\epsilon^1|^2) dx + \int v_\epsilon^1 v_\epsilon^2 dx = \int v_\epsilon^1 d\mu_\epsilon.$$

Thus H^1-estimates for v_ϵ^2 follow from Lemma 4.4 provided $\mu_1/\xi - 1 - \lambda_2/\xi > 0$, which is again condition (ii) of our theorem. Again, H^1-estimates for v_ϵ^1 now follow from the first equation of (9) and Lemma 4.4.

Finally, we briefly consider the case of complex eigenvalues: $\lambda_1 = \lambda + i\sigma$, $\lambda_2 = \lambda - i\sigma$. Again from (9), we can write

$$\int (|\nabla v_\epsilon^2|^2 + \lambda_2 |v_\epsilon^2|^2) dx = -\xi \int \bar{v}_\epsilon^2 d\mu_\epsilon$$

$$= -\xi \int |v_\epsilon^2|^2 + \xi(\bar{\lambda}_2 - \lambda_1) \int \bar{v}_\epsilon^2 v_\epsilon^1 + |\xi|^2 \int v_\epsilon^1 d\mu_\epsilon.$$

Thus

(20) $$\int |\nabla v_\epsilon^2|^2 dx + (\lambda_2 + \xi) \int |v_\epsilon^2|^2 dx = |\xi|^2 \int v_\epsilon^1 d\mu_\epsilon.$$

The real part of (20) is

$$\int |\nabla v_\epsilon^2|^2 dx + \frac{B_{22}}{A_{22}} \int |v_\epsilon^2|^2 dx = |\xi|^2 \int v_\epsilon^1 d\mu_\epsilon.$$

H^1-estimates for v_ϵ^2 follow from this, condition (ii), and Lemma 4.4. An estimate for $||v_\epsilon^1||_{H^1}$ is now obtained in a manner similar to earlier arguments. This completes our proof of Theorem 4.1. □

Proof of Theorem 4.2. Again, we begin by assuming that λ_1, λ_2 are real. Now system (5) becomes

(21) $$\begin{cases} -\Delta v_\epsilon^1 + \lambda_1 v_\epsilon^1 + v_\epsilon^2 = 0 \\ -\Delta v_\epsilon^2 + \lambda_2 v_\epsilon^2 = -\frac{B_{22}}{\det(A)} \cdot \frac{1}{\epsilon}\eta(v_\epsilon^1 - \psi) \\ v_\epsilon^1 = v_\epsilon^2 = 0, \quad \text{on } \partial\Omega. \end{cases}$$

Hence

(22) $$\begin{aligned} \textstyle\int |v_\epsilon^2|^2 dx &= \textstyle\int v_\epsilon^2 (\Delta v_\epsilon^1 - \lambda_1 v_\epsilon^1) dx \\ &= (\lambda_2 - \lambda_1) \textstyle\int v_\epsilon^1 v_\epsilon^2 dx + \frac{B_{22}}{\det(A)} \frac{1}{\epsilon} \textstyle\int \eta(v_\epsilon^1 - \psi) v_\epsilon^1 dx \end{aligned}$$

So if we assume, as we may, that $\lambda_1 \leq \lambda_2$, then by the maximum principle we have $v_\epsilon^2 \leq 0, v_\epsilon^1 \geq 0$ on Ω, so that L^2-estimates for v_ϵ^2 follow from (22) and the following

LEMMA 4.5. *Suppose v_ϵ^1 and η are as above, then there is a constant Q (independent of ϵ) such that*

$$-\frac{1}{\epsilon} \int \eta(v_\epsilon^1 - \psi) v_\epsilon^1 dx \leq Q \left(\int \psi_+^2 dC_2 \right)^{1/2} \cdot ||v_\epsilon^2||_2.$$

Proof of lemma. Choose a $w \in H_0^2(\Omega)$ such that $w \geq \psi_+$ on Ω, then

$$-\tfrac{1}{\epsilon}\int \eta \cdot v_\epsilon^1 dx \leq \tfrac{\det(A)}{B_{22}} \int w(-\Delta v_\epsilon^2 + \lambda_2 v_\epsilon^2)dx$$

$$\leq Q\|v_\epsilon^2\|_2 \ \|\Delta w\|_2$$

and the result follows by Theorem 2.1.

Returning to the proof of our theorem, we easily see that the L^2-estimates above on v_ϵ^2 and the first equation of (21) give H^2-estimates for v_ϵ^1. Finally, the case of complex eigenvalues is handled in a similar manner as before. This completes our proof of Theorem 4.2. □

For our next results, we will assume that $\psi \in W^{2,p}(\Omega)$, $\psi \in W^{3,2}(\Omega) = H^3(\Omega)$, respectively.

THEOREM 4.6. *There is a constant Q_3, independent of ϵ, such that if*

(i) conditions (i) - (iv) of Theorem 4.1 hold,

(ii) $\|\Delta\psi\|_p < \infty$,

then any solution u_ϵ to (4) satisfies

$$\|u_\epsilon\|_{W^{2,p}(\Omega)\times W^{2,p}(\Omega)} \leq Q_3\left(\int \psi_+^2 dC_1 + \|\Delta\psi\|_p^2\right)$$

for any finite p.

THEOREM 4.7. *There is a constant Q_4, independent of ϵ, such that if*

(i) conditions (i) - (iii) of Theorem 4.2 hold,

(ii) $\int_\Omega [-\Delta\psi + \lambda_1^+\psi]_+^2 dC_1 < \infty$,

then any solution u_ϵ to (4) satisfies

$$\|u_\epsilon\|_{H^3(\Omega)\times H^1(\Omega)}^2 \leq Q\int [-\Delta\psi + \lambda_1^+\psi]_+^2 dC_1$$

where $\lambda_1^+ = \max(Re\,\lambda_1, 0)$.

Remark 4.8. For smooth ψ, there are two solution classes - the one of Theorem 4.6 and the one of Theorem 4.7. The first case contains some strongly elliptic systems well treated in the literature; the second case contains the version of the biharmonic obstacle problem referred to earlier. For $N \times N$ weakly elliptic systems, $N > 2$, there are many more possible distinct solution classes. Such systems will be treated in Part II.

Proof of Theorem 4.6. The approach is standard and we shall be brief (cf. [C]). Set $F_p(t) = |t|^p t$, for $t \in \boldsymbol{R}$, $p \geq 2$, then $F_p'(t) = (p-1)|t|^{p-2}$. Thus from (5) with

$\eta_\epsilon(t) = \frac{1}{\epsilon}\eta(t)$,

$$\int |\eta_\epsilon(v_\epsilon^1 - \psi)|^p dx = \int F_p(\eta_\epsilon(v_\epsilon^1 - \psi)) \cdot \eta_\epsilon(v_\epsilon^1 - \psi) dx$$

$$= \frac{1}{A^1}\int F_p(\eta_\epsilon(\cdot))[-\Delta v_\epsilon^1 + \lambda_1 v_\epsilon^1 + v_\epsilon^2] dx$$

$$= -\frac{1}{A^1}\int F_p(\eta_\epsilon)\Delta(v_\epsilon^1 - \psi)dx - \frac{1}{A^1}\int F_p(\eta_\epsilon)\Delta\psi dx$$

$$+\frac{1}{A^1}\int F_p(\eta_\epsilon)(\lambda_1 v_\epsilon^1 + v_\epsilon^2)dx$$

$$\leq Q||\eta_\epsilon(v_\epsilon^1 - \psi)||_p^{p-1}(||\Delta\psi||_p + ||v_\epsilon^1||_p + ||v_\epsilon^2||_p).$$

With this and (5), we can boot strap our way from L^2-estimates on Δv_ϵ to L^p-estimates via the Sobolev inequality. Thus the result of Theorem 4.6 follows. □

Proof of Theorem 4.7. We start by assuming that λ_1, λ_2 are real. From (21), we have

$$\int (|\nabla v_\epsilon^2|^2 + \lambda_2 |v^2|^2)dx = \frac{A^2}{\epsilon}\int \eta(v_\epsilon^1 - \psi)v_\epsilon^2 dx \tag{23}$$

$$= \frac{A^2}{\epsilon}\int \eta(v_\epsilon^1 - \psi)\Delta(v_\epsilon^1 - \psi)dx + A^2\int \eta(v_\epsilon^1 - \psi)\Delta\psi dx$$

$$-\lambda_1\frac{A^2}{\epsilon}\int \eta(v_\epsilon^1 - \psi)v^1 dx.$$

The first term of (23) is ≤ 0 (integrate by parts), the third is ≤ 0 when $\mu_1 < \lambda_1 \leq 0$ and less than or equal to

$$-\lambda_1\frac{A^2}{\epsilon}\int \eta(v_\epsilon^1 - \psi)\psi dx \tag{24}$$

when $\lambda_1 > 0$. Thus (23) and (24) give

$$\int (|\nabla v_\epsilon^2|^2 + \lambda_2 |v_\epsilon^2|^2)dx \leq -\frac{A^2}{\epsilon}\int \eta(v_\epsilon^1 - \psi)[-\Delta\psi + \lambda_1^+\psi]_+ dx,$$

where $\lambda_1^+ = \max(\lambda_1, 0)$. The rest of the argument proceeds as before, especially with regard to the proof of Lemma 4.4.

In case the eigenvalues are complex, $\lambda_1 = \lambda + i\sigma$, $\lambda_2 = \bar{\lambda}_1$, then we proceed as we did in (23) and then take the real parts of both sides. The previous argument now works except as a replacement for what we learn from the maximum principle, that $v^1 \geq 0$ on Ω, we notice that the imaginary part produces

$$\int |v_\epsilon^2|^2 dx = -\frac{A^2}{\epsilon}\int \eta(v_\epsilon^1 - \psi)v_\epsilon^1 dx$$

when $\sigma \neq 0$. Thus we still have $\int \eta v_\epsilon^1 dx \leq 0$. This completes the proof of Theorem 4.7. □

5. Existence theorems. We begin by applying the estimates of Theorem 4.1.

THEOREM 5.1. *Suppose that conditions (i) - (iv) of Theorem 4.1 hold, then there exists a solution to problem (2).*

Also with regard to the estimates of Theorem 4.2, we prove the following revised existence result, revised since we need to modify the definition of the convex set $\boldsymbol{K}$. Set

$$\boldsymbol{K}' = \{v \in H^2(\Omega) \cap H_0^1(\Omega) \times L^2(\Omega) : \ \ v^1 \geq \psi, \ \ \text{a.e. on } \Omega\}.$$

We then prove

THEOREM 5.2. *Suppose that conditions (i) - (iv) of Theorem 4.2 hold, there exists a solution to the problem*

$$(25) \qquad \begin{cases} \textit{find } u \in \boldsymbol{K}' \textit{ such that} \\ \langle \mathcal{L}u, V - u \rangle \geq 0, \quad \textit{for all } V \in \boldsymbol{K}'. \end{cases}$$

Here the brackets $\langle \cdot, \cdot \rangle$ now denote the duality paring between $H^2(\Omega) \cap H_0^1(\Omega)$ and its dual.

Remark 5.3. If we assume the conditions of Theorem 4.7 instead of those of 4.2, then clearly we will have enough regularity to conclude that problem (2) has a solution rather than the modified version (25). However, it should be pointed out that Theorem 5.2 gives the most general existence theorem for H^2-solutions of the biharmonic obstacle problem: cf. [F2] and [CF].

Proof of Theorem 5.1. From (9), we can write

$$(26) \qquad \int_\Omega (-\Delta v_\epsilon^1 + \lambda_1 v_\epsilon^1 + v_\epsilon^2)(V^1 - v_\epsilon^1)dx \geq 0$$

for all V^1 such that $(V^1, V^2) \in \boldsymbol{K}$. Now since $||v_\epsilon||_{H^1 \times H^1} \leq Q < \infty$, for all $\epsilon > 0$, there exists a subsequence $\{v_{\epsilon'}\}$ which converges weakly in $H_0^1(\Omega) \times H_0^1(\Omega)$ and strongly in $L^2(\Omega) \times L^2(\Omega)$ to $v \in H_0^1(\Omega) \times H_0^1(\Omega)$ as $\epsilon' \to 0$. Thus from (26), we must have

$$(27) \qquad \int_\Omega (\lambda_1 v^1 + v^2)(V^1 - v^1)dx + \int_\Omega \nabla v^1 \nabla V^1 dx \geq \int_\Omega |\nabla v^1|^2 dx$$

by the lower semi-continuity of the H_0^1-norm with respect to weak H^1-convergence. But (27) implies

$$(28) \qquad \langle -\Delta u^1 + M_{1,1}u^1 + M_{1,2}u^2, \ V^1 - u^1 \rangle \geq 0.$$

Also we easily get

$$(29) \qquad A_{2,1}\Delta u^1 + A_{2,2}\Delta u^2 - B_{2,1}u^1 - B_{2,2}u^2 = 0$$

in the sense of $\mathcal{D}'(\Omega)$. Hence using (29) in (28) yields

$$-\frac{A_{2,2}}{\det (A)} \langle A_{1,1}\Delta u^1 + A_{1,2}\Delta u^2 - B_{1,1}u^1 - B_{1,2}u^2, V^1 - u^1 \rangle \geq 0$$

which with (29) gives our result provided we can show $(u^1, u^2) \in \boldsymbol{K}$. But this is easy too, since $\eta(u^1_{\epsilon'} - \psi) \to \eta(u^1 - \psi)$ strongly in $L^2(\Omega)$ as $\epsilon' \to 0$ and hence in $\mathcal{D}'(\Omega)$. But clearly

$$|\int \eta(u^1_{\epsilon'} - \psi)\phi dx| \leq Q \cdot \epsilon'$$

for some Q independent of ϵ'. Thus $\eta(u^1 - \psi) = 0$ a.e. or $u^1 \geq \psi$ a.e. □

Proof of Theorem 5.2. There is a subsequence $\{u_{\epsilon'}\}$ for which $u^1_{\epsilon'}$ converges to u^1 weakly in $H^2(\Omega)$, strongly in $H^1(\Omega)$ and $L^2(\Omega)$, and for which $u^2_{\epsilon'}$ converges to u^2 weakly in $L^2(\Omega)$. Thus from (21), we deduce

$$\Delta u^1_\epsilon = M_{1,1}u^1_\epsilon + M_{1,2}u^2_\epsilon \tag{30}$$

and hence $\Delta u^1 = M_{1,1}u^1 + M_{1,2}u^2$, in $\mathcal{D}'(\Omega)$. But (30) together with

$$\int_\Omega (A_{1,1}\Delta u^1_\epsilon + A_{1,2}\Delta u^2_\epsilon - B_{1,1}u^1_\epsilon - B_{1,2}u^2_\epsilon)(V^1 - u^1_\epsilon)dx \geq 0$$

for all V^1 such that $(V^1, V^2) \in \boldsymbol{K}'$, implies

$$\int_\Omega (A_{1,1}\Delta u^1_{\epsilon'} - B_{1,1}u^1_{\epsilon'} - B_{1,2}u^2_{\epsilon'})(V^1 - u^1_{\epsilon'})dx + \int_\Omega A_{1,2}u^2_{\epsilon'}\Delta V^1 dx$$

$$\geq -\frac{A^2_{1,2}B_{2,1}}{\det (A)}\int_\Omega u^1_{\epsilon'}u^2_{\epsilon'}dx - \frac{A^2_{1,2}B_{2,2}}{\det (A)}\int |u^2_{\epsilon'}|^2 dx.$$

Now passing to the limit as $\epsilon' \to 0$ yields our desired result, this time due to the lower semi-continuity of the L^2-norm with respect to weak L^2-convergence. The rest of the proof proceeds as in our earlier argument. □

6. Uniqueness. Here we briefly comment on the uniqueness of the solution to problem (2).

THEOREM 6.1. *Suppose that ψ and $\mathcal{L}$ satisfy the conditions of either Theorem 4.6 or Theorem 4.7, then if both eigenvalues λ_1, λ_2 of M belong to the interval $(\mu_1, +\infty)$, the solution to problem (2) is unique.*

Remark 6.2. The additional regularity assumptions on ψ in the above theorem are needed to guarantee that problem (2) can be written out in a distributional sense in terms of a measure $\mu \in H^{-1}$ concentrated on the set in Ω where $u^1 = \psi$, or equivalently, where $v^1 = \psi$. These reduced forms are systems (31) and (39) below corresponding to whether or not $A_{2,2} \neq 0$ or $A_{2,2} = 0$; they are limiting versions of (9) and (21). This additional regularity probably is not necessary for the uniqueness of problem (2) or (25).

Proof of Theorem 6.1. By passing to the limit in $\mathcal{D}'(\Omega)$ as $\epsilon \to 0$ in (9), we get when $A_{2,2} \neq 0$, the following system equivalent to problem (2): $v = (v^1, v^2) \in H^1_0(\Omega) \times H^1_0(\Omega)$,

$$\begin{cases} -v^1 + \lambda_1 v^1 + v^2_\epsilon = \mu \\ -\Delta v^2 + \lambda_2 v^2 = (\lambda_2 - \sigma)\mu, \end{cases} \tag{31}$$

where $\sigma = B_{2,2}/A_{2,2}$ with $\mu \in H^{-1}(\Omega)$ satisfying supp $\mu \subset \{x \in \Omega : \ v^1(x) = \psi(x)\}$. We now rewrite (31) as a single equation with a nonlocal term. This is always possible when $\sigma > \mu_1$.

$$(32) \qquad -\Delta v^1 + (\lambda_1 + \lambda_2 - \sigma)v^1 + (\lambda_1 - \sigma)(\lambda_2 - \sigma)G_\sigma v^1 = \mu$$

where G_σ is the Green's function for the operator $(-\Delta + \sigma)$ on Ω. Now setting $U = v - \tilde{v}$, where v and $\tilde{v}$ are solutions to (31) with corresponding measures μ and $\tilde{\mu}$ respectively, we can write, from (32),

$$\textstyle\int_\Omega |\nabla U^1|^2 dx + (\lambda_1 + \lambda_2 - \sigma)\int_\Omega |U^1|^2 dx + (\lambda_1 - \sigma)(\lambda_2 - \sigma)\int_\Omega G_\sigma U^1 \cdot U^1 dx \leq 0$$

or

$$(33) \qquad \int_\Omega (-\Delta + \sigma)U^1 \cdot U^1 dx + (\lambda_1 + \lambda_2 - 2\sigma) \int_\Omega |U^1|^2 dx + (\lambda_1 - \sigma)(\lambda_2 - \sigma)\int_\Omega G_\sigma U^1 \cdot U^1 dx \leq 0$$

because

$$(34) \qquad \int_\Omega U^1 d\mu - \int_\Omega U^1 d\tilde{\mu} \leq 0.$$

Next, we note that

$$(35) \qquad \int_\Omega (-\Delta + \sigma)U^1 \cdot U^1 dx + 2(\mu_1 - \sigma) \int_\Omega |U^1|^2 dx + (\mu_1 - \sigma)^2 \int_\Omega G_\sigma U^1 \cdot U^1 dx \geq 0.$$

To see (35), we need only notice that in general $\int (-\Delta + \sigma)\phi \cdot \phi =$ $\int |\nabla\phi|^2 + \sigma\phi^2 \geq (\sigma - \mu_1)\int \phi^2$ for any $\phi \in H_0^1(\Omega)$, hence applying this to $\phi = U^1 + (\mu_1 - \sigma)G_\sigma U^1$ achieves the desired result. Thus (33) and (35) combined, imply

$$(36) \qquad [(\lambda_1 + \lambda_2 - 2\sigma) - 2(\mu_1 - \sigma)] \int_\Omega |U^1|^2 dx + [(\lambda_1 - \sigma)(\lambda_2 - \sigma) - (\mu_1 - \sigma)^2]\int_\Omega G_\sigma U^1 \cdot U^1 dx \leq 0.$$

Next, we use the inequality

$$(37) \qquad \int_\Omega |U^1|^2 dx \geq (\sigma - \mu_1)\int_\Omega G_\sigma U^1 \cdot U^1 dx$$

to conclude that (36) implies

$$(38) \qquad (\lambda_1 - \mu_1)(\lambda_2 - \mu_1)\int_\Omega G_\sigma U^1 U^1 dx \leq 0$$

which gives $U^1 = 0$ and then $U^2 = 0$ from (31). To see (37), we use a very standard argument to show that $\delta \geq \sigma - \mu_1$, where

$$\delta = \inf \frac{\int_\Omega [(-\Delta + \sigma)w]^2 dx}{\int_\Omega w \cdot (-\Delta + \sigma)w dx},$$

the infimum being taken over all $w \in H^2(\Omega) \cap H_0^1(\Omega)$ for which $\int_\Omega w \cdot (-\Delta + \sigma)w dx > 0$.

When $A_{2,2} = 0$, our reduced system is

(39)
$$\begin{cases} -\Delta v^1 + \lambda_1 v^1 + v^2 = 0 \\ \\ -\Delta v^2 + \lambda_2 v^2 = \frac{B_{2,2}}{\det(A)} \cdot \mu. \end{cases}$$

Again setting $U = v - \tilde{v}$, we can rewrite (39), using (34), as

(40)
$$\begin{cases} \int_\Omega (\nabla U^1 \nabla U^2 + \lambda_2 U^1 U^2) dx \geq 0 \\ \\ -\Delta U^1 + \lambda_1 U^1 + U^2 = 0. \end{cases}$$

Multiplying the second equation in (40) by U^2, integrating by parts and then comparing the two, leads to

(41)
$$\int_\Omega |U^2|^2 dx \leq (\lambda_2 - \lambda_1) \int_\Omega U^1 U^2 dx.$$

But we clearly have $\int U^1 U^2 dx \leq 0$ by (40), hence if we choose $\lambda_1 \leq \lambda_2$, we get $U^2 = 0$ and then $U^1 = 0$. □

Remark 6.3. As in [A2], we could rework the proofs above to include $H^1 \times H^1$ estimates for the difference of two solutions in terms of a Choquet integral of the difference of the two obstacle; see [A2], theorem 7. The key is to replace (34) by

$$\int_\Omega U^1 d\mu - \int_\Omega U^1 d\tilde{\mu} \leq \int (\psi - \tilde{\psi}) d(\mu - \tilde{\mu}).$$

We leave the rest to the interested reader.

7. Further regularity. We conclude this note with a short summary of additional regularity that can be established for the solution to problem (2) with arbitrarily smooth ψ.

In the case $A_{2,2} \neq 0$, assuming that the eigenvalues λ_1, λ_2 of M are real, $A_{2,2}\det(A) < 0$, $\mu_1 < B_{2,2}/A_{2,2}$, both λ_1 and λ_2 are greater than μ_1, and one of the eigenvalues (call it λ_2) belongs to the interval $(\mu_1, B_{2,2}/A_{2,2}]$, then the argument of Theorem 4.6 can be modified to yield an L^∞-estimate for both Δv_ϵ^1 and Δv_ϵ^2, independent of ϵ. In fact, the maximum principle implies $v_\epsilon^2 \leq 0$ and hence

$$||\eta_\epsilon(v_\epsilon^1 - \psi)||_p \leq Q||(-\Delta\psi + \lambda_1^+ \psi)||_p$$

with Q independent of ϵ. Now just let $p \to \infty$. A consequence of this estimate is that $W^{2,\infty}$ estimates for solutions to problem (2) follow by familiar arguments in this case; cf. [C].

When $A_{2,2} = 0$, we can again obtain $W^{2,\infty}$-estimates on u^1, the first component of the solution vector for problem (2), assuming λ_1, λ_2 in (μ_1, ∞), and $B_{2,2}/\det(A) < 0$. Again this result follows a familiar pattern: $\Delta u^1 \geq \Delta\psi$, C_1-a.e. on supp μ, μ the measure of (35); thus $G_2\mu$ is bounded on supp μ and the Evans' maximum principle implies that $G_2\mu$ is bounded in Ω; hence so is Δv^1 or equivalently, Δu^1; now one resorts to the method of Freshe [F2], see also [CF], to bound arbitrary second partials.

These are the main regularity results for 2×2 weakly elliptic systems under a single obstacle constraint in one component direction. Some open questions remain, in particular concerning the optimal $W^{2,\infty}$-regularity with regard to the size of the eigenvalues, and the uniqueness under the weakest existence hypotheses.

REFERENCES

[A1] D.R. ADAMS, *L^p-capacitary integrals with some applications*, Proc. Symp. Pure Math., AMS **35** (1979),359-367.

[A2] D.R. ADAMS, *Capacity and the obstacle problem*, Appl. Math. Optim. **8** (1981), 39-57.

[CF] L.A. CAFFERELLI, A. FRIEDMAN, *The obstacle problem for the biharmonic operator*, Ann. Sc. Norm. Sup. Pisa **6** (1979), 151-184.

[C] M. CHIPOT, *Variational inequalities and flow in porous media*, App. Math. Sciences **52**, Springer-Verlag, 1984.

[F1] J. FRESHE, *On systems of second order variational inequalities*, Israel J. Math. **15** (1973), 421-429.

[F2] J. FRESHE, *On the regularity of the solution of the biharmonic variational inequality*, Manuscripta Math., **9** (1973), 91-103.

[H] K. HANSSON, *Imbeddings theorems of Sobolev type in potential theory*, Math. Scand. **45** (1979), 77-102.

[HW] S. HILDEBRANDT, K.O. WIDMAN, *Variational inequalities for vector valued functions*, J. für Math. **309** (1979), 191-220.

[KS] D. KINDERLEHRER, G. STAMPACCHIA, An Introduction to Variational Inequalities and their Applications, **88** Pure and Applied Math., Academic Press, 1980.

SOME REMARKS ON WIDDER'S THEOREM AND UNIQUENESS OF ISOLATED SINGULARITIES FOR PARABOLIC EQUATIONS

A. ANCONA AND J.C. TAYLOR*

Abstract. An elementary proof is given of Widder's Theorem and of the uniqueness of isolated singularities for parabolic differential equations $Lu = u_t$. It applies equally well to operators L that are Hölder continuous, in divergence form, or of Hörmander type.

Key words. Widder's Theorem, isolated singularities, parabolic partial differential equations

AMS(MOS) subject classifications. 35K05,35K99, 58G11,

Introduction. The purpose of this elementary note is to show that Widder's Theorem and the uniqueness of isolated singularities are both straightforward consequences of the existence of a fundamental solution for a parabolic partial differential equation $Lu = u_t$ on a manifold M. Assuming (i) the solvability of the first boundary value problem (including the minimum principle), (ii) Harnack's inequality, and (iii) the continuity and strict positivity of the fundamental solution Γ, both results are obtained without the use of Gaussian or other estimates for Γ. However, Gaussian estimates on Γ enter the picture as one way to obtain Harnack's inequality.

As a consequence, Kryzanski's results in $\mathbf{R}^n$ [7] are extended to operators with only Hölder continuous coefficients, Aronson's result [1] for operators in divergence form on $\mathbf{R}^n$ is obtained simply, and for operators in Hörmander form on a manifold (e.g. the Kohn Laplacian on the Heisenberg group) these (new) results are obtained all by the same argument. According to Saloff-Coste, our arguments also prove these theorems for operators L of the form $Lf = \sum X_i a_{ij} X_j$, with (a_{ij}) uniformly elliptic and the X_i left invariant vector fields on a Lie group M that satisfy Hörmander's condition [9]. In other words, all the hard work in proving these theorems in particular cases may be done by verifying the above assumptions.

1. Background information. Let M denote a smooth manifold equipped with a positive measure σ whose support is M and such that all points have measure zero, and let L denote a second order differential operator on M with possibly time dependent coefficients. Consider the sheaf $\mathcal{H}$ of continuous solutions of the equation $Lu = u_t$ on $M \times \mathbf{R}$. It will be assumed to have the following properties. To begin with, there is a base for the topology of M consisting of open relatively compact sets O such that the first boundary value problem on $D = O \times (c, d)$ is solvable: that is for each continuous function f on the parabolic boundary $\partial_p D = \partial O \times [c, d] \cup O \times \{c\}$ there is a continuous function u on $\overline{D}$ that agrees with f on the parabolic boundary and is a solution of the equation on D; furthermore, this function u will be assumed to be ≥ 0 if $f \geq 0$; as a result, for each $\dot{x} = (x, t) \in D$, there is a unique **parabolic measure** $\mu_{\dot{x}}^D = \mu_{\dot{x}}$ on the parabolic boundary of D such that $u(\dot{x}) = \int u d\mu_{\dot{x}}$. Note that if $\dot{x} = (x, t)$, the support of $\mu_{\dot{x}}^D$ is $\partial_p D \cap M \times [c, t]$

*Materially supported by NSERC Operating Grant #A3108

Consequently, a continuous function u on an open subset U of M$\times$**R** is a solution if and only if it is averaged by all the parabolic measures determined by any family of sets of the form D that covers U, where each $\overline{\mathrm{D}} \subset \mathrm{U}$.

It will be assumed that there is a (Hadamard-Moser type) parabolic Harnack inequality for the nonnegative solutions u:

$$u(x_0, t+b) \geq C(x_0, a, b, \mathrm{K}, \mathrm{U})u(y, s), y \in \mathrm{K}, t \leq s \leq t+a,$$

if u is defined on a neighbourhood U of $\mathrm{K} \times [t, t+b]$, $0 < a < b$, where K is a compact neighbourhood of x_0.

Finally, the equation is assumed to have a fundamental solution Γ — defined below — on $\mathrm{M} \times [S, T)$ for some interval $[S, T) \subset \mathbf{R}$.

DEFINITION 1.1. *Let u be a continuous function u on $M \times (s, t)$ that is a solution of the equation $Lu = u_t$ on $M \times (s, t)$. It will be said to have* **initial value** *f if it has a continuous extension u to $M \times [s, t)$ and $f(x) = u(x, s), \forall x \in M$.*

DEFINITION 1.2. *A* **fundamental solution** Γ *for the equation $Lu = u_t$ on $M \times [S, T)$ is a continuous function $\Gamma(x, t; y, s), x, y \in M$ and $S \leq s < t < T$ such that*

(1) *For all (y,s), as a function of (x, t) it is a solution of the equation on $M \times (s, T)$;*
(2) *For all $f \in C_0^+(M)$, $\Gamma f(x, t) = \Gamma_s f(x, t) = \int \Gamma(x, t; y, s) f(y) \sigma(dy)$ is a solution of the equation on $M \times (s, T)$ with initial value f; and*
(3) *$\Gamma_s f(x, t) \leq u(x, t)$, $s < t < s+\delta$ for any solution $u \geq 0$ on $M \times (s, s+\delta), \delta > 0$ with initial value $\geq f \in C_0^+(M)$.*

REMARK. Condition (3) is automatically verified if $\Gamma = \sup_n \Gamma_n$, where Γ_n is a fundamental solution on $\mathrm{U}_n \times [S, T)$, U_n open relatively compact and $\Gamma_n(x, t; y, s) = 0$ if $x \in \partial \mathrm{U}_n, y \in \mathrm{U}_n$.

The fundamental solution will assumed to be strictly positive. This, for example is the case if, for all $\varphi \in C_0^+(M)$,

$$(*) \qquad \int \varphi(x) \Gamma(x, t; y, s) \sigma(dx) \to \varphi(y) \text{ as } t \to s.$$

These hypotheses are satisfied for at least three broad categories of operator L:

(1) $\mathrm{M} = \mathbf{R}^n$ and $Lu = \sum_{i,j=1}^{n} a_{ij}(x, t) \frac{\partial^2 u}{\partial x_i \partial x_j} + \sum_{i=1}^{n} b_i(x, t) \frac{\partial u}{\partial x_i} + c(x, t)u$, where (a_{ij}) is symmetric positive definite and $(1/\lambda)||\xi||^2 \leq \sum_{i,j=1}^{n} a_{ij}(x, t) \xi_i \xi_j \leq \lambda ||\xi||^2, c \leq 0$, the coefficients are uniformly α-Hölder continuous in x and the a_{ij} are $\alpha/2$-Hölder continuous in t (cf. Friedman [5] for the first and

third hypothesis and the method in Fabes and Stroock [4] for Harnack's inequality);

(2) $M = \mathbf{R}^n$ and $Lu = \sum_{i=1}^n \frac{\partial}{\partial x_i}\left\{\sum_{j=1}^n a_{ij}(x,t)\frac{\partial u}{\partial x_j} + a_j(x,t)u\right\} + \sum_{i=1}^n b_i(x,t)\frac{\partial u}{\partial x_i} +$ $c(x,t)u$ where the coefficients are all bounded, the matrix (a_{ij}) is uniformly elliptic and some technical $L^{p,q}$-conditions are verified (see Aronson [1]); and

(3) M a smooth manifold and $L = (1/2)\sum_{i=1}^m Y_i^2 + Y_0$, where Y_i and Y are smooth vector fields on M and the $Y_i, 1 \le i \le m$ satisfy Hörmander's condition: at each point of M the rank of the Lie algebra generated by $Y_1, Y_2, \ldots, Y_m$ is n, the dimension of M (Bony [2]) . For the first hypothesis see Appendix 1. Presumably the vector fields may be time dependent but that is not covered by Bony's work.

Since the first two hypotheses are local, L may be any second order elliptic operator on a manifold, that in local coordinates is of type (1) provided the global fundamental solution exists.

2. Some properties of fundamental solutions

LEMMA 2.1. *Let f be a continuous non-negative function on M and let O be an open subset. Then,*

$$\lim_{t\to s}\int_O \Gamma(x,t;y,s)f(y)\sigma(dy) = f(x), \forall x \in O,$$

provided one of the following conditions holds:

(1) *O is open relatively compact; or*

(2) *f is the initial value of a non-negative solution u on $M \times (s, s+\delta)$.*

Proof. (1) is obvious given (2) of Definition 1.2. For (2) note that (3) of this definition implies that $\int_O \Gamma(x,t;y,s)u(y,s)\sigma(dy) \le u(x,t)$. The result follows by using (2) of the above definition to get a lower estimate. □

LEMMA 2.2. *Let $y_0 \in M$ and let u be a positive solution on $M \times (s, s+\delta)$ that extends continuously to $(U\backslash\{y_0\}) \times \{s\}$. Let K be a compact neighbourhood of $y_0 \subset U$. Then*

(1) *the measures $\nu_\varepsilon(dy) = 1_K(y)u(y, s+\varepsilon)\sigma(dy)$ are uniformly bounded; and*

(2) *the weak limit points ν as $\varepsilon \to 0$ are of the form $\nu(dy) = 1_{K\backslash\{y_0\}}(y)u(y,s)\sigma(dy) +$ $c\,\varepsilon_{y_0}(dy)$.*

Proof. If $\varepsilon < \delta/2$, $\int_K \Gamma(y_0, s+\delta/2; y, s+\varepsilon)\mathrm{u}(y, s+\varepsilon)\sigma(dy) \le \mathrm{u}(y_0, s+\delta/2)$. The continuity and strict positivity of Γ ensures that there is a constant C> 0 such that
$\Gamma(y_0, s+\delta/2; y, s+\varepsilon) > C, y \in K$ and $\varepsilon < \delta/4$. This proves (1). For (2), note that the continuity of u on the lower boundary except at y_0 implies that a weak limit point ν has density u(y,s) on $K\backslash\{y_0\}$. □

PROPOSITION 2.3. *Let u be a solution on $M\times(s,s+\delta)$ which vanishes continuously $O\times\{s\}$, O an open subset of M. Then the function $\tilde{u}$ defined by*

$$\tilde{u}(x,t)=\begin{cases} u(x,t), & \text{for } s<t<s+\delta\\ 0, & \text{for } x\in O,\ t=s\\ 0, & \text{for } t<s,\end{cases}$$

is a solution on $M\times(-\infty,s+\delta)\cap\complement(\complement O\times\{s\})$.

Proof. The function $\tilde{u}(x,t)$ is averaged by the parabolic measures for those $D=O'\times(c,d)$ with $\overline{D}\subset M\times(-\infty,s+\delta)\cap\complement(\complement O\times\{s\})$. This is obvious for $s<c$ or $d\leq s$. In case $c\leq s<d$, $\overline{O'}\subset O$. Let $D_\varepsilon=O'\times(s+\varepsilon,d)$. For $s<t, \tilde{u}(x,t)=u(x,t)=\lim_{\varepsilon\to 0}\int u d\mu_x^{D_\varepsilon}=\int\tilde{u}d\mu_x^D$. If $t\leq s$ the averaging property is obvious. □

LEMMA 2.4. *Let v_n be a sequence of uniformly bounded non-negative solutions on $O\times(s-\varepsilon,s+\delta/2)$ that converges to ψ and is such that each v_n vanishes below s. Let D= $O_1\times(s-\varepsilon/2,s+\delta/4)$, $O_1\subset O$ be such that the first boundary value problem is solvable. Then*

(1) $\int\psi\mu_{\dot x}^D=\psi(\dot x)$, *and*
(2) $\lim_{t\to s}\psi(x,t)=0, \forall x\in O_1$.

COROLLARY 2.5. $\lim_{t\to s}\Gamma(x,t;y,s)=0, x\neq y$.

Proof. Let f_n be a sequence of smooth functions with $\int f_n d\sigma=1$ and compact support K_n decreasing to $\{y_0\}$. Then, if $u_n(x,t)=\Gamma_s f_n(x,t)$, the continuity of Γ implies that $\lim_{n\to\infty}u_n(x,t)=\Gamma(x,t;y_0,s)$. Let $v_n(x,t)u_n(y_0,t_0)=u_n(x,t), s<t_o<T$. By proposition 2.3, the functions v_n extend as solutions to $M\times(-\infty,T)\cap\complement(K_n\times\{s\})$. It follows from Harnack's inequality that the functions v_n are locally uniformly bounded on $M\times(-\infty,\frac{s+t_0}{2})\cap\complement(K_n\times\{s\})$. Let

$$\psi(x,t)=\begin{cases}\Gamma(x,t;y_0,s)/\Gamma(y_0,t_0;y_0,s), & \text{for } 0<t<s+\delta\\ 0, & \text{for } x\neq y_0,\ t=s\\ 0, & \text{for } t<s.\end{cases}$$

It follows from Lemma 2.4 that $\lim_{t\to s}\psi(x,t)=0$ on $M\times(-\infty,\frac{s+t_0}{2})\cap\complement\{(y_0,s)\}$.
□

COROLLARY 2.6. *The function $\Gamma_{(y,s)}$ defined by*

$$\Gamma_{(y,s)}(x,t)=\begin{cases}\Gamma(x,t;y,s), & \text{for } S<s<t<T\\ 0, & \text{for } x\neq y,\ t=s\\ 0, & \text{for } t<s,\end{cases}$$

is a solution on $M\times(-\infty,T)\cap\complement\{y,s)\}$.

Finally, here is a standard observation.

PROPOSITION 2.7. *Let O be an open relatively compact subset of M such that the first boundary value problem on* $D = O \times (s_0, t_0)$ *is solvable. Let* Γ *be a fundamental solution for the equation* $Lu = u_t$ *on* $M \times [S,T)$, $S \le s_0 < t_0 \le T$. *Define* $\Gamma_D(x,t;y,s) = \Gamma(x,t;y,s) - H(x,t;y,s)$, *where* $H(x,t;y,s)$ *is the value at (x,t) of the solution of the first boundary value problem on D with boundary value* $\Gamma_{(y,s)}$. *Then* Γ_D *is a fundamental solution for the equation* $Lu = u_t$ *on D.*

Proof. Property (1) of Definition 1.2 is obvious. For (2) note that $H(x,t;y,s)$ is continuous on

D and that $H(x,s;y,s) = 0$ for all $x \in O$ as $H(x,t';y,s) = 0$ for all $x \in O, s_0 < t' < s$. Property (3) follows from the minimum principle since if $f \in C_0^+(O)$, then $\Gamma_D f$ vanishes on the lateral boundary $\partial O \times [s_0, t_0]$ of D. □

3. Isolated singularities

THEOREM 3.1. *Let* $u \ge 0$ *be a solution on* $M \times (s, s+\delta)$ *which vanishes continuously on* $(M \backslash \{y_0\}) \times \{s\}$. *Then there is a constant* $c \ge 0$ *and a solution v on* $M \times (-\infty, s+\delta)$ *which vanishes below s such that*

$$u = c\Gamma_{(y_0,s)} + v.$$

Proof. It follows from Lemma 2.1 and Proposition 2.3, that for any $\varepsilon > 0$, there is a solution v_ε on $M \times (-\infty, s+\delta) \cap \complement(\complement B \times \{s+\varepsilon\})$, such that for $t > s+\varepsilon$,

$$u(x,t) = \int_B \Gamma(x,t;y,s+\varepsilon)u(y,s+\varepsilon)\sigma(dy) + v_\varepsilon(x,t),$$

where B is the open ball $B(y_0,r)$ about y_0 of radius r. Note that v_ε vanishes below s.

Choose $\varepsilon_n \to 0$ so that the measures $\nu_{\varepsilon_n}(dy) = 1_B(y)u(y,s+\varepsilon_n)\sigma(dy)$ have a weak limit $\nu(dy) = c\varepsilon_{y_0}(dy)$. Then $\int_B \Gamma(x,t;y,s+\varepsilon_n)\sigma(dy) \to c\Gamma_{(y_0,s)}(x,t)$ for $t > s$. Hence, $v_{\varepsilon_n}(x,t) \to v(x,t) = u(x,t) - c\Gamma_{(y_0,s)}(x,t), s < t < s+\delta$. Since $0 \le v \le u$, v vanishes continuously on $(M\backslash\{y_0\}) \times \{s\}$.

Let $B_1 = B(y_0, \theta r), 0 < \theta < 1$ and $D = B_1 \times (s-1, s+\delta/2)$. The solutions v_{ε_n} restricted to $\partial_p D$ are (i) uniformly bounded and (ii) converge to a continuous function f. Let $\tilde{v}$ be the solution to the first boundary value problem with boundary value f. It agrees with v on $B_1 \times (s, s+\delta)$. As in Lemma 2.4, this implies that v vanishes continuously on $M \times \{s\}$. The result follows from Proposition 2.3. □

THEOREM 3.2 (UNIQUENESS OF ISOLATED SINGULARITIES). *Let* $u \ge 0$ *be a solution on a punctured neighbourhood* $U \subset M \times (S,T)$ *of* (y_0, s_0). *Then*

$$u = c\Gamma_{(y_0,t_0)} + h, \; h \text{ a solution on } U.$$

If $U = M \times (S,T) \backslash \{(y_0,s_0)\}$ *then* $h \ge 0$.

Assume the first boundary value problem is solvable for $D = O\times(s-\delta,s+\delta), \delta < (T-s)\wedge(s-S)$, *O open relatively compact. If* $U = D\backslash\{(y_0,s_0)\}$ *and* $\Gamma = \Gamma_D$ *then h is the solution of the first boundary value problem with initial value u.*

Proof. In view of Corollary 2.6 and Proposition 2.7, it will suffice to prove this for $U = D\backslash\{(y_0,s_0)\}$, where $D = O\times(s-\delta,s+\delta), \delta < (T-s)\wedge(s-S)$ and $O = B(y_0,r)$. To simplify notation it will be assumed that $s = 0$. Let $\Gamma = \Gamma_D$ denote the fundamental solution for D. Then $\Gamma u(x,t) = \int \Gamma_D(x,t;y,0)u(y,0)\sigma(dy)$. Let $f \in C_0(O), 0 \le f \le 1$, with support disjoint from $\{y_0\}$. Then $u(x,t) \ge \int \Gamma_D(x,t;y,s)f(y)u(y,s)\sigma(dy), 0 < s < t$. Continuity implies that $u(x,t) \ge \int \Gamma_D(x,t;y,0)f(y)u(y,0)\sigma(dy) = u'(x,t)$. Hence, $u(x,t) \ge \int \Gamma_D(x,t;y,0)u(y,0)\sigma(dy)$ as σ has no atoms. Also, since u' has initial value $f(x)u(x,0)$, it follows that $u-\Gamma u$ satisfies the hypotheses of Theorem 3.1.

Hence, there is a constant c and a non-negative solution v on $O\times(-\infty,\delta)$ that vanishes below zero and such that $u = c\Gamma_{(y_0,0)} + v + \Gamma u$ on $O\times(0,\delta)$.

Let h be the solution of the first boundary value problem for D with initial value u. It agrees with u below zero and so $h(x,0) = u(x,0), x \ne y_0$. Hence, $\Gamma h = \Gamma u$ on $O\times(0,\delta)$. Since the Green function $\Gamma = \Gamma_D$ for D has the property that for any bounded continuous function f on O, $\lim_{t\to 0}\Gamma f(x,t) = f(x)$, it follows that $h = \Gamma u + v$ for $t > 0$. The formula relating Γ to Γ_D shows that $h \ge 0$ if $U = M\times(S,T)\backslash\{(y_0,s_o)\}$. □

4. Widder's Theorem.

The following result is an easy consequence of Lemma 2.2.

PROPOSITION 4.1. *Let* $u \ge 0$ *be a solution of the equation* $Lu = u_t$ *on* $M\times(s,s+\delta)$. *Let O be a relatively compact open subset of M. Then,*

$$u = \Gamma_s\mu_O + h_O,$$

where h_O *is a solution of the equation on* $M\times(-\infty,s+\delta)\cap \complement(M\backslash O\times\{s\})$ *that vanishes below zero and* μ_O *is a positive measure on* $\overline{O}$.

Proof. In view of Lemma 2.2 (2) it follows that there is a sequence $\varepsilon_n \to 0$ such that $\nu_{\varepsilon_n}(dy)$ has a weak limit μ_O. Let v_{ε_n} be the function defined in the proof of theorem 3.1. Then $\lim_{n\to\infty} v_{\varepsilon_n}(x,t) = h_O(x,t) = u(x,t) - \Gamma_s\mu_O(x,t), s < t < s+\delta$. Further, if $O_1 \subset \overline{O_1} \subset O$ as in Lemma 2.4, it follows from this Lemma that $\lim_{t\to s} h_O(x,t) = 0, \forall x \in O_1$ since $v_{\varepsilon_n}(x,s+\delta/2) \le u(x,s+\delta/2)$. □

THEOREM 4.2 (WIDDER'S THEOREM). *Let* $u \ge 0$ *be a solution of the equation* $Lu = u_t$ *on* $M\times(s,s+\delta)$. *Then,*

$$u = \Gamma_s\mu + h,$$

where h is a solution of the equation on $M\times(-\infty,s+\delta)$ *that vanishes below zero and* μ *is a positive measure on M. Further, the measure is unique if, for example, the fundamental solution satisfies (*).*

Proof. Let $(\mathrm{O}_n)_{n\geq 0}$ be an exhaustion of M by a sequence of relatively compact open sets $\mathrm{O}_n, \overline{\mathrm{O}_n} \subset \mathrm{O}_{n+1}$.

Let $\mathrm{h}_0 = \mathrm{h}_{\mathrm{O}_0}$ and $\mu_0 = \mu_{\mathrm{O}_0}$. Apply Proposition 4.1 to h_0 and O_1 : $\mathrm{h}_0 = \Gamma_s\mu_{\mathrm{O}_1} + \mathrm{h}_{\mathrm{O}_1}$. The measure $\mu_1 = \mu_{\mathrm{O}_1}$ has no mass in O_0, $\mathrm{h}_1 = \mathrm{h}_{\mathrm{O}_1}$ vanishes on O_1 and $\mathrm{h}_1 \leq \mathrm{h}_0$. Continuing in this way one obtains a sequence $(\mu_n)_{n\geq 0}$ of measures μ_n and solutions h_n that vanish on O_n such that for each n

$$\mathrm{u} = \Gamma_s(\mu_0 + \mu_1 + \cdots + \mu_n) + \mathrm{h}_n.$$

Let $\mu = \sum_{n\geq 0}\mu_n$ and $\mathrm{h} = \lim_{n\to\infty 0} \mathrm{h}_n$.

The uniqueness follows easily from (*). $\square$

REMARK 1. The classical form of Widder's theorem [12] gives the form of the positive solutions u of the heat equation $\frac{1}{2}\Delta u = u_t$ on $\mathbf{R}^n \times (0,T)$:

$$u(x,t) = \frac{1}{(2\pi t)^{n/2}} \int e^{-\frac{(x-y)^2}{2t}} \mu(dy) = \int \Gamma(x,t;y,0)\mu(dy),$$

where μ is a unique positive Borel measure on $\mathbf{R}^n$.

From the point of view of abstract potential theory, this theorem is simply the Riesz representation of a particular superharmonic function as a potential plus a harmonic function. The Gauss kernel is in effect an abstract Green kernel in view of the uniqueness of isolated singularities established above. In other words, if

$$\Gamma_{(y,s)}(x,t) = \begin{cases} \frac{1}{(2\pi(t-s))^{n/2}} e^{-\frac{(x-y)^2}{2(t-s)}}, & \text{for } t > s; \text{ and} \\ 0, & \text{for } t \leq s, \end{cases}$$

then $\Gamma_{(y,s)}$ is a potential on $\mathbf{R}^n \times (-\infty, T)$ with point support $\{(y,s)\}$ and the function Γ is a Green function on $\mathbf{R}^n \times (-\infty, T)$.

Extend the positive solution u of the heat equation below the strip $\mathbf{R}^n \times (0,T)$ by zero and the resulting function ũ is lower-semicontinuous and even superharmonic. The Riesz representation theorem guarantees that ũ is a Green potential — given by a unique measure μ supported by the complement of the set on which it is harmonic, namely, $\mathbf{R}^n \times \{0\}$ — plus a harmonic function h. This function vanishes below zero and so in the case of $\mathbf{R}^n$ is identically zero.

The trouble with this proof is that the abstract functional analytic machinery used to prove a general Riesz representation theorem in axiomatic potential theory is not particularly attractive, especially to analysts used to a priori estimates and the systematic use of inequalities and estimates.

REMARK 2. In general, the function h in Widder's theorem is not zero (cf. Mair and Taylor [10] for examples involving the classical heat equation). In Widder's original theorem [12] h is identically zero. This is because of the well known fact that, for $\mathrm{M} = \mathbf{R}^n$, a non-negative solution of the heat equation on $\mathrm{M} \times (0,T)$ which vanishes at $t = 0$ is identically zero. This " Principle of Uniqueness for the Positive Cauchy Problem" is always valid when the so-called minimal solutions of the heat equation on $M \times (-\infty, T)$ never vanish [6]. This is the case for a Riemannian manifold with $L = \Delta$ when the Ricci curvature of M is bounded below since Li and Yau [8] showed the Harnack constant to be independent of $x \in M$ (cf. Davies [3]). It is also true when L is the Ornstein-Uhlenbeck operator on $\mathbf{R}^n$ as shown in [11] Lemma 1.1.

Appendix 1: the first boundary value problem for Hörmander type operators

Assume that L is an operator on M of the form $L = (1/2)\sum_{i=1}^m Y_i^2 + Y_0$, where Y_i and Y are smooth vector fields on M and the $Y_i, 1 \leq i \leq m$ satisfy Hörmander's condition: at each point of M the rank of the Lie algebra generated by $Y_1, Y_2, \ldots, Y_m$ is n, the dimension of M.

Let $\mathcal{H}$ denote the sheaf of solutions of the equation $Lu = u_t$ on $M \times \mathbf{R}$. A relatively compact open set $U \subset M \times \mathbf{R}$ will be said to be **regular** if the following Dirichlet problem has a unique solution (cf. Bony [2] théorème 5.2)): for each continuous function f on $\overline{U}$ and continuous function φ on ∂U, there is a unique continuous function u on $\overline{U}$ such that:

$$Lu - u_t = -f \text{ as a distribution on U, and}$$
$$u = \varphi \text{ on } \partial\, U.$$

This implies that, for each continuous function φ on ∂U, there is a unique solution $u = u_\varphi$ on U with boundary value φ i.e. u extends continuously to $\overline{U}$ and equals φ on ∂U.

Bony's general result ([2] théorème 5.2) implies that $M \times \mathbf{R}$ has a base for its topology consisting of regular sets. Consequently, a continuous function u on an open set W is a solution if and only if it is averaged by the **harmonic measures** for a base of the topology of W consisting of regular subsets.

Finally, since the question of parabolic measures for small sets is a local question, assume that $M = \mathbf{R}^n$. Bony [2] p 294 gives an explicit formula for a regular open set U. Let h be a unit vector such that $\sum a_{ij}(x_o)h_i h_j > 0$, where $Lu = \sum a_{ij}(x)\frac{\partial^2 u}{\partial x_i \partial x_j} + \sum a_i(x)\frac{\partial u}{\partial x_i}$. Set

$$U(M, \varepsilon, x_0) = U(x_o) = \{x \mid ||x - (x_o + Mh)|| < M + \varepsilon, ||x - (x_o - Mh)|| < M + \varepsilon\}.$$

This set is regular for M sufficiently large.

Proposition A.1. *If $U = U(x_o)$ is as above, then the first boundary value problem is solvable for $U \times (c, d)$.*

Proof. If $\Omega \subset \mathbf{R}^n \times \mathbf{R}$, let $\Omega(t) = \{x \in \mathbf{R}^n \mid (x, t) \in \Omega\}$. Define Ω by setting

$$\Omega(t) = \begin{cases} U(x_o), & \text{for } c \leq t \leq d \\ s\{U(x_o) - x_o\} + x_o, & \text{for } t = d + (1-s), 0 \leq s \leq 1 \\ s\{U(x_o) - x_o\} + x_o, & \text{for } t = c - (1-s), 0 \leq s \leq 1 \\ \emptyset & \text{otherwise.} \end{cases}$$

Then Ω is regular for $Lu - u_t$, Bony [2] (théorème 5.2). Hence, if f is a continuous function on $\partial\Omega$ which vanishes below c, the solution u of the corresponding Dirichlet problem when restricted to $U \times (c, d)$ solves the first boundary value problem with boundary value f on the lateral boundary $\partial U \times [c, d]$ and zero on the bottom $U \times \{c\}$. The fundamental solution Γ' on $U \times \mathbf{R}$ solves the first boundary value problem for zero boundary value on the lateral boundary and given continuous function g on the bottom. □

By using stochastic calculus, this result is easily extended to arbitrary sets U that are regular for L.

Bibliography

[1] D. G. Aronson, *Non-negative solutions of linear parabolic equations*, Annali della Scuola Norm. Sup. — Pisa, 22 (1968), pp. 607-694.

[2] J.- M. Bony, *Principe du maximum, inégalité de Harnack et unicité du problème de Cauchy pour les opérateurs elliptiques dégénérés*,, 19 (1969), pp. 277-304.

[3] E.B. Davies, *Heat kernels and spectral theory*, Cambridge University Press, Cambridge, 1989.

[4] E.B. Fabes and D.W. Stroock, *A new proof of Moser's parabolic Harnack inequality via the old ideas of Nash*, Arch. Rat. Mech. Anal., 96 (1986), pp. 327-338.

[5] A. Friedman, *Partial differential equations of parabolic type*, Prentice-Hall, Englewood Cliffs, N.J., 1964.

[6] A. Koranyi and J.C. Taylor, *Minimal solutions of the heat equation and uniqueness of the positive Cauchy problem on homogeneous spaces*, Proc. Amer. Math. Soc., 94 (1985), pp. 273-278.

[7] M. Kryzyanski, *Sur les solutions non négative de l'équation linéaire normale parabolique*, Revue Roumaine de Mathématiques Pures et Appliquées, 9 (1964), pp. 393-408.

[8] P. Li and S.T. Yau, *On the parabolic kernel of theSchrödinger operator*, Acta Math., 156 (1986), pp. 153-201.

[9] L. Saloff-Coste and D.W.Stroock, *Opérateurs uniformément sous-elliptiques sur les groupes de Lie*, Journal of Functional Analysis, (to appear), pp. 273-278.

[10] B. Mair and J.C. Taylor, *Integral representation of positive solutions of the heat equation*, Lecture Notes in Math., 1096 (1984), pp. 419-433.

[11] J.C.Taylor, *The minimal eigenfunctions characterize the Ornstein-Uhlenbeck process*, The Annals of Probability, 17 (1989), pp. 1055-1062.

[12] D. V. Widder, *Positive temperatures on the infinite rod*, Trans. Amer. Math. Soc., 55 (1944), pp. 85-95.

GENERALIZED DERIVATIVES

J. MARSHALL ASH, JONATHAN COHEN, CHRIS FREILING, A.E. GATTO, AND DAN RINNE

0. Introduction. Generalized differential operators are ones which agree with differential operators when applied to sufficiently smooth functions but have special symmetry properties which allow them to be defined on less smooth functions. Such operators were used by Cantor [4] in his proof of the uniqueness of representation by trigonometric series and have been an integral part of all extensions of Cantor's theorem to higher dimensions.

In this article we introduce some generalized mixed partial derivative operators and a generalized biharmonic operator. We give theorems describing properties of the solutions of the related homogeneous equations and describe their connection with problems in the uniqueness of multiple trigonometric series.

1. The mixed partials. For functions $f \in C^2(\mathbf{R}^2)$, we know that $u_{xy} = 0$ for all $(x, y) \in \mathbf{R}^2$ implies that $u(x, y) = A(x) + B(y)$ 0 for all (x, y). We are interested in weakening the limiting process in taking the mixed partials and seeing if the splitting of the function u into a sum of functions of one variable still holds. We define two such limiting processes. The first is a non-centered limiting process and the second is a symmetric one.

DEFINITION 1.1. Let

$$\Delta u(x, y; h, k) = u(x + h, y + k) - u(x + h, y) - u(x, y + k) + u(x, y)$$

We then define the operator:

$$D^2_{xy} u = Du(x, y) = \lim_{\substack{h,k \to 0 \\ h,k \neq 0}} \frac{\Delta u(x, y; h, k)}{hk}$$

DEFINITION 1.2. Let

$$\Delta_s u(x, y; h, k) = u(x + h, y + k) - u(x + h, y - k) - u(x - h, y + k) + u(x - h, y - k).$$

We define:

$$D^s u(x, y) = \lim_{\substack{h,k \to 0 \\ h,k \neq 0}} \frac{\Delta_s u(x, y; h, k)}{4hk}.$$

If $f \in C^2(\mathbf{R}^2)$, then both of these agree with the mixed partial derivative. That is $f_{xy}(x, y) = Df(x, y) = D^s f(x, y)$ for all (x, y). However, these need not always coincide. Our concern in this paper is to whether one can still conclude that $u(x, y) = A(x) + B(y)$ if either $Du \equiv 0$ or $D^s u \equiv 0$. A theorem due to Bögel [3] says that if the unsymmetric generalized mixed partial $Du(x, y) = 0$ for all $(x, y) \in \mathbf{R}^2$

then $u(x,y) = A(x) + B(y)$. A simple proof of this can be found in [1]. This theorem requires no regularity assumptions on u of any kind, not even measurability.

One then would like to know if the same result will hold for the symmetrically defined generalized mixed partial. The following example shows that the two operators are different.

Example. The function $S(x,y) = \text{sgn}\ (|y| - |x|)$ is easily seen to satisfy $D^s S \equiv 0$ but doesn't decompose to a sum of a function of x and a function of y. (See [1] for details.)

It is clear from the example function that one needs additional assumptions on u to have $D^s u \equiv 0$ imply $u(x,y) = A(x) + B(y)$. We can now state our result:

THEOREM I. *If u is continuous in $\mathbf{R}^2$ and $D^s u \equiv 0$ then $u(x,y) = A(x) + B(y)$.*

Proof. In this proof a box will be a nondegenerate closed rectangle with sides parallel to the axes. Let B be a box with vertices (x,y), $(x+h,y)$, $(x+h,y+k),(x,y+k)$ where $h > 0, k > 0$ and define $\Delta B = \Delta u(x,y;h,k)$. It will be of great importance in the proof that ΔB satisfies the finite additivity property

$$\Delta B = \sum \Delta B' \tag{1.3}$$

whenever $B = \cup B'$ where the $B's$ are boxes with mutually disjoint interiors.

To prove theorem I it then suffices to prove $\Delta B = 0$ for all rectangles B in $\mathbf{R}^2$ since if that is the case $u(x,y) = u(x,0) + [u(0,y) - u(0,0)]$ which is the splitting of u into functions of x and of y. We will show,

LEMMA 1.4. *u continuous and $D^s u \geq c > 0$ for all (x,y) implies $\Delta B \geq 0$ for all boxes B in $\mathbf{R}^2$.*

Remark. This will prove our theorem since if $D^s u = 0$ for all (x,y) then $D^s[u(x,y) + \epsilon xy] = \epsilon > 0$ which implies $\Delta B \geq 0$. On the other hand $D^s[-u(x,y) + \epsilon xy] = \epsilon > 0$ which implies $\Delta B \leq 0$. Hence $\Delta B = 0$ which as we have just pointed out proves the theorem.

We now prove lemma 1.4. Let S be the set of all open subsets O of $\mathbf{R}^2$ which are dense and satisfy

$$\Delta B \geq 0 \text{ for all boxes } B \subset O. \tag{1.5}$$

We will show that S is non empty and that the union of all the subsets of S is itself in S and is all of $\mathbf{R}^2$.

Let $A_n = \{(x,y) : \Delta_s u(x,y) \geq 0 \text{ for all } (h,k) \in (0,\frac{1}{n}] \times (0,\frac{1}{n}]\}$. The $A_n's$ are closed, $A_n \subset A_{n+1}$ for all n and $\cup_{n=1}^{\infty} A_n = \mathbf{R}^2$. Let A_n^o denote the interior of A_n, ∂A_n the boundary of A_n and $A = \cup_{n=1}^{\infty} A_n^o$. We show that $A \in S$.

A is clearly open. Furthermore A is dense since $\mathbf{R}^2 = A \cup (\cup \partial A_n)$ implies that $A \supset (\cup \partial A_n)^c$ and $(\cup \partial A_n)^c$ is dense in $\mathbf{R}^2$ by the Baire Category Theorem. Finally, since $A_n^o \subset A_{n+1}^o$ and any box $B \subset A$ is compact, $B \subset A_N^o$ for some N. Subdivide B

into the union of subboxes B' with non-overlapping interiors and with sides of length less than $1/N$. Then $\Delta B' \geq 0$ and by finite additivity, (1.3), $\Delta B = \sum \Delta B' \geq 0$. Thus $A \in S$ and hence $S \neq \emptyset$.

Let $T = \cup_{O \in S} O$. Clearly T is open and dense in $\mathbf{R}^2$. Property (1.5) follows from a compactness argument and so T is also in S. It remains to show that $T = \mathbf{R}^2$.

Let $C = \mathbf{R}^2 \backslash T$. Assume that $C \neq \emptyset$. Then C is a complete metric space and $C = \cup_{n=1}^{\infty}(A_n \cap C)$ The sets $A_n \cap C$ are closed in C and so by the Baire Category Theorem there is an open set $O \subset \mathbf{R}^2$ with $\emptyset \neq O \cap C \subset A_n \cap C$. One can then find a box D with side lengths less than $1/n$, $D \subset O$ and $D^o \cap C \neq \emptyset$. We will show that $D^o \cup T \in S$ thus contradicting the maximality of T.

Clearly $D^o \cup T$ is open and dense. It remains to show that for any box $B \subset D^o \cup T$, $\Delta B \geq 0$. For B any such box, B can be written as the union of $B \cap D$ with at most four boxes which are entirely contained in T. Furthermore the union is disjoint except for some of the edges so that by additivity we need only show that $\Delta D \cap B \geq 0$.

We now let B denote $D \cap B$. It is clear that if $B' \subset B$ is a subbox and $B' \cap C = \emptyset$ then $\Delta B' \geq 0$. Furthermore, if B' is a subbox of B and is centered at a point $c \in C$, then $\Delta B' \geq 0$ because the sidelength of B' is less than $1/n$ and $c \in A_n$. We are then done once we prove the following lemma:

LEMMA 1.6. *Let B be a box which meets the closed nowhere dense set C such that for every box $B' \subset B$*

(1.7) $$B' \text{ disjoint from } C \text{ implies } \Delta B' \geq 0$$

and

(1.8) $$\text{center of } B' \text{ in } C \text{ implies } \Delta B' \geq 0,$$

then $\Delta B \geq 0$.

Proof. For ease of calculation we assume that B is centered at (0,0) and has first quadrant vertex (a, b). Let Y be the y-axis. We distinguish three cases.
Case 1: $Y \cap B \cap C = \emptyset$

For $0 < c \leq a$, let $N(c)$ be the box with vertices $(0, -b)$, $(c, -b)$, (c, b), and $(0, b)$. Since $Y \cap B$ and C are disjoint closed sets, they are separated by some positive distance. Thus, for some $c' > 0$, $N(c') \cap C = \emptyset$, and, for any $B' \subset N(c')$ we have $\Delta_s B' \geq 0$. Let $R = \{c | 0 < c \leq a \text{ and } \Delta B' \geq 0 \text{ for any } B' \subset N(c) \text{ with left edge in } Y\}$ and let $\tau = \sup R$. By the selection of c' above, $R \neq \emptyset$, and the continuity of f implies that $\tau \in R$. Suppose that $\tau < a$ and let $t := \min\{\tau, (a - \tau)/2\}$. We show that this gives $\tau + t \in R$, contradicting the choice of τ. It suffices to show $\Delta N(\tau + t) \geq 0$, since the same argument will apply to any box contained in $N(\tau + t)$ with left edge contained in Y.

Let V_1 be the box with opposite corners $(\tau, -b)$ and $(\tau + \frac{t}{2}, b)$. If $V_1^o \cap C \neq \emptyset$, pick $(x_1, y_1) \in V_1 \cap C$ closest to the horizontal bisector of V_1. Let B_1 be the largest box centered at (x_1, y_1) that is contained in $N(\tau+t)$ and let B_1' be the box that has the same right edge as B_1, and has left edge in Y. By (1.3) $\Delta B_1' \geq 0$ since $\Delta B_1 \geq 0$ and the left edge of B_1 is in $N(\tau)$. Note that the right edge of B_1 is contained in the right edge of $N(\tau+t)$. Now suppose V_k, (x_k, y_k), B_k, and B_k', $k = 1, ..., n-1$ have been selected. We define the box $V_n = V_{n-1} \backslash B_{n-1}$. If $V_n^o \cap C \neq \emptyset$, pick $(x_n, y_n) \in V_n \cap C$ closest to the horizontal bisector of V_n. Let B_n be the largest box centered at (x_n, y_n) that is contained in the closure of $N(\tau+t) \backslash \cup_{k=1}^{n} B_k$ and let B_n' be the box that has the same right edge as B_n, and has left edge in Y. If some $V_n^o \cap C = \emptyset$, we stop and obtain a finite sequence of boxes $\{B_n'\}$. Otherwise, the sequence is infinite. If $\cup B_n' = N(\tau+t)$, then, by the continuity of f, we are done. Otherwise, $N(\tau+t)$ is the union of at most four nonoverlapping boxes O_1, U_1, L_1, and R_1, of the following form. We have O_1 and U_1 as the closures of unions of adjacent elements of $\{B_n'\}$, L_1 has left edge in Y and right edge in the vertical line $x = \tau + \frac{t}{2}$ and R_1 has left edge equal to the right edge of L_1 and right edge in the right edge of $N(\gamma+t)$. (O is for "over", U for "under", L for "left", and R for "right".) Note that ΔO_1, ΔU_1, and ΔL_1 are nonnegative and $\Delta B' \geq 0$ for any box $B' \subset L_1$ with left edge in Y. If $\Delta R_1 \geq 0$ we are done by additivity.

Otherwise we iterate the process described in the preceding paragraph to obtain a nested sequence of boxes $\{R_n\}$, each R_n having right edge contained in the right edge of $N(\tau+t)$ as follows: having constructed R_{n-1}, perform the process inside the box $L_{n-1} \cup R_{n-1}$ with V_1 chosen to be the left half of R_{n-1}. If $\Delta R_n \geq 0$, we are done by additivity. If for each $n = 1, 2, ...,$ $\Delta R_n < 0$; then $\cap R_n$ is a line segment (possibly degenerate) contained in the right edge of $N(\tau+t)$, so that $\Delta N(\tau+t) \geq 0$ by the continuity of f.

Applying the same argument to the left half of B and using additivity gives $\Delta B \geq 0$.

Case 2: $Y \cap B^o \cap C = \emptyset$.

For $0 < \epsilon < b$, let B_ϵ be the box centered at (0,0) with first quadrant vertex $(a, b-\epsilon)$. We apply Case 1 to get $\Delta B_\epsilon \geq 0$. Let $\epsilon \to 0$ and, by the continuity of f, we get $\Delta B \geq 0$.

Case 3: $Y \cap B^o \cap C \neq \emptyset$.

Let $V_1 = B$ and pick $(0, y_1) \in Y \cap V_1^o \cap C$ closest to the horizontal bisector of V_1. Let B_1 be the largest box centered at $(0, y_1)$ contained in V_1. Suppose V_k, $(0, y_k)$, and B_k, $k = 1, ..., n-1$ have been selected. We let $V_n = V_{n-1} \backslash B_{n-1}$ and pick $(0, y_n) \in Y \cap V_n^o \cap C$ closest to the horizontal bisector of V_n. Let B_n be the largest box centered at $(0, y_n)$ and contained in V_n. This generates the sequence $\{B_n\}$, which is finite if some $Y \cap V_n^o \cap C = \emptyset$. If $\cup B_n = B$, then $\Delta B \geq 0$ by the continuity of f. Otherwise, $B' = B \backslash \cup B_n$ is a box to which we apply Case 1 or 2. Again, the continuity of f, along with the fact that $\Delta B' \geq 0$, gives $\Delta B \geq 0$.

2. A generalized biharmonic operator. We now study a generalized version of the biharmonic operator Δ^2. To avoid any confusion we point out that in this section we are using the symbol Δ to denote the Laplace operator as opposed to denoting a second difference operator as it did in §1.

For $f \in C^4(\mathbf{R}^n)$, $(n \geq 2)$, it can be shown that

$$\Delta^2 u(x_o) = (a_n/R^4)[M_R u(x_o) - u(x_o) - \frac{\Delta u(x_o)}{2n} R^2] + o(1), \tag{2.1}$$

where $a_n = \Delta^2|x|^4 = 16n + 8n^2$,

$$M_R u(x_o) = \frac{1}{\omega_n R^{n-1}} \int\limits_{\partial B_R(x_o)} u(t)dt$$

the average of u over the surface of the ball $B_R(x_o)$ of radius R centered at x_o, $R > 0$, ω_n is the area of the boundary of the unit sphere in $\mathbf{R}^n$, and $\partial B_R(x_o)$ is the boundary of $B_R(x_o)$. We then define the generalized biharmonic operator by:

$$G\Delta^2 u(x) = a_n \lim_{R\to 0}(1/R^4)[M_R(x) - u(x) - \frac{\Delta u(x)}{2n} R^2]. \tag{2.2}$$

We remark that the function $x^2 \operatorname{sgn} x$ satisfies $G\Delta^2 u \equiv 0$ but obviously does not satisfy $\Delta^2 u = 0$ at the origin. Our main theorem is then:

THEOREM II. *If $G\Delta^2 u(x) = 0$ for all $x \in \mathbf{R}^n$ and $\Delta u(x)$ is continuous, then $\Delta^2 u(x) = 0$ for all $x \in \mathbf{R}^n$.*

Proof. We begin by giving the following formulae:

(2.3)
$$\frac{1}{(n-2)\omega_n} \int\limits_{B_R(0)} [\frac{1}{|t|^{n-2}} - \frac{1}{R^{n-2}}][\Delta f(x+t) - \Delta f(x)]dt = M_R f(x) - f(x) - \frac{\Delta f(x)}{2n} R^2,$$

when $n \geq 3$, and when $n = 2$,

$$\frac{1}{2\pi} \int\limits_{B_R(0)} [\log R - \log|t|][\Delta f(x+t) - \Delta f(x)]dt = M_R f(x) - f(x) - \frac{\Delta f(x)}{4} R^2. \tag{2.4}$$

This can be verified for $n \geq 3$ by applying Green's theorem

$$\int\limits_{\Omega} u\Delta v - v\Delta u = \int\limits_{\partial\Omega} u\frac{\partial v}{\partial n} - v\frac{\partial u}{\partial n}$$

over the annulus $\Omega_\epsilon = \{\epsilon < |t| < R\}$ with $u(t) = |t|^{2-n} - R^{2-n}$ and $v(t) = f(x + t) - (\Delta f(x)|t|^2/2n)$. For $n = 2$, use the same argument with $u(t) = \log R - \log|t|$. We are led to define, for $n \geq 3$:

$$D^2 g(x) = \lim_{R\to 0} \frac{c_n}{R^4} \int\limits_{B_R(0)} [|t|^{2-n} - R^{2-n}][g(x+t) - g(x)]dt \tag{2.5}$$

where $c_n = \frac{8n^2+16n}{(n-2)\omega_n}$, with a similar definition if $n = 2$.

We will show that $D^2g(x) = 0$ and $g(x)$ continuous will imply that g is harmonic. Hence when $g = \Delta F$, $D^2g = 0$ and ΔF is continuous, we have F is biharmonic. For $f(x)$ a function continuous on the closed ball $\overline{B_R(x_o)}$ we let $PI(f, x_o, R)$ denote the Poisson integral of f on $B_R(x_o)$. We next establish the following lemma:

LEMMA 2.6. *If $g(x)$ is continuous in $\mathbf{R}^n$ and $D^2g(x) > 0$ for all $x \in \mathbf{R}^n$, then $g(x) \leq PI(g, x_o, R)(x)$ on any ball $B_R(x_o)$.*

Proof of lemma. Assume the contrary. Then there is a ball $B_R(x_o)$ and a point $x_1 \in B_R(x_o)$ such that $g(x_1) > PI(g, x_o, R)(x_1)$.

Let $w(x) = g(x) - PI(g, x_o, R)(x)$ for $x \in \overline{B_R(x_o)}$. Then $w(x)$ is zero on $\partial B_R(x_o)$ and $w(x_1) = g(x_1) - PI(g, x_o, R)(x_1) > 0$. Hence w attains its maximum at a point c in the open ball $B_R(x_o)$.

Choose $R_1 < \frac{1}{2}\text{dist}(c, \partial B_R(x_o))$. Let h be the Newtonian potential of $PI(g, x_o, R)$ for the ball $B_{R_1}(c)$. Then inside $B_{R_1}(c)$, $\Delta h(x) = PI(g, x_o, R)(x)$. Hence,

$$\begin{aligned} D^2w(c) =& D^2g(c) - D^2(PI(g, x_o, R))(c) \\ =& D^2g(c) - D^2(\Delta h)(c) \\ =& D^2g(c) - G\Delta^2 h(c) \\ =& D^2g(c) - \Delta^2 h(c) \\ =& D^2g(c) - \Delta PI(g, x_o, R)(c) \\ =& D^2g(c) > 0. \end{aligned}$$

On the other hand,

$$D^2w(c) = \lim_{R\to 0} \frac{1}{R^4} \int_{B_R(0)} k(t, R)[w(c+t) - w(c)]dt \leq 0 \tag{2.7}$$

since $w(c+t) - w(c) \leq 0$ and $k(t, R) \geq 0$, where $k(t, R) = [(n-2)\omega_n]^{-1}[|t|^{(2-n)} - R^{(2-n)}]$ if $n \geq 3$ or $(2\pi)^{-1}[\log R - \log|t|]$ if $n = 2$. Hence the assumption that $g(x_1) > PI(g, x_o, R)(x_1)$ for some x_o, R, x_1 leads to a contradiction.

Proof of theorem. We assume that $G\Delta^2 f(x) = 0$. Then for $\epsilon > 0$,

$$D^2[\Delta(f(x) + \epsilon|x|^4)] = G\Delta^2 f(x) + 8n(n+2)\epsilon = 8n(n+2)\epsilon > 0. \tag{2.8}$$

This tells us that the function $\Delta(f(x) + \epsilon|x|^4)$ lies below its Poisson integral on any ball in $\mathbf{R}^n$ and for any $\epsilon > 0$. Letting $\epsilon \to 0$ we conclude that if $G\Delta^2 f(x) = 0$ for all $x \in \mathbf{R}^n$, and if $\Delta f(x)$ is continuous in $\mathbf{R}^n$, then on any ball in $\mathbf{R}^n$, Δf is less than or equal to its Poisson integral over that ball.

By considering $D^2[\Delta(-f(x) + \epsilon|x|^4)]$, the same argument shows that on any ball in $\mathbf{R}^n$, $-\Delta f$ is less than or equal to its Poisson integral over that ball. Hence the conditions $G\Delta^2 f(x) = 0$ and Δf continuous will imply that Δf is harmonic which shows $\Delta^2 f(x) = 0$.

3. Concluding remarks. Our theorems were motivated in part by trying to extend Cantor's [4] theorem on the uniqueness of multiple trigonometric series:

THEOREM III. (Cantor) *If* $\lim_{R\to\infty}\sum_{|n|\le R} c_n e^{inx} = 0$ *for all* $x \in \mathbf{R}$, *then* $c_n = 0$ *for all* $n \in Z$.

The proof of Cantor's theorem involves looking at the formal second integral $F(x) = c_o x^2/2 - \sum_{n\neq 0}(c_n/n^2)e^{inx}$. The pointwise convergence of the original series implies that the $c_n s$ go to zero (Cantor-Lebesgue) and this control of the $c_n s$ implies that $F(x)$ is continuous. Term by term application of the second derivative to the series defining F suggests that $F''(x) \equiv 0$ from which it would immediately follow that the $c_n s$ are all zero. However, the convergence of the original series is not uniform and so one cannot interchange summation and differention. Instead, Cantor used the following:

LEMMA 3.2. *If* $F(x)$ *is continuous and its Schwartz derivative*

$$DF(x) := \lim_{h\to 0} \frac{F(x+h) - 2F(x) + F(x-h)}{h^2} = 0$$

then $F(x)$ *is a line.*

Our Theorems I and II are analogs of Lemma 3.2. Our ultimate goal is to refine these results to where they can be applied to the appropriate formally integrated trigonometric series in d dimensions and thus obtain uniqueness theorems.

Shapiro proved [7] that if $c_n = o(R^{2-d})$ and $\sum c_n e^{inx}$ Abel sums to zero for all x where $c_n \in \mathbf{C}$, $n \in Z^d$ and $nx = n_1x_1 + \ldots + n_dx_d$, then $c_n = 0$ for all $n \in Z^d$. Roger Cooke [6] later proved that the circular convergence of a two dimensional trigonometric series for all x implies that the coefficients c_n satisfy Shapiro's hypothesis and since circular convergence implies Abel summability one combines Shapiro and Cooke's works to obtain:

THEOREM IV. *If* $\lim_{r\to\infty}\sum_{|n|\le r} c_n e^{inx} = 0$ *for all* $x \in \mathbf{R}^2$, *then* $c_n = 0$ *for all* $n \in Z^2$. *Later, estimates for the coefficients of spherically convergent trigonometric series were obtained by Bernard Connes [5] in all dimensions but they only satisfy Shapiro's requirements in dimension two.*

We can apply our theorem II in two dimensions to the series $\sum(c_m/|m|^4)e^{imx}$ where it is assumed that $\sum c_m e^{imx}$ circularly sums to zero everywhere. Using Shapiro's result that $\lim_{t\to 0}\sum(c_m/|m|^2)e^{imx-|m|t}$ is continuous, and Cooke's coefficient bounds we can obtain another (more complicated) proof of uniqueness.

For the case of unrestricted rectangular convergence Ash and Welland[2] proved:

THEOREM V. *If*

$$\lim_{\min(M,N)\to\infty} \sum_{|m|\le M} \sum_{|n|\le N} c_{mn} e^{i(mx+ny)} = 0 \qquad \textit{for all } (x,y) \in \mathbf{R}^2,$$

then $c_{mn} = 0$ *for all* $(m,n) \in Z^2$.

The proof relied in a crucial way on Shapiro's results and did not generalize to higher dimensions. Our theorem I is part of a program to develop an inductive procedure to show that in d-dimensions the assumption

$$\lim_{\min(N_1,\dots,N_d)\to\infty} \sum_{|n_1|\le N_1} \cdots \sum_{|n_d|\le N_d} c_{n_1..n_d} e^{i(n_1x_1+\dots+n_dx_d)} = 0$$

$\forall x \in \mathbf{R}^d$ implies that the formal d^{th} integral $F(x)$ obtained by termwise integration in each variable separately is a sum of functions of $d-1$ variables.

In two dimensions we have shown [1] that $D^2_{L^2,s}F(x) = 0$ ($D^2_{L^2,s}$ is a square integrably averaged version of the operator D^2_s of this paper), where F is the twice formally integrated trigonometric series first in x and then in y. Our counter-example shows that additional regularity (such as continuity) is needed to obtain $F(x,y) = A(x) + B(y)$. We have not been able to show directly that $F(x,y)$ is continuous. (Continuity of course follows from the uniqueness theorem [2] since in that case the coefficients are all zero). However we are able to show in one dimension that $c_n = o(\frac{1}{n})$ implies that $\sum c_n e^{inx} \in VMO$. Our conjecture is then that in d-dimensions if the suitably defined generalized mixed partial operator $D^d_{L^2,s}F(x) = 0$ for all $x \in \mathbf{R}^d$, and F is in some suitably defined product VMO, then F can be written as the sum of functions of $d-1$ variables.

REFERENCES

[1] J.M. ASH, J. COHEN, C. FREILING, D. RINNE, *Generalizations of the wave equation*, To appear, Transactions of the American Math Society.

[2] J.M. ASH, G. WELLAND, *Convergence, uniqueness, and summability of multiple trigonometric series*, Trans. Amer. Math Soc., 163 (1972), pp. 401–436.

[3] K. BÖGEL ÜBER DIE MEHRDIMENSIONALE DIFFERENTIATION, Jber. Deutsch. Math.-Verein., 65 (1962), pp. 45–71.

[4] G. CANTOR, *Beweis, das eine für jeden reellen Wert von x durch eine trigonometrische Reihe gegebene Funktion f(x) sich nur auf eine einzige Weise in dieser Form darstellen lässt*, Crelles J. für Math., 72 (1870), pp. 139–142; also in Gesammelte Abhandlungen, Georg Olms, Hildesheim (1962), pp. 80–83.

[5] B. CONNES, *Sur les coefficients des séries trigonométriques convergentes sphériquement*, C.R. Acad. Sci. Paris, Ser A, 283 (1976), pp. 159–161.

[6] R. COOKE, *A Cantor-Lebesgue theorem in two dimensions*, Proc. Amer. Math. Soc., 30 (1971), pp. 547–550.

[7] V. SHAPIRO, UNIQUENESS OF MULTIPLE TRIGONOMETRIC SERIES, Ann. of Math., (2) 66 (1957), pp. 467–480.

ON NULL SETS OF P-HARMONIC MEASURES*

PATRICIO AVILÉS** AND JUAN J. MANFREDI†

1. Introduction. In a Lipschitz domain $\Omega \subset \mathbf{R}^n$, associated to the p-Laplace equation

$$\operatorname{div}(|\nabla u|^{p-2}\nabla u) = 0 \tag{1}$$

one can define a notion of p-harmonic measure on subsets $E \subset \partial\Omega$ by solving the Dirichlet problem for (1) with boundary values χ_E. Denote by $w_p(x;E)$ the p-harmonic function with boundary values 1 on E and 0 on $\partial\Omega\backslash E$. In the linear case $p = 2$ for each $x \in \Omega$ we do obtain a Borel measure on $\partial\Omega$. This is no longer true in the nonlinear case $p \neq 2$. Yet the monotonicity properties of classical harmonic measures extend to their nonlinear counterparts. Many applications of this principle are in [GLM], [HM].

Given two compact sets on the boundary of p-harmonic measure zero, it remains an open problem to decide whether their union has p-harmonic measure zero or not, that is, is $w_p(x,\cdot)$ additive on null sets? If the compact sets are disjoint that is indeed the case [GLM]. We will prove below that for arbitrary compact set F

$$w_p(x, E \cup F) = w_p(x, F), \quad x \in \Omega \tag{2}$$

holds if the Haussdorff dimension of the compact set E is smaller than a nonnegative number β which depends only on p, n and the Lipschitz character of $\partial\Omega$. This condition on E implies $w_p(x, E) \equiv 0$ (see below). The validity of (2) when E has zero p-harmonic measure remains in doubt even if $E \cap F = \phi$.

2. Definitions and statements. Let Ω be a bounded domain in $\mathbf{R}^n$ and $1 < p < \infty$. A function $u \in W^{1,p}_{loc}(\Omega)$ is p-harmonic if

$$\int_\Omega |\nabla u|^{p-2} < \nabla u, \nabla \phi > dx = 0$$

for all $\phi \in C_0^\infty(\Omega)$. It is well known [DB] that $u \in C^{1,\alpha}_{loc}(\Omega)$ for some $\alpha = \alpha(p,n) \in (0,1]$.

A lower semicontinuous function $v : \Omega \longrightarrow \mathbf{R} \cup \{+\infty\}$ is p-superharmonic if for every domain $D \subset\subset \Omega$ and p-harmonic function u in D such that $u \in C(\bar{D}), u \leqq v$ on ∂D implies $u \leqq v$ in D. This notion turns out to be more general than the notion of weak supersolution. See [L] and [K].

*Research partially supported by NSF Grants DMS-8901695 and DMS-8901524.
**Department of Mathematics, University of Illinois, Urbana, Illinois 61801.
†Department of Mathematics and Statistics, University of Pittsburgh, Pittsburgh, PA 15260.

Let $E \subset \Omega$. The upper class U consists of all superharmonic functions $v : \Omega \longrightarrow \mathbf{R} \cup \{+\infty\}$ such that

$$\lim_{x \to y \in \partial\Omega} \inf v(x) \geq \chi_E(y)$$

for all $y \in \partial\Omega$. The following Perron solution

$$w_p(x, E) = \inf\{v(x) : v \in U\}$$

turns out to be p-harmonic and it satisfies

$$0 \leqq w_p(x, E) \leqq 1$$

for all $x \in \Omega$. The set E has p-harmonic measure zero if $w_p(x, E) = 0$ for some $x \in \Omega$. Harnack's inequality for nonnegative p-harmonic functions implies then that $w_p(x, E) = 0$ for all $x \in \Omega$. For the details of this construction (in the more general case of quasilinear elliptic equations) see [GLM], [HKR] and for more information on sets of zero harmonic measure for quasilinear operators see [M].

We shall restrict ourselves to the case $\Omega = \mathbf{B}^n$ the unit ball. It will be clear from the proofs that if Ω is a Lipschitz domain our arguments remain valid with minor modifications and the help of [FGMS].

THEOREM. *There exists $\beta = \beta(p, n) > 0$ such that if Hauss-dim $(E) \leqq \beta$, then*

$$w_p(x, E \cup F) = w_p(x, F), \quad x \in \mathbf{B}^n$$

for compact subsets E, F of $\partial\mathbf{B}^n$.

We choose β such that Hass-dim $(E) \leqq \beta$ implies $w_L(x, E) = 0$ for any second order linear differential operator in general form whose ellipticity constants are as in (3) below. See [MW] and [FGMS]. It will follow then from (6) below that Hauss-dim $(E) \leqq \beta$ implies $w_p(x, E) \equiv 0$.

3. Proof. The key observation in the proof is to linearize the p-Laplacian at a particular p-harmonic function along the lines of [MW].

Set $u(x) = w_p(x, E \cup F)$ and

$$a^u_{ij}(x) = \begin{cases} \delta_{ij} + (p-2)\dfrac{u_{x_i}(x) u_{x_j}(x)}{|\nabla u(x)|^2}, & \text{if } \nabla u(x) \neq 0 \\ \delta_{ij} & \text{otherwise.} \end{cases}$$

Observe that $a^u_{ij} \in L^\infty(\Omega)$ and for $\xi \in \mathbf{R}^n$

$$\min(1, p-1)|\xi|^2 \leqq \sum_{i,j=1}^{n} a^u_{ij}(x)\xi^i\xi^j \leqq \max(p-1, 1)|\xi|^2. \tag{3}$$

We now define L_u, a linear uniformly elliptic operator with bounded measurable coefficients and ellipticity constants depending only on p, by

$$L_u(v)(x) = \sum_{i,j=1}^{n} a^u_j(x) v_{x_i x_j}(x) \tag{4}$$

for $v \in W^{2,2}_{loc}$. Differentiation of (1) leads to

$$L_u u(x) = 0 \tag{5}$$

at least formally. When $1 < p < 3 + \frac{2}{n-2}$, L_u is a Cordes operator. It follows that the Dirichlet problem for the operator L_u can be uniquely solved, $u \in W^{2,2}_{loc}(\Omega)$ and (5) holds a.e. see [MW]. In particular we can talk about the elliptic harmonic measure associated to L_u, denoted w_{L_u}.

LEMMA 1.

$$\boxed{u(x) \leq w_{L_u}(x, E \cup F)}$$

Proof. Both functions are solutions of the equation $L_u v = 0$, certainly $\overline{\lim_{x \to y}} u(x) \leqq 1$ for $y \in E \cup F$ and $\lim_{x \to y} u(x) = 0$ for $y \in \partial\Omega \backslash E \cup F$ by the boundary regularity of p-harmonic functions. The lemma now follows from the comparison principle.

Indeed, we have proved

$$w_p(x, C) \leqq w_{L_c}(x, C) \tag{6}$$

for any compact set $C \subset \partial\Omega$, where L_c is the operator defined by (4) with $u(x)$ replaced by $w_p(x, c)$. Whether equality actually hold in (6) or not remains an open question. Let us observe that by taking complements and using $L_C = L_{\partial\Omega \backslash C}$ for C compact it follows that

$$w_p(x, G) \geqq w_{LG}(x, G) \tag{7}$$

for any open set $G \subset \partial\Omega$.

Since L_u is a linear operator we have

$$u(x) \leqq w_{L_u}(x, E) + w_{L_u}(x, F). \tag{8}$$

It was shown in [MW] that we can find β such that Hauss-dim $(E) \leq \beta$ implies $w_{L_u}(x, E) = 0$. From (8) we obtain

$$u(x) \leqq w_{L_u}(x, F).$$

Thus

$$\limsup_{x \to y} u(x) = 0 \quad \text{for } y \notin F$$

The comparison principle gives then

$$w_p(x, E \cup F) = u(x) \leqq w_p(x, F).$$

Since the opposite inequality is obvious, we have proved the Theorem for $1 < p < 3 + \frac{2}{n-2}$.

When $p \geq 3 + \frac{2}{n-2}$ the operator L_u is no longer Cordes and we can not talk about w_{L_u}. We apply an approximation technique as in [FGMS] to write

$$u(x) = \int_{\partial\Omega} K(x,y)dv(y),$$

where K is a limit of kernel functions of operators with ellipticity bounds as in (3) and dv is a nonnegative measure on $\partial\Omega$. Since u is bounded it follows from the proof of Theorem 4.5 in [CFMS] that $dv(y) = f(y)dW(y)$ where dW is a weak limit of harmonic measures at zero of approximating operators. Since $u(y) = 0$ in $\partial\Omega\backslash E \cup F$ one can easily check that $\operatorname{supp} dv \subset E \cup F$. We can now mimic the argument in the Cordes case,

$$u(x) \leq \int_F K(x,y)f(y)dW(y) + \int_E K(x,y)f(y)dW(y)$$

Our choice of β gives $W(E) = 0$, so

$$u(x) \leq \int_F K(x,y)f(y)dW(y).$$

From the properties of the kernel function see [CFMS] we easily conclude

$$\limsup_{x \to y} u(x) = 0 \quad \text{for} \quad y \notin F.$$

REFERENCES

[CFMS] Caffarelli, L. Fabes, E. Mortola, S. and Salsa, S., *Boundary behavior on nonnegative solutions of elliptic operators in divergence form*, Indiana Univ. Math. J. 30 (1981), pp. 621-640.

[DB] DiBenedetto, E., *$C^{1+\alpha}$-local regularity of weak solutions of degenerate elliptic equations*, Nonlinear Anal., Theory, Methods, Appl. 7 (1983), pp. 827-850.

[FGMS] Fabes, E. Garofalo, N. Marín-Malave, S., and Salsa, S., *Potential Theory and Fatou theorems for some nonlinear elliptic operators*, to appear in the Revista Iberoamericana de Matemáticas.

[GLM] Granlund, S. Linqvist, P. and Martio. O., *F-harmonic measure in space*, Ann. Acad. Sci. Fenn. A I Math 7 (1982), pp. 233-247.

[HKR] Heinonen, J. Kilpeläinen, T. and Rossi, J., *The Growth of A-subharmonic Functions and Quasiregular Mappings along Asymptotic Paths*, Indiana Univ. Math. J. 38 (3) (1989), pp. 581-601.

[HM] Heinonen, J. and Martio, O., *Estimates for F-harmonic measures and Oksendal's Theorem for quasiconformal mappings*, Indiana Univ. Math. J. 36, No. 3 (1987), pp. 659-683.

[L] Linqvist, P., *On the definition and properties of p-superharmonic functions*, J. Reine Angew. Math. 365 (1986), pp. 67-79.

[M] Martio, O., *Sets of Zero Elliptic Harmonic Measures*, Ann. Acad. Sci. Fenn. A I. Math 14 (1989), pp. 47-55.

[MW] Manfredi, J. and Weitsman, A., *On the Fatou Theorem for p-harmonic functions*, Comm. in PDE 13(6) (1988), pp. 651-688.

LIFETIME AND HEAT KERNEL ESTIMATES IN NON-SMOOTH DOMAINS

RODRIGO BAÑUELOS*

0. Introduction

Let D be a domain in $\mathbb{R}^n$, $n \geq 2$, and let B_t be Brownian motion in D with lifetime τ_D. The transition probabilities for this motion are given by the Dirichlet heat kernel $P_t^D(x,y)$ for $\frac{1}{2}\Delta$ in D. If h is a positive harmonic function in D the Doob h-process, the Brownian motion conditioned by h, is determined by the following transition functions:

$$P_t^h(x,y) = \frac{1}{h(x)} P_t^D(x,y)h(y).$$

We will write P_x^h for the measure on path space induced by the transition densities $P_t^h(x,y)$ and denote the corresponding expectation by E_x^h. When $h = 1$, the case of Killed Brownian motion in D, we simply write P_x and E_x. With this notation we have

$$P_x^h\{\tau_D > t\} = \frac{1}{h(x)} \int_D P_t^D(x,y)h(y)dy \tag{0.1}$$

and integrating in time we have

$$E_x^h(\tau_D) = \frac{1}{h(x)} \int_D G_D(x,y)h(y)dy \tag{0.2}$$

where $G_D(x,y)$ is the Green's function for $\frac{1}{2}\Delta$ in D. In answer to a question of K. L. Chung, M. Cranston and T. McConnell [10] proved the following

Theorem A. *Let $H^+(D)$ be the set of positive harmonic functions in D. If $D \subset \mathbb{R}^2$, then*

$$\sup_{\substack{x\in D \\ h\in H^+(D)}} E_x^h(\tau_D) \leq C \text{ area}(D) \tag{0.3}$$

where C is an absolute constant. For any $n \geq 3$, there exists a bounded domain $D \subset \mathbb{R}^n$ and $h \in H^+(D)$ such that $P_x^h\{\tau_D = \infty\} = 1$ for all $x \in D$.

We will often refer below to the estimate $\sup_{\substack{x\in D \\ h\in H^+(D)}} E_x^h(\tau_D) < \infty$ as the Cranston-McConnell or lifetime estimate.

The Cranston-McConnell proof of (0.3), as well as its refinement given by Chung [7], is probabilistic. We feel that the result has been of sufficient interest to analysts that an analytical proof is warranted. In §1 of this paper we present such a

*Department of Mathematics, Purdue University. Supported in part by NSF.

proof which we learned from T. Wolff in the spring of 1985. However, any errors in presentation are due to the author. As T. Wolff himself has pointed out, the idea behind the analytic proof is still based on that of the probabilistic proof of Cranston and McConnell. In §1, we also discuss extensions of Theorem A to two classes of domains in several dimensions. The first class is what we call **Hölder domains of order** 0 and which we denote by $H(0)$. This class of domains has been studied previously in connection with the Poincaré inequality. It includes the class of Lipschitz, NTA, uniform (BMO-extension), and John domains. The second class of domains is what we call **uniformly Hölder of order** α and which we denote by $UH(\alpha)$. This class of domains was introduced in Bañuelos [3]. Our results for the class $UH(\alpha)$ are sharp in the sense that if $D \in UH(\alpha)$, $0 < \alpha < 2$, then D satisfies the Cranston-McConnell estimate and for every $\alpha \geq 2$ there exists $D \in UH(\alpha)$ for which the estimate fails. In §2, we discuss the notion of intrinsic ultracontractivity, (IU), which was first introduced by E. B. Davies and B. Simon [11]. Any domain whose Dirichlet semigroup is (IU) satisfies the Cranston-McConnell estimate. Our result is that if $D \in H(0)$ or if $D \in UH(\alpha)$, $0 < \alpha < 2$, then the semigroup of the Dirichlet Laplacian plus a potential in the Kato class is (IU). For every $\alpha \geq 2$ there is a $D \in UH(\alpha)$ for which the Dirichlet Laplacian is not (IU). Our proofs, which we only briefly outline here, are via Dirichlet forms and logarithm Sobolev inequalities.

1. Lifetime Estimates

It is well known, with no assumptions in D, that (case $h = 1$)

$$\sup_{x\in D} E_x(\tau_D) \leq \frac{1}{n}\left(\frac{\text{vol}(D)}{C_n}\right)^{2/n}, \tag{1.1}$$

where C_n is the volume of the unit ball in $\mathbf{R}^n$. Perhaps not as well known is the following Hayman-Pommerenke-Stegenga type result first proved in $\mathbf{R}^2$ by C. Muller [17] and subsequently extended to all dimensions (and to elliptic diffusions) by R. Bañuelos and Øksendal [5]:

(1.2) $\sup_{x\in D} E_x(\tau_D) < \infty$ iff there exist constants R_0 and C_0 such that for all $x \in D$, $Cap(B(x,R_0)\cap D^C) \geq C_0 R_0^{n-2}$, where Cap denotes the Newtonian capacity, (in $\mathbf{R}^2$ this has to be modified). In fact, whenever the capacity condition holds we have $\sup_{x\in D} E_x(\tau_D) \leq CR_0^2$ with $C = C(n, C_0)$.

The case of conditioned Brownian motion is much more complicated and one does not have such satisfactory results. Let us first consider a simple case. Suppose $D \subset \mathbf{R}^2$ is simply connected. Let φ map the unit disc $D(0,1)$ conformally onto D. We may assume that $\varphi(0) = x$. By the conformal invariance of the Green's function,

$$E_x^h(\tau_D) = \frac{1}{h(x)}\int_D G_D(x,y)h(y)dy = \frac{1}{h(x)}\int_{D(0,1)} \log\left(\frac{1}{|z|}\right) h(\varphi(z))|\varphi'(z)|^2 dz.$$

Since $h(\varphi)$ is a positive harmonic function in D we have, with $T = \partial D(0,1)$,

$$h(\varphi(z)) = \int_T \frac{1-|z|^2}{|z-e^{i\theta}|} d\mu(\theta)$$

for some positive measure μ on T with $\mu(T) = h(\varphi(0)) = h(x)$. From this and Fubini's Theorem,

$$E_x^h(\tau_D) = \frac{1}{h(x)} \int_T g_*^2(\varphi)(\theta) d\mu(\theta)$$

where g_*^2 is the classical Littlewood-Paley square function

$$g_*^2(\varphi)(\theta) = \frac{1}{\pi} \int_{D(0,1)} \log\left(\frac{1}{|z|}\right) \frac{1-|z|^2}{|z-e^{i\theta}|^2} |\varphi'(z)|^2 dz.$$

It follows from the fact that $\log\left(\frac{1}{|z|}\right) \sim (1-|z|)$ for $|z|$ near 1 that for all $\theta \in T$,

$$g_*^2(\varphi)(\theta) \leq C \int_{D(0,1)} |\varphi'(z)|^2 dz = C \text{ area}(D).$$

Thus,

$$E_x^h(\tau_D) \leq \frac{C}{h(x)} \text{ area}(D)\mu(T) = C \text{ area}(D)$$

and (0.3) is proved for simply connected domains.

If the domain is not simply connected the above argument breaks down. However, it may be possible to adapt the proof to the case of multiply connected regions by introducing universal covering maps. We believe such a proof would be of interest.

We now prove (0.3) for any planar domain. The basic estimate holds in fact in any dimensions and with no restrictions on the domain.

Basic Estimate. Let D be a domain in $\mathbb{R}^n$, $n \geq 2$. For $h \in H^+(D)$ set $D_m = \{z \in D : 2^{m-1} < h(z) < 2^{m+2}\}$, $m = \pm 1, \pm 2, \ldots$. Then for all $x \in D$,

$$\frac{1}{h(x)} \int_D G_D(x,y)h(y)dy \leq 8 \sum_{m=-\infty}^{\infty} \left(\sup_{z \in D_m} \int_{D_m} G_{D_m}(z,y)dy \right).$$

This basic estimate is inequality (2.7), p.6, in Cranston and McConnell [10]. Before we give the analytic proof we show how Theorem A, and many extensions to higher dimensions, immediately follow from it. First, if $D \subset \mathbb{R}^2$, then by (1.1),

$$\sup_{z \in D_m} \int_{D_m} G_{D_m}(z,y)dy \leq \frac{1}{2\pi} \text{ area}(D_m)$$

and we have

$$\frac{1}{h(x)} \int_D G_D(x,y)h(y)dy \leq \frac{8}{2\pi} \sum_{m=-\infty}^{\infty} \text{ area}(D_m) = C \text{ area}(D)$$

which gives (0.3).

If we follow the above argument in $\mathbb{R}^n$, $n \geq 3$, we have to show that $\sum_{m=1}^{\infty} (vol(D_m))^{2/n} < C(D) < \infty$ where $C(D)$ depends only on the domain and not

on the function h and starting point x. This will ensure that $\sup_{\substack{x\in D\\ h\in H^+(D)}} E_x^h(\tau_D) < \infty$.
We next show that this is the case for a large class of domains, including Lipschitz. First, if $x \in D$ let $d_D(x)$ be the distance from x to ∂D. Also fix a point $x_0 \in D$, called the centered of D, and assume $d_D(x_0) = 1$. For any $x \in D$, let $\rho_D(x)$ be quasi-hyperbolic distance from x_0 to x, (see Bañuelos [2] for definitions). We shall say that D is a **Hölder domain of order** 0, and denote this class of domains by $H(0)$, if there exists constants C_1 and C_2 such that for all $x \in D$,

$$\rho_D(x) \le C_1 \log\left(\frac{1}{d_D(x)}\right) + C_2. \tag{1.3}$$

For any $\alpha > 0$, we shall say that D is a **uniformly Hölder domain of order** α, and denote this class by $UH(\alpha)$, if there exists constants C_1 and C_2 such that

$$\rho_D(x) \le C_1 \frac{1}{d_D(x)^\alpha} + C_2 \tag{1.4}$$

and in addition, if there exists a constant C_3 such that for all $Q \in \partial D$ and all $r > 0$,

$$\mathrm{Cap}(B(Q,r) \cap D^C) \ge C_3 r^{n-2}. \tag{1.5}$$

The class of domains $H(0)$ has been studied by several authors in connection with the Poincaré inequality. In [20], W. Smith and D. Stegenga proved that $\rho_D \in L^p(D)$ for any $0 < p < \infty$. Using the equivalent formulation for ρ_D in terms of the Whitney distance, (see Bañuelos [2], [3]), it follows by the Harnack inequality that $D_m \subseteq \{x \in D : \rho_D(x) > C|m|\}$ where C is a universal constant. Thus if $D \subset H(0)$, we have, by the Smith-Stegenga result, that $\sum_{m=-\infty}^{\infty} (vol(D_m))^{2/n} < C(D) < \infty$. Therefore,

$$\sup_{\substack{x\in D\\ h\in H^+(D)}} \left(\frac{1}{h(x)} \int_D G_D(x,y)h(y)dy\right) < \infty$$

for any $D \in H(0)$ in any dimension.

The class of domain $H(0)$ includes, as an easy exercise shows once the reader has looked up the relevant definitions, Lipschitz, NTA, BMO-extension (uniform), and John domains, (see Bañuelos [3] for definitions). We should mention, however, that if the domain is Lipschitz the (difficult) result of Smith and Stegenga is not needed. Indeed, $\{x \in D : \rho(x) > C|m|\} \subseteq \{x \in D : d_D(x) < C_1 e^{-C_2|m|}\}$ and a simple exercise, (by hand or using the co-area formula), shows that for a Lipschitz domain $vol\{x \in D : d(x) < t_0\} \le Ct_0$, as soon as t_0 is small enough. This implies the result.

The first extension of the lifetime estimate to several dimensions was done by Cranston [8] for Lipschitz domains. It was subsequently extended by Bañuelos [2] to uniform domains and to elliptic diffusions, both in divergence and non-divergence form. Several other people have obtain (0.3) for Lipschitz domains including Cranston, Fabes and Zhao [9], Kenig and Pipher [16], and N. Falkner [15].

The class of domains $UH(\alpha)$ was introduced in Bañuelos [3]. Because of the uniform capacity condition (1.5) the estimate (1.2) may be used to obtain the lifetime estimate. More precisely, the Harnack inequality together with (1.4) immediately imply that $D_m = \{x \in D : \rho_0(x) > C|m|\} \subseteq \{x \in D : d_D(x) \leq \frac{C_1}{|m|^{1/\alpha}}\} = \widetilde{D}_m$. By (1.2) we have,

$$\sup_{x\in\tilde{D}_m} \int_{\tilde{D}_m} G_{\tilde{D}_m}(x,y)dy \leq \frac{C_2}{|m|^{2/\alpha}}.$$

This together with the Basic Estimate imply that if $0 < \alpha < 2$, then the Cranston-McConnell estimate holds for D. The sharpness of this result follows from the examples in Theorem 9.6 of Davies and Simon [11] together with the results of Xu [21] concerning the lifetime between two Lipschitz curves. The only thing one has to do is compute the quasi-hyperbolic metric for the Davies-Simon domains.

In [6], R. Bass and K. Burdzy introduced a class of domains which they called "twisted Hölder of order α" and proved the lifetime estimate for these domains for $\frac{1}{3} < \alpha \leq 1$. (The proof for uniformly Hölder domains extends easily to their class of domains.) They also obtained the lifetime estimate for domains whose boundary is given locally by the graph of an L^p-function for $p > n-1$. The case of bounded functions was also done independently by B. Davis [13]. We refer the reader to these paper for more details on their results.

Proof of Basic Estimate. Assume $x = 0$ and $h(0) = 1$. Define $C_m = \{x \in D : 2^m < h(x) < 2^{m+1}\}$, $m = \pm 1, \pm 2, \ldots$. Then $C_m \subset D_m$. Define the functions

$$U_m(x) = \int_{C_m} G_D(x,y)dy$$

and

$$u_m(x) = \int_{C_m} G_{D_m}(x,y)dy.$$

The function U_m is harmonic in $D \setminus C_m$ and zero on ∂D. Thus

$$U_m(0) \leq W_0^{D\setminus C_m}(\partial C_m) \sup_{x\in\partial C_m} U_m(x)$$

where $W_0^{D\setminus C_m}(\partial C_m)$ denotes the harmonic measure of ∂C_m relative to the domain $D \setminus C_m$ evaluated at 0. Since $h(0) = 1$ and $h(x) \geq 2^m$ for $x \in \partial C_m$ we have $W_0^{D\setminus C_m}(\partial C_m) \leq 2^{-m}$ and

$$U_m(0) \leq 2^{-m} \max_{x\in\partial C_m} U_m(x). \tag{1.6}$$

Next, for $x_0 \in \partial C_m$ write

$$U_m(x_0) = u_m(x_0) + (U_m(x_0) - u_m(x_0)). \tag{1.7}$$

$U_m - u_m$ is nonnegative and harmonic in D_m and u_m is zero on ∂D_m. Therefore, (1.8)

$$U_m(x_0)-u_m(x_0) \le \left(\max_{x\in A_m} U_m(x_0)\right) W_{x_0}^{D_m}(A_m)+\left(\max_{x\in\partial D_m\setminus A_m} U_m(x)\right) W_{x_0}^{D_m}(\partial D_m\setminus A_m)$$

where $A_m = \{x \in \partial D_m : h(x) = 2^{m+2}\}$ and $\partial D_m \setminus A_m = \{x \in \partial D_m : h(x) = 2^{m-1}\{\cup\{x \in \partial D_m : x \in \partial D\}$. By the maximum principle for the domain $D \setminus C_m$, we have

$$\max_{x\in A_m} U_m(x) \le \max_{x\in\partial C_m} U_m(x). \tag{1.9}$$

Set $M = \max_{x\in\partial C_m} U_m(x)$ and define $V_m : D\setminus C_m \to \mathbb{R}$ by $V_m(x) = U_m(x) - \frac{M}{2^m}h(x)$. V_m is negative on $\partial(D\setminus C_m)$ and hence negative on $D\setminus C_m$. Thus evaluating V_m at points x with $h(x) = 2^{m-1}$ and using the fact that $U_m = 0$ on ∂D we have

$$\max_{x\in\partial D_m\setminus A_m} U_m(x) \le \frac{1}{2}\max_{x\in\partial C_m} U_m(x). \tag{1.10}$$

Let $\alpha = W_{x_0}^{D_m}(A_m)$. Then $\alpha \le \frac{1}{2}$ for otherwise $h(x_0) \ge 2^{m+2}\alpha > \frac{1}{2}\cdot 2^{m+2} = 2^{m+1}$ which is false. From this, (1.8), (1.9) and (1.10) we have for any $x_0 \in \partial C_m$,

$$U_m(x_0) - u_m(x_0) \le (\frac{1}{2}(1-\alpha)+\alpha)\max_{x\in\partial C_m} U_m(x) \le \frac{3}{4}\max_{x\in\partial C_m} U_m(x).$$

It follows from (1.7) that

$$\max_{x\in\partial C_m} U_m(x) \le 4\max_{x\in\partial C_m} u_m(x) \le 4\sup_{x\in D_m}\int_{D_m} G_{D_m}(x,y)dy. \tag{1.11}$$

The Basic Estimate follows from (1.6) and (1.11):

$$\begin{aligned}\int_D G_D(0,y)h(y)dy &\le \sum_{m=-\infty}^{\infty} 2^{m+1}U_m(0) \le 2\cdot 4\sum_{m=-\infty}^{\infty}\max_{x\in\partial C_m} u_m(x)\\ &\le 8\sum_{m=-\infty}^{\infty}\sup_{x\in D_m}\int_{D_m} G_{D_m}(x,y)dy.\end{aligned}$$

In the case of unconditioned Brownian motion, ($h = 1$), it follows trivially from (1.2) that there are domains of infinite volume where the expected lifetime is finite. This is not as clear for conditioned Brownian motion. J. Xu [21] was the first to give an example of a domain of infinite area where the Cranston-McConnell estimate holds. His domain is simply connected and his result can be stated analytically as follows: There exists a univalent function in the unit disc whose Littlewood-Paley g_*-function is in $L^\infty(T)$ but whose Dirichlet norm is infinite. His proof does not use the Basic Estimate. Instead, he uses an estimate of B. Davis [12] on the probability of hitting whitney cubes.

2. Heat Kernel Estimates: Intrinsic Ultracontractivity

It follows from the fact that $P_t^D(x,y) \le \frac{1}{(2\pi t)^{n/2}}$ for all x, $y \in D$ and $t > 0$ that the Dirichlet semigroup is ultracontractive. That is, if $T_t f(x) = \int_D P_t(x,y)f(y)dy$ then $\|T_t f\|_{L^\infty} \le C_t \|f\|_2$ for all $t > 0$. Here the norms are with respect to the Lebesgue measure in D. In [11], E. B. Davies and B. Simon introduced a stronger notion of ultracontractivity for symmetric Markovian semigroup. We now define this for our Dirichlet semigroup. Let φ_0 be the lowest eigenfunction for D normalized by $\|\varphi_0\|_2 = 1$. Let λ_0 be the corresponding eigenvalue. Define the new semigroup on $L^2(\varphi_0^2)$ by

$$\widetilde{T}_t f(x) = \int_D \frac{e^{\lambda_0 t} P_t^D(x,y)}{\varphi_0(x)\varphi_0(y)} f(y)\varphi_0^2(y)dy$$

for $f \in L^2(\varphi_0^2)$. A simple exercise shows that this is indeed a symmetric Markovian semigroup. It is the semigroup of Brownian motion conditioned to stay forever in D. We will say that D is **intrinsic ultracontractive**, (IU), if the new semigroup is ultracontractive. That is, D is (IU) if

$$\|\widetilde{T}_t f\|_\infty \le \widetilde{C}_t \|f\|_{L^2(\varphi_0^2)} \tag{2.1}$$

for all $t > 0$ and all $f \in L^2(\varphi_0^2)$. The constant $\widetilde{C}_t$ depends only on t and the dimension. Using the definition of $\widetilde{T}_t$ and applying (2.1) to f/φ_0 for any $f \in L^2(dx)$ we can rewrite (IU) in terms of the original semigroup T_t. Namely, D is (IU) iff

$$|T_t f(x)| \le \varphi_0(x) e^{-\lambda_0 t} \widetilde{C}_t \|f\|_{L^2(dx)} \tag{2.2}$$

for all $x \in D$, all $t > 0$, and all $f \in L^2(dx)$.

Remark. There are several other equivalent definitions of (IU); see Davies and Simon [11], Theorem 3.2, and Davis [13].

Proposition 1. *Suppose D is (IU). Then*

(a). For any $\epsilon > 0$ there exists $t(\epsilon)$ depending only on ϵ such that for all $t > t(\epsilon)$ and all x, $y \in D$,

$$(1-\epsilon)e^{-\lambda_0 t}\varphi_0(x)\varphi_0(y) \le P_t^D(x,y) \le (1+\epsilon)e^{-\lambda_0 t}\varphi_0(x)\varphi_0(y), \tag{2.3}$$

(b).

$$\sup_{\substack{x\in D\\ h\in H^+(D)}} \left(\frac{\varphi_0(x)}{h(x)} \int_D \varphi_0(y)h(y)dy\right) < \infty \tag{2.4}$$

and for any $h \in H^+(D)$,

$$\lim_{t\to\infty} e^{\lambda_0 t} P_x^h\{\tau_D > t\} = \frac{\varphi_0(x)}{h(x)} \int_D \varphi_0(y)h(y)dy. \tag{2.5}$$

In particular,

$$\lim_{t\to\infty}\frac{1}{t}\log P_x^h\{\tau_D > t\} = -\lambda_0. \tag{2.6}$$

Proof. Let φ_n be the n-th eigenfunction with eigenvalue λ_n, $n = 1, 2, \ldots$, and normalized by $\|\varphi_n\|_2 = 1$. Applying (2.2) with $f = \varphi_n$ we have $|\varphi_n(x)| \le \widetilde{C}_t e^{(\lambda_n-\lambda_0)t}\varphi_0(x)$ and it follows that

$$\begin{cases} |\varphi_n(x)| \le C_n\varphi_0(x), \\ C_n = \inf\{\widetilde{C}_t e^{(\lambda_n-\lambda_0)t} : 0 < t < \infty\}. \end{cases} \tag{2.7}$$

From the eigenfunction expansion of the heat kernel we have

$$\frac{e^{\lambda_0 t}P_t^D(x,y)}{\varphi_0(x)\varphi_0(y)} = 1 + \sum_{n=1}^{\infty} e^{-(\lambda_n-\lambda_0)t}\frac{\varphi_n(x)\varphi_n(y)}{\varphi_0(x)\varphi_0(y)}$$

and (2.3) follows from this and the estimate (2.7).

From (0.1),

$$\frac{1}{h(x)}\int_D P_t(x,y)h(y)dy = P_x^h\{\tau_D > t\} \le 1$$

and (2.4) follows from (2.3). Also, (2.3) gives

$$\lim_{t\to\infty}\frac{e^{\lambda_0 t}P_t^D(x,y)}{\varphi_0(x)\varphi_0(y)} = 1$$

uniformly in x, $y \in D$. This immediately implies (2.5) and (2.6).

The above proposition shows that whenever D is (IU) not only does the Cranston-McConnell estimate holds but we also have a very precise estimate on the tail of the distribution of the lifetime. R. D. De Blassie [14] was the first to investigate the possibility of having (2.6) for conditioned Brownian motion and proved the result for Lipschitz domains with sufficiently small Lipschitz constant. C. Kenig and J. Pipher [16] extended De Blassie's result to all Lipschitz and NTA domains and to reversible and some non-reversible diffusions; R. Pinsky [19] also has some extensions to some nonreversible diffusions. In [11], Davies and Simon proved intrinsic ultracontractivity for Lipschitz domains and gave an example of a simply connected domain of finite area which is not (IU). Since Theorem A holds for any planar domain of finite area we have that (IU) is strictly stronger than the lifetime estimate. The connection between (IU) and conditioned Brownian motion seem to have been first explicitly noticed in Bañuelos and Davis [4]. In this paper, which was inspired by the results of [14] and [16], it is proved that even though an arbitrary planar domain of finite area may not be (IU) it is what one may call "one half (IU)" in the following sense: Fix $y \in D$. then

$$\lim_{t\to\infty}\frac{e^{\lambda_0 t}P_t^D(x,y)}{\varphi_0(x)\varphi_0(y)} = 1 \quad \text{uniformly in } x \in D. \tag{2.8}$$

This is enough to prove (2.4), (2.5) and (2.6). The argument in [4] shows in fact that whenever, (notation as in the Basic Estimate),

$$(2.9) \qquad \sum_{m=-\infty}^{\infty} \sup_{x\in D_m} \int_D G_{D_m}(x,y)dy \leq C(D,n) < \infty$$

$C(D,n)$ depending only on D and n, we have (2.8). In §1 we showed that if $D \in H(0)$ *or* $D \in UH(\alpha)$, $0 < \alpha < 2$, then (2.9) holds. However, these domains are actually (IU). The following result is proved in Bañuelos [3].

THEOREM 1. *Let $V \in K_n$, the Kato class of potentials in $\mathbf{R}^n$.*

(a) *Suppose $D \subseteq \mathbf{R}^n$, $n \geq 2$, has the property that $\rho_D \in L^p(D)$ for some $p > \frac{n}{2}$, (this is the case in particular if D is Hölder of order 0), then the semigroup of $H = -\frac{1}{2}\Delta_0 + V$, where Δ_0 is the Laplacian with Dirichlet boundary conditions in D, is (IU).*

(b) *Suppose $D \in UH(\alpha)$ for $0 < \alpha < 2$. Then H is (IU).*

(c) *For every $\alpha \geq 2$ there exists a $D \in UH(\alpha)$ for which Δ_0 is not (IU).*

From the special case of part (a) when D is an NTA domain and the conditional gauge theorem of Cranston, Fabes and Zhao [9] we obtain the following corollary which answers a question raised by E. B. Davies and B. Simon [11].

COROLLARY. *Let D be a Lipschitz (or NTA) domain. Let $V \in K_n$ and let $P_t^V(x,y)$ be the heat kernel for $-\frac{1}{2}\Delta_0 + V$. There are constants a_t and b_t depending only on t such that*

$$a_t P_t^D(x,y) \leq P_t^V(x,y) \leq b_t P_t^D(x,y)$$

for all $t > 0$, x, $y \in D$.

Remark. If $D \subset \mathbf{R}^n$ is a Hölder domain of order zero, then $|\partial D| = 0$ where $|\cdot|$ is the n-dimensional Lebesgue measure, (see [20]). In contrast, if $p = n-1$ there exists a D with $\rho_D \in L^p(D)$ and $|\partial D| = \infty$. Thus boundary smoothness is not the determining factor for (IU). What is more relevant is the rate of growth of the quasi-hyperbolic metric ρ_D, (for the lifetime estimate it was already noticed in Bañuelos [2], p.321, that $\rho_D \in$ weak $L^p(D)$, $p > \frac{n}{2}$ will suffice).

There are other recent results on I.U. In [13], B. Davis proved I.U. for $-\frac{1}{2}\Delta_0$ in domains above the graph of a bounded function, (no other assumptions), and gave examples of domains of infinite volume which are (IU), also answering a question of Davies and Simon [11]. Subsequently R. Bass and K. Burdzy [6] extended his first result to domains whose boundary is given locally by the graph of an L^p-function for $p > n-1$. In this paper Bass and Burdzy also give a new proof of part (b) of our Theorem 1 when $V \equiv 0$ and observed that there are uniformly Hölder domains of infinite volume.

The proof of Theorem 1 is via Logarithmic Sobolev inequalities. It is well known nowadays that for any symmetric Markovian semigroup T_t on $L^2(X,dx)$,

(X a locally compact second countable Hausdorff space with a Borel measure dx) with Dirichlet form Q, the estimate

$$\int_X u^2 \log u \, dx \le \epsilon Q(u) + \beta(\epsilon)\|u\|_2^2 + \|u\|_2^2 \log \|u\|_2^2 \tag{2.10}$$

for all $\epsilon > 0$, $0 \le u \in$ Dom$(Q) \cap L^1 \cap L^\infty$, implies

$$\|T_t f\|_\infty \le C_t \|f\|_2$$

provided $\beta(\epsilon)$ is monotonically decreasing and we have some information on the rate, ($\beta(\epsilon) = \epsilon^{-\gamma}$ for $\gamma > 0$ or $\beta(\epsilon) = \log\left(\frac{1}{\epsilon}\right)$ will do). Let us briefly outline the proof of (2.10) for our particular semigroup when $D \in UH(\alpha)$, $0 < \alpha < 2$, and $V \equiv 0$, (case $V = 0$ in part (b) of Theorem 1). After some work (2.10) can be reduced to proving

$$\int_D u^2 \log\left(\frac{u}{\varphi_0}\right) dx \le \epsilon \int_D |\nabla u|^2 dx + \beta(\epsilon)\|u\|_2^2 + \|u\|_2^2 \log \|u\|_2^2, \tag{2.11}$$

for all $u \in C_0^\infty(D)$, $u \ge 0$, where $C_0^\infty(D)$ are the C^∞ function of compact support in D. Once (2.11) is proved we apply it to $u\varphi_0 \in C_0^\infty(D)$ and obtain the inequality (2.10) for our semigroup $\widetilde{T}_t$ on $L^2(\varphi_0^2)$, (one first has to compute the Dirichlet form of $\widetilde{T}_t$).

To prove (2.11) we prove

$$\text{(2.12)} \quad \begin{cases} \text{(i)} & \int_D u^2 \log u \, dx \le \epsilon \int_D |\nabla u|^2 dx + \beta_1(\epsilon)\|u\|_2^2 + \|u\|_2^2 \log \|u\|_2^2 \\ \text{and} & \\ \text{(ii)} & \int_D u^2 \log\left(\frac{1}{\varphi_0}\right) dx \le \epsilon \int_D |\nabla u|^2 dx + \beta_2(\epsilon)\|u\|_2^2 + \|u\|_2^2 \log \|u\|_2^2. \end{cases}$$

For all $u \in C_0^\infty(D)$, $u \ge 0$. To prove (2.12) (i) assume $\|u\|_2 = 1$. Let $p = \frac{2n}{n-2}$. Then by the Jensen's and Sobolev's inequalities,

$$\begin{aligned} \int_D u^2 \log u \, dx &\le \frac{1}{p-2} \log\left(\int_D |u|^p dx\right) \le \frac{1}{p-2} \log\left(\int_D |\nabla u|^2 dx\right)^{p/2} \\ &= \frac{p}{2(p-2)} \log \int_D |\nabla u|^2 dx \\ &\le \frac{p}{(p-2)} \epsilon \int_D |\nabla u|^2 dx + \frac{p}{2(p-2)} \log\left(\frac{1}{\epsilon}\right) + C \end{aligned}$$

where in the last inequality we have used $\log x \le \epsilon x^2 + \frac{1}{2}\log\left(\frac{1}{\epsilon}\right) + C$. (2.12) (i) follows from this with $\beta_1(\epsilon) = c_1 \log\left(\frac{1}{\epsilon}\right) + C_2$, C_1 and C_2 constants independent of ϵ.

Recall that $x_0 \in D$ is fixed and $\rho_D(x) = \rho_D(x_0, x)$. Assume $\varphi_0(x_0) = 1$. By the Harnack inequality $\varphi_0(x) \ge C_1 e^{-C_2 \rho_D(x)}$. Thus for $D \in UH(\alpha)$ we have

(2.13)

$$\begin{aligned} \int_D u^2 \log\left(\frac{1}{\varphi_0}\right) dx &\le C_1 \int_D \frac{u^2}{(d_D(x))^\alpha} dx + C_2 \int_D u^2 dx \\ &\le C_1 \epsilon \int_D \frac{u^2}{(d_D(x))^2} dx + C_1 \epsilon^{-\frac{\alpha}{2-\alpha}} \int_D u^2 dx + C_2 \int_D u^2 dx. \end{aligned}$$

where we have also used the elementary inequality $ab \le a^p + b^{p'}$ with $p = \frac{2}{\alpha}$. Because of the uniform regularity of our domains, (1.5), we have the following inequality, (known as Hardy's inequality), due to A. Ancona [1]: For all $u \in C_0^\infty(D)$,

$$\int_D \frac{u^2}{(d_D(x))^2}dx \le C\int_D |\nabla u|^2 dx$$

and (2.12) (i) follows from this and (2.13). This gives (2.11) with $\beta(\epsilon) = C_1 \log\left(\frac{1}{\epsilon}\right) + C_2\epsilon^{-\frac{\alpha}{2-\alpha}} + C_3$ which is good enough to show $\|\widetilde{T}_t f\|_\infty \le C_t \|f\|_{L^2(\varphi_0^2)}$ with $C_t = t^{-\nu}e^{t^{-\gamma}}$, with ν and γ depending on n and α. Part (c) of the theorem is already contained in Theorem 9.6 of Davies and Simon [11]. Again, the only thing one has to do is compute the quasi-hyperbolic metric for those domains. We refer the reader to Bañuelos [3] for details and for the other cases of Theorem 1 as well as the proof of the Corollary.

Remark. The above argument can be modified to relax the growth condition in the quasi-hyperbolic metric and still get (IU). More precisely, suppose D still satisfies the uniform capacity condition but that $\rho_D(x) \le \varphi(d_D(x)) + C$ where $\varphi(t) = \frac{\eta(t)}{t^2}$ and C is a constant. If $\eta(t) = \frac{1}{(\log \frac{1}{t})^\beta}$ for $1 < \beta < \infty$, then again we have (IU) and in general we can give an integral test on η which guarantees (IU). However, if we only assume $\eta(t) \to 0$ as $t \to 0$ we can only prove what Davies and Simon [11] call "intrinsic super contractivity".

Since the above proof is via Dirichlet form techniques we can obtain the same results if the Laplacian is replaced by a second order divergence form operator with bounded measurable coefficients.

Intrinsic ultracontractivity is a very strong mixing condition. This has some other interesting consequences. The following is a functional central limit Theorem which was noticed in conversations with B. Davis. The proof is essentially trivial; it follows immediately from Theorem 4.1 in W. Philipp and W. Stout [18].

PROPOSITION 2. *Suppose D is (IU). Let $d\mu = \varphi_0^2 dx$ and let $g \in L^\infty(D)$ with $\int_D g\,d\mu = 0$. Define*

$$\sigma_g^2 = 2\int_D\int_D \widetilde{G}(x,y)g(x)g(y)d\mu(x)d\mu(y)$$

where

$$\widetilde{G}(x,y) = \int_0^\infty \left(\frac{e^{\lambda_0 t}P_t^D(x,y)}{\varphi_0(x)\varphi_0(y)} - 1\right)dt.$$

Let X_t be Brownian motion conditioned to stay forever in D. Let $L_t(g) = \int_0^t g(X_s)ds$ and set $S_n(t) = \frac{L_{nt}(g)}{\sigma_g\sqrt{n}}$, $0 \le t \le 1$. Then S_n converges weakly to Brownian motion.

Similarly we have functional laws of the iterated logarithm of the Kolmogorov and Chung type.

REFERENCES

[1] A. Ancona, *On strong barriers and an inequality of Hardy for domain in* $\mathbf{R}^n$, J. London Math. Soc., 34 (1986), pp. 274–290.

[2] R. Bañuelos, *On an estimate of Cranston and McConnell for elliptic diffusions in uniform domains*, Prob. Th. Rel. Fields, 76 (1987), pp. 311–323.

[3] ———, *Intrinsic ultracontractivity and eigenfunction estimates for Schrödinger operators*, to appear, J. Func. Anal.

[4] R. Bañuelos and B. Davis, *Heat Kernel, eigenfunctions, and conditioned Brownian motion in planar domains*, J. Func. Anal., 84 (1989), pp. 188–200.

[5] R. Bañuelos and B. Øksendal, *Exit times for elliptic diffusions and BMO*, Proc. Edinburgh Math. Soc., 30 (1987), pp. 273–289.

[6] R. Bass and K. Burdzy, *Lifetime of conditional diffusions*, (preprint).

[7] K. L. Chung, *The lifetime of conditional Brownian motion in the plane*, Ann. Inst. Henri Poincaré, 20 (1984), pp. 349–351.

[8] M. Cranston, *Lifetime of conditional Brownian motion in Lipschitz domains*, Z. Wahrscheinlichkeitstheorie Verw. Geb., 70 (1985), pp. 335–340.

[9] M. Cranston, G. Fabes, and Zhao, *Conditional gauge and potential theory for the Schrödinger operator*, Trans. Amer. Math. Soc., 20 (1988), pp. 171–194.

[10] M. Cranston and T. McConnell, *The lifetime of conditioned Brownian motion*, Z. Wahrscheinlichkeitstheorie Verw. Geb., 65 (1983), pp. 1–11.

[11] E. B. Davies and B. Simon, *Ultracontractivity and the heat kernel for Schrödinger operators and Dirichlet Laplacians*, J. Func. Anal., 59 (1984), pp. 335–395.

[12] B. Davis, *Conditioned Brownian motion in planar domains*, Duke Math. J., 59 (1988), pp. 397–421.

[13] ———, *Intrinsic ultracontractivity for the Dirichlet Laplacian*, to appear, J. Func. Anal.

[14] R. D. De Blassie, *The lifetime of conditioned Brownian motion in certain Lipschitz domains*, Prob. Th. Rel. Fields, 75 (1987), pp. 55–65.

[15] N. Falkner, *Conditioned Brownian motion in rapidly exhaustible domains*, Ann. Probab., 17 (1987), pp. 1501–1514.

[16] C. Kenig and J. Pipher, *The h-paths distribution of the lifetime of conditioned Brownian motion for non-smooth domains*, Prob. Th. Rel. Fields, 82 (1989), pp. 615–623.

[17] C. Muller, *A characterization of* BMO *and* BMO_ρ, Studia Math., 72 (1982), pp. 47–57.

[18] W. Philipp and W. Stout, *Almost sure invariance principles for partial sums of weakly dependent random variables*, Memoirs of Amer. Math. Soc., 161 (1975).

[19] R. G. Pinsky, *The lifetimes of conditioned diffusion processes*, Ann. Inst. Henri Poincaré, 26 (1990), pp. 87–99.

[20] W. Smith and D. Stegenga, *Holder domains and Poincaré domains*, Trans. Amer. Math. Soc., 319 (1990), pp. 67–100.

[21] J. Xu, *The lifetime of conditioned Brownian motion in planar domains of infinite area*, (Preprint).

ON THE POISSON KERNEL FOR NONDIVERGENCE ELLIPTIC EQUATIONS WITH CONTINUOUS COEFFICIENTS

TOMEU BARCELO*

Let us consider the class of elliptic equations

$$Lu(x) = \sum_{i,j=1}^{n} a_{ij}(x)u_{x_i x_j} + \sum_{i=1}^{n} b_i(x)u_{x_i} + c(x)u \tag{1}$$

defined on a bounded C^2 domain D in $\mathbf{R}^n$.

We assume that the coefficients $a_{ij}(x)$ are continuous on $\overline{D}$, symmetric

$$a_{ij}(x) = a_{ji}(x) \qquad i,j = 1,2,\ldots,n$$

and that they satisfy the ellipticity condition

$$\frac{1}{\lambda}|\xi|^2 \leq \sum_{i,j=1}^{n} a_{ij}(x)\xi_i\xi_j \leq \lambda|\xi|^2 \tag{2}$$

for some $\lambda > 1$, every $x \in D$ and every vector $(\xi_1, \xi_2, \ldots, \xi_n) \in \mathbf{R}^n$.

If $\delta(x) = \operatorname{dist}(x, \partial D)$ is the distance to the boundary and if $\eta(s)$ is an increasing function such that $\eta(0) = 0$ and $\lim_{s\to 0}\eta(s) = 0$ we assume also that the coefficients $b_i(x)$ are measurable functions bounded as

$$|b_i(x)| \leq \frac{\eta(\delta(x))}{\delta(x)} \tag{3}$$

and $c(x) \leq 0$ and

$$|c(x)| \leq \frac{\eta(\delta(x))}{\delta(x)^2} \tag{4}$$

We will denote by $\mathcal{L}_\lambda$ this class of equations.

Our aim is to do some potential theory for $\mathcal{L}_\lambda$, in particular to solve the Dirichlet problem and study its harmonic measure.

Given a continuous function φ on ∂D the classical Dirichlet problem is to find a continuous solution u to $Lu = 0$ in D such that $u = \varphi$ on ∂D. Due to the nature of the lower order coefficients is not at first clear that we can solve the Dirichlet problem for $\mathcal{L}_\lambda$ as the following example shows:

Example. The upper half disc

$$\Omega = \{(x,r) \in \mathbf{R}^2_+ : x^2 + r^2 \leq 1\}$$

*Partially supported by Dirección General de Investigación Científica y Técnica (DGICYT).

comes from the unit ball

$$B_1 = \{(x, y, z) \in \mathbf{R}^3 \mid x^2 + y^2 + z^2 \leq 1\}$$

under the transformation

$$(x, y, z) \to (x, \sqrt{y^2 + z^2}) = (x, r).$$

Let us assume v satisfies

$$\Delta v = v_{xx} + v_{yy} + v_{zz} = 0$$

in B_1, then if

$$u(x, r) = v(x, y, z)$$

is well defined, $u(x, r)$ satisfies the equation

$$u_{xx} + u_{rr} + \frac{1}{r} u_r = 0 \tag{5}$$

in Ω.

Now observe that the piece of the boundary of Ω $\gamma_1 = \partial\Omega \cap \mathbf{R}^2_+$ comes from the unit sphere ∂B_1, while the other part $\gamma_2 = \partial\Omega \cap \{r = 0\}$ comes from the line $B_1 \cap \{y = 0, z = 0\}$ which lies inside B_1. Since the values of v inside B_1 are determined by the values at the boundary ∂B_1, the corresponding values of u at γ_2 are determined by the values of u at γ_1 and therefore the Dirichlet problem is not solvable for (5) in Ω.

Equation (5) is a particular case of the well known GASP (generalized axially symmetric potential theory) or Weinstein equations

$$Lu = \sum_{i,j=1}^{n} a_{ij}(x) u_{x_i x_j} + \frac{1}{x_n} \sum_{i=1}^{n} b_i(x)$$

which are defined on the upper half-space $\mathbf{R}^+_n = \{(x_1, x_2, \ldots, x_n) \in \mathbf{R}^n | x_n > 0\}$.

Although Weinstein's equations do not belong to our class $\mathcal{L}_\lambda$, the methods developed for $\mathcal{L}_\lambda$ can be applied to them as well. For more references about GASP equations see [4], [38].

Let $W^{2,p}_{\text{loc}}(D)$ denote the Sobolev space $W^{2,p}_{\text{loc}}(D) = \{u : D^\alpha u \in L^p_{\text{loc}}(D), |\alpha| \leq 2\}$.

By a solution u to $Lu = 0$ in D we mean a continuous $u \in C(\overline{D}) \cap W^{2,p}_{\text{loc}}(D)$ for some $p > 1$ such that $Lu(x) = 0$ almost everywhere.

In Section 1 we solve the Dirichlet problem for $\mathcal{L}_\lambda$

THEOREM A (SOLUTION TO THE DIRICHLET PROBLEM). *Given any $\varphi \in C(\partial D)$ there exists a unique $u \in C(\overline{D})$ such that $u \in W^{2,p}_{\text{loc}}(D)$ for all $p \in (1, \infty)$ and $u = \varphi$ at ∂D.*

The proof of Theorem A comes from two main properties for $\mathcal{L}_\lambda$, which are basic ingredients for the posterior development of its potential theory: the weak maximum principle and the existence of appropriate barriers.

Once the Dirichlet problem is solved, for a fixed $x \in D$ we can consider the linear functional

$$T_x : C(\partial D) \to \mathbf{R}$$

given by

$$T_x(\varphi) = u(x)$$

where u is the solution to $Lu = 0$ in D with boundary values φ.

By the maximum principle T_x is positive and bounded and therefore by the Riesz representation theorem there exists a positive measure w_L^x such that

$$u(x) = \int_{\partial D} \varphi(Q) dw_L^x(Q) \tag{6}$$

w_L^x is the harmonic measure associated to L and D.

The harmonic measure w_L^x depends on x, but by Harnack's inequality $w_L^{x_1}$ and $w_L^{x_2}$ are mutually absolutely continuous for two different $x_1, x_2 \in D$. Therefore we will fix $x_0 \in D$ and denote $w = w_L^{x_0}$.

Another important piece in the potential theory for $\mathcal{L}_\lambda$ is the Comparison Theorem, Theorem 2.6 in Section 2, which states that if two nonnegative solutions u, v to $Lu = 0$ vanish continuously at a portion of the boundary then they vanish at the same rate.

With the representation (6) given by the harmonic measure and the Comparison Theorem we can apply the techniques given by L. Caffarelli - E. Fabes - S. Mortola and S. Salsa in [11] to obtain a Fatou Theorem for $\mathcal{L}_\lambda$: If $u \geq 0$ is a solution to $Lu = 0$ in D then u has nontangential limits to the boundary almost everywhere with respect to ω. We omit the details.

Therefore a natural question comes in: Is ω absolutely continuously with respect to the surface Lebesgue measure $d\sigma$?

And since the Poisson kernel $k(x, Q)$ is defined as the density

$$d\omega_L^x(Q) = K(x, Q) d\sigma(Q)$$

the above question can be rephrased as: Does the Poisson kernel exists? The answer is negative since given any continuous increasing function $\eta(s)$ for $s \geq 0$ such that $\eta(0) = 0$, $s^{-1/4}\eta(s)$ decreasing for $s > 0$ and

$$\int_0^1 \frac{\eta(s)^2}{s} ds = +\infty$$

we can construct an elliptic operator in $\mathcal{L}_\lambda$ for some $\lambda > 1$ on the unit ball B_1 of the form

$$L = \sum a_{ij}(x) D_{x_j x_i} + \sum b_i(x) D_{x_i},$$

for which $\eta(s)$ measures the modulus of continuity of the coefficients $a_{ij}(x)$ at the boundary, that is

$$\eta(s) = \sup_{\substack{0\leq t\leq s \\ Q\in\partial D}} \{|a_{ij}(Q+tn_Q) - a_{ij}(Q)| : i,j = 1,2,\ldots,n\}$$

and such that its harmonic measure w is completely singular with respect to Lebesgue measure.

Here n_Q denotes the unit inward normal at $Q \in \partial D$. We sketch the proof in Section 3.

But the Poisson kernel exists if the modulus of continuity $\eta(s)$ satisfies the Dini condition

$$\int_0^{\epsilon_0} \frac{\eta(s)^2}{s} ds < \infty \tag{7}$$

for some $\epsilon_0 > 0$.

In fact, by assuming also

$$|c(x)| \leq \frac{\eta(\delta(x))}{\delta(x)^{2-\alpha}} \tag{8}$$

for some $\alpha > 0$ we can prove

THEOREM 1. *If L is an elliptic operator in $\mathcal{L}_\lambda$ as in (1) with condition (8), and if the modulus of continuity $\eta(s)$ of the coefficients $a_{ij}(x)$ at the boundary ∂D satisfies the Dini condition (7) then*

(a) *Harmonic measure ω and surface measure σ on ∂D are mutually absolutely continuous.*

(b) *If $K(Q) = \frac{d\omega}{d\sigma}(Q)$ denotes the Poisson kernel then $K \in L^2(d\sigma)$ and*

$$\left(\frac{1}{\sigma(\Delta)}\int_\Delta K^2 d\sigma\right)^{1/2} \leq c\frac{1}{\sigma(\Delta)}\int_\Delta K d\sigma$$

for every surface ball $\Delta \subset \partial D$.

To prove Theorem 1 we first obtain it for a small perturbation of a constant coefficients operator in the upper half-space $\mathbf{R}^{n+1}_+ = \mathbf{R}^n \times (0,\infty)$. This is done using the Fourier transform and treating the corresponding singular integrals with some multilinear Littlewood-Paley estimates. Due to the continuity of the coefficients L is a small perturbation of a constant coefficients operator at a neighborhood of every point $x \in D$. Therefore, locally, L is as the model case in $\mathbf{R}^{n+1}_+$ and the corresponding local estimates are then patched together using the Comparison Theorem. We sketch the details in Section 4.

As a consequence of Theorem 1 any nonnegative solution to $Lu = 0$ has nontangential limits for almost every $Q \in \partial D$ with respect to Lebesgue measure. Theorem 4 also allows to solve the Dirichlet problem with boundary values in L^2.

The Comparison Theorem, once proved for the Laplacian on a Lipschitz domain, see [27], has been extended by A. Ancona [2] to solutions of elliptic equations in nondivergence form with Hölder continuous coefficients. By L. Caffarelli, E. Fabes, S. Mortola, S. Salsa in [11] for solutions of elliptic equations in divergence form and bounded measurable coefficients and later, by P. Bauman in [8], for elliptic equations in divergence form with continuous coefficients plus a bounded first order term.

Our proof is modeled on those approaches, mainly from the one given by P. Bauman in [8]. She proved the Comparison Theorem on Lipschitz domains with constants depending on the continuity of the coefficients. Our approach gives constants depending only on ellipticity λ, η and the geometry of D.

A Comparison and Fatou Theorem for GASP equations with Hölder coefficients was given by A. Ancona in [3].

L. Caffarelli, E. Fabes, C. Kenig constructed in [12] a divergence form elliptic operator with continuous coefficients up to the boundary whose corresponding harmonic measure is completely singular with respect to Lebesgue measure. Another previous example, is given by Modica-Mortola-Salsa in [34] for a nondivergence form elliptic in the upper half-plane $\mathbb{R}^2_+$. Our example improves that one in the sense that we have a control on the moduli of continuity of the coefficients.

The positive result, Theorem 1, is modeled after the similar ones given for divergence form elliptic equations and for parabolic equations by E. Fabes, D. Jerison, C. Kenig in [21] and [22].

A different proof for divergence form with measurable coefficients was given by B. Dahlberg in [16] and some later refinements about the Dini condition [3] can be found in R. Fefferman, C. Kenig, J. Pipher [25].

Some interesting open questions still remain in this subject. A important one is the study of the Poisson kernel for nondivergence elliptic equations with just measurable coefficients, which is far of being understood since we don't even know how to solve its Dirichlet problem.

Acknowledgement. I am very grateful to the constant help and encouragement and human atmosphere received by E. Fabes during the whole development of this work.

1. The Dirichlet problem. We say that D is a C^2 domain if for any $Q \in \partial D$ there exists a ball with center at Q, a coordinate system $x' = (x_1, x_2, \dots, x_{n-1}), x_n$ with origin at Q, which can be obtained as a translation and rotation of the usual coordinate system, and a C^2 function $\varphi : \mathbb{R}^{n-1} \to \mathbb{R}$ such that $\varphi(Q) = 0$ and

$$D \cap B = \{x = (x', x_n) : x_n > \varphi(x')\} \cap B \tag{9}$$

in this coordinate system.

If D is bounded we may assume that the radius of the ball B is independent of $Q \in \partial D$ and that there exist two positive constants a, r_0 depending only on the geometry of D, such that for $Q_0 = (x_0', \varphi(x_0'))$ at the boundary and $r \leq r_0$. The cylinder

$$\Omega(r, Q_0) = \{(x', x_n) | |x' - x_1'| < r, |x_n - \varphi(x_0')| < ar\}$$

has the property

$$\Omega(r, Q_0) \cap D = \{(x', x_n) | |x' - x_0'| < r, |x_n - \varphi(x_0')| < ar, x_n > \varphi(x')\}$$

We denote by $A_r(Q_0)$ the point in $\Omega(r, Q_0)$ with coordinates $(x_0', \varphi(x_0') + \frac{ar}{2})$ and we call it the center of the cylinder $\Omega(r, Q_0)$.

As usual $B_r(Q)$ is the open ball with center Q and radius r and $\Delta(r, Q)$ is the surface ball

$$\Delta(r, Q) = \Omega(r, Q) \cap \partial D = B_r(Q) \cap \partial D$$

If L has a bounded drift term then the Schauder estimates give

$$\|u\|_{W^{2,p}(D')} \leq c\{\|Lu\|_{L^p(D)} + \|u\|_{L^p(D)}\}$$

for any subdomain D' such that $\overline{D}' \subset D$.

LEMMA 1.1 (WEAK MAXIMUM PRINCIPLE). *If D is a bounded C^2 domain in $\mathbf{R}^n$, $L \in \mathcal{L}_\lambda$, $u \in W^{2,p}_{\text{loc}}(D) \cap C(\overline{D})$ for some $p > n$, $Lu \geq 0$ in D and $u \leq 0$ on ∂D then $u \leq 0$ in D.*

Proof. Was given by J. Bony in [10]. □

Notice that if u is a solution, that is $Lu = 0$, then Lemma 1.1 gives

$$\|u\|_{L^\infty(D)} \leq \|u\|_{L^\infty(\partial D)}.$$

Let us consider now the tubular neighborhood of ∂D

$$T_\epsilon = \{x \in D | \delta(x) = \text{dist}(x, \partial D) \leq \epsilon\}$$

LEMMA 1.2 (EXISTENCE OF BARRIERS). *Given $L \in \mathcal{L}_\lambda$ there exists a function $w \in C^2(T_{\epsilon_0}) \cap C(\overline{T}_{\epsilon_0})$ and positive constants $\epsilon_0, c_1, c_2, c_3$ depending only on λ, η and domain D such that*

1) $c_1\delta(x)^\alpha \leq w(x) \leq c_2\delta(x)^\alpha$ *for all* $x \in T_{\epsilon_0}, 0 < \alpha < 1$
2) $Lw(x) \leq -c_3\delta(x)^{\alpha-2}$ *for all* $x \in T_{\epsilon_0}$

Proof. Consider a finite number of balls $B_1, B_2, \ldots B_N$ covering ∂D for which the boundary is given as the graph of C^2 functions $\varphi_1, \varphi_2, \ldots, \varphi_N$.

Let B stand for one of these balls and take φ such that

$$D \cap B = \{(x = (x', x_n) : x_n > \varphi(x')\} \cap B.$$

Notice that we can take B, by shrinking it if necessary, such that

$$(10)\qquad \begin{aligned} &|\nabla \varphi| \leq M \\ &|\varphi_{x_i x_j}| \leq M \qquad i,j = 1,2,\ldots,n-1 \end{aligned}$$

for some constant $M > 0$ on $\partial D \cap B$.

Notice also that

$$(11)\qquad c_4(x_n - \varphi(x')) \leq \delta(x) \leq c_5(x_n - \varphi(x'))$$

for all $x = (x', x_n) \in T_\epsilon \cap B$ and some $\epsilon > 0$.

Consider $w(x) = (x_n - \varphi(x'))^\alpha \quad 0 < \alpha < 1$ we get

$$\begin{aligned}
w_{x_i} &= -\alpha(x_n - \varphi(x'))^{\alpha-1}\varphi_{x_i} \quad i = 1,2,\ldots,n-1 \\
w_{x_n} &= \alpha(x_n - \varphi(x'))^{\alpha-1} \\
w_{x_i x_j} &= \alpha(\alpha-1)(x_n - \varphi(x'))^{\alpha-2}\varphi_{x_i}\varphi_{x_j} - \alpha(x_n - \varphi(x'))^{\alpha-1}\varphi_{x_i x_j} \\
&\qquad\qquad i,j = 1,2,\ldots,n-1 \\
w_{x_i x_n} &= -\alpha(\alpha-1)(x_n - \varphi(x'))^{\alpha-2}\varphi_{x_i} \quad i = 1,2,\ldots,n-1 \\
w_{x_n x_n} &= \alpha(\alpha-1)(x_n - \varphi(x'))^{\alpha-2}.
\end{aligned}$$

Therefore using the ellipticity condition (2) the bounds for the lower order terms (3) and (4) and the bounds in (10) we obtain

$$\begin{aligned}
Lw(x) \leq \alpha(x_n - \varphi(x'))^{\alpha-2}&\left[\frac{\alpha-1}{\lambda}(|\nabla\varphi|^2 + 1) + (n-1)\lambda M(x_n - \varphi(x'))\right] + \\
&+ \frac{1}{c_4}\alpha(M + n - 1)\eta(\delta(x))(x_n - \varphi(x'))^{\alpha-2} + \frac{1}{c_4^2}\eta(\delta(x))(x_n - \varphi(x')))^{\alpha-2}
\end{aligned}$$

which gives

$$Lw(x) \leq -c_3\delta(x)^{\alpha-2} \text{ for } x \in T_\epsilon \cap B, \ \epsilon \text{ small enough }.$$

This proves property 2 for $w(x)$ locally in $T_\epsilon \cap B$.

Property 1) comes from (11).

To find the global $w(x)$ let us assume that the balls $B_1, B_2, \ldots, B_N$ that cover ∂D have associated a partition of unity, that is, a family of functions $\varphi_1, \varphi_2, \ldots, \varphi_N$ such that $\varphi_i \in C_o^\infty(B_i)$ for $i = 1,2,\ldots,N$ and such that

$$\sum_{i=1}^{N} \varphi_i = 1.$$

If $w_i(x)$ is the function satisfying 1) and 2) in $B_i \cap T_{\epsilon_i}$ then the function

$$w(x) = \sum_{i=1}^{N} \varphi_i(x) w_i(x)$$

will satisfy 1) and 2) in T_{ϵ_0} where $\epsilon_0 = \min\{\epsilon_1, \epsilon_2, \ldots, \epsilon_N\}$ □

Comment. A similar construction could be carried out if D is a C^1 or a convex Lipschitz domain. We could have then barriers for such domains which would allow to solve the Dirichlet problem for them.

Proof of Theorem A. (solution to the Dirichlet problem). The weak maximum principle implies uniqueness, therefore we only need to show existence of the solution.

Let us assume for the moment that φ is smooth, say of class C^2, on ∂D. Construct a sequence $\{D_m\}$ of smooth domains such that

$$D_m \subset D_{m+1}$$

where

$$\frac{c_4}{m} \leq \operatorname{dist}(x, \partial D) \leq \frac{c_5}{m}$$

for every $x \in \partial D_m$ and for two fixed positive constants c_4, c_5.

In order to solve the problem

$$\text{(a)} \qquad \begin{cases} Lu & = 0 \quad \text{in } D \\ u & = \varphi \quad \text{at } \partial D \end{cases}$$

if $\bar{\varphi}$ is a C^2 extension of φ to D, by using $v = u - \bar{\varphi}$ it is enough to solve the homogeneous problem

$$\text{(b)} \qquad \begin{cases} Lu & = -L\bar{\varphi} \quad \text{in } D \\ v & = 0 \quad \text{at } \partial D \end{cases}$$

To solve (b) let v_m be the unique function in $C(\overline{D}_m) \cap W^{2,p}_{\text{loc}}(D_m)$ for all $1 < p < \infty$ given by the Schauder theory such that

$$Lv_m = -L\bar{\varphi} \text{ in } D_m \text{ and } v_m = 0 \text{ at } \partial D_m$$

Let

$$\bar{v}_m(x) = \begin{cases} v_m(x) & \text{if } x \in D_m \\ 0 & \text{if } x \in \overline{D} \backslash D_m \end{cases}.$$

We will prove that $\bar{v}_m$ converges uniformly on $\overline{D}$ and that the limit function is a solution to (b).

Let $w(x)$ be the barrier constructed in Lemma 1.2 such that

$$\text{(12)} \qquad Lw(x) \leq -c_3 \delta^{\alpha-2}(x)$$

for all $x \in \overline{D} \backslash D_{m_0}$, for some m_0.

If $m \geq m_0$ since $\bar{v}_m + \bar{\varphi}$ solves $L(\bar{v}_m + \bar{\varphi}) = 0$ in D_m with $\bar{v}_m = \varphi$ at ∂D_m, we have by the maximum principle

$$|\bar{v}_m(x) + \bar{\varphi}(x)| \leq \max_{x \in \partial D_m} \bar{\varphi}(x) \leq \max_{x \in D} \bar{\varphi}(x) = k_1$$

for every $x \in D_m$. Therefore

$$|\bar{v}_m(x)| \leq |\bar{v}_m(x) + \bar{\varphi}(x)| + |-\bar{\varphi}(x)| \leq 2k_1$$

for every $x \in D_m$.

Since on ∂D_{m_0} $w(x) \geq c_1\delta^\alpha(x) \geq c_1 \left(\frac{c_4}{m_0}\right)^\alpha = c_6$ we get

$$|\bar{v}_m(x)| \leq \frac{2k_1}{c_6} w(x)$$

for $x \in \partial D_{m_0}$. The same inequality holds also on ∂D_m since $\bar{v}_m(x) = 0$ there.

Taking now c large enough we can get

$$\pm\bar{v}_m(x) - cw(x) \leq 0$$

on $\partial(\overline{D}_{m_0} \backslash D_m)$ and

$$L(\pm\bar{v}_m(x) - cw(x)) = \mp L\bar{\varphi} - cLw(x) \geq 0$$

where we have used (12).

Maximum principle gives then

$$|\bar{v}_m(x)| \leq cw(x)$$

for all $x \in \overline{D}_{m_0} \backslash D_m$ and $m \geq m_0$.

This inequality holds also for $x \in \overline{D} \backslash D_m$ since $\bar{v}_m(x) = 0$ there. Since $w(x) \to 0$ as x approaches the boundary ∂D the same thing happens to $\bar{v}_m(x)$.

Therefore, given $\in > 0$, $|\bar{v}_m(x)| \leq \in$ for all $x \in \overline{D} \backslash D_{m_1}$ $m \geq m_0$, and then

$$|\bar{v}_p(x) - \bar{v}_q(x)| \leq 2 \in$$

for $x \in \overline{D} \backslash D_{m_1}$ and $p, q \geq m_0$. By the maximum principle the inequality extends to all $\overline{D}$ for $p, q \geq m_0$ and then $\bar{v}_m$ converges uniformly to some function $v \in C(\overline{D})$ such that $v = 0$ on ∂D. The L^p-Schauder interior estimates give now $u \in W^{2,p}_{\text{loc}}$ for all $p \in (1, \infty)$ and $Lv = -L\bar{\varphi}$ in D.

To remove the original assumption that φ is smooth on ∂D consider just φ continuous on ∂D and take a sequence of C^2 functions φ_m such that

$$\varphi_m \to \varphi$$

uniformly on ∂D. Let u_m be the solution to the problem

$$\begin{cases} Lu_m & = 0 \text{ in } D \\ u & = \varphi_m \text{ at } \partial D \end{cases}$$

By the maximum principle

$$\|u_p - u_q\|_\infty \leq \max_{x \in \partial D} |\varphi_p(x) - \varphi_q(x)|$$

Therefore u_m converges uniformly to a function $u \in C(\overline{D})$ such that $u = \varphi$ on ∂D. The interior Schauder estimates give now $u \in W^{2,p}_{\text{loc}}$ for all $p \in (1, \infty)$ and $Lu = 0$ in D. □

2. The comparison theorem

LEMMA 2.1 (HARNACK'S INEQUALITY). *If u is a nonnegative solution to $Lu = 0$ in $B_{4r}(P) \subset D$ then there exists a constant $c > 0$ depending only on λ, η and D such that*

$$\max_{B_r(P)} u \leq c \min_{B_r(P)} u$$

Proof. Harnack's inequality for a uniformly elliptic second order operator with bounded lower order terms appears in [30]. Here the lower order coefficients of L blow up as we approach the boundary, but using the dilation $v(y) = u(ry)$ we are in the preceeding case. The Lemma follows by this change of variable. □

LEMMA 2.2 (CARLESSON ESTIMATE). *Suppose $Q \in \partial D, 0 < 2r < r_0$ and u is a positive solution to $Lu = 0$ in $\Omega(2r, Q)$ which vanishes continuously on $\Delta(2r, Q)$. Then, there exists a constant $c > 0$ depending only on λ, η, D such that*

$$u(x) \leq cu(A_r(Q))$$

for every $x \in \Omega(r, Q) \cap D$.

Proof. Let us consider first a continuous function $h(x)$ on $\partial(\Omega(2r, Q) \cap D)$ such that $0 \leq h \leq 1, h = 0$ on $\Delta\left(\frac{3r}{2}, Q\right)$ and $h = 1$ on $\partial\Omega(2r, Q) \cap D$. Let us call again h the solution to the Dirichlet problem $Lu = 0$ in $\Omega(2r, Q) \cap D$ such that $u = h$ at $\partial(\Omega(2r, Q) \cap D)$. (We can smooth out the corners of $\Omega(2r, Q) \cap D$ if we want to make it a C^2 domain).

By the maximum principle we get

$$u(x) \leq M(u)h(x)$$

for every $x \in \Omega(2r, Q) \cap D$, where $M(u)$ is the maximum of u on $\Omega(2r, Q) \subset D$.

This proves that u is continuous at the piece of the boundary $\Delta\left(\frac{3r}{2}, Q\right)$. With this, maximum principle and Harnack's inequality we can adapt closely the proof for the Carlesson estimate given in [8]. □

DEFINITION. A nonnegative weak solution of L is a function $v \in L^1_{\text{loc}}(D), v \geq 0$, such that

$$\int vL\varphi = 0$$

for every $\varphi \in C_0^2(D)$.

Notice that we can take as a test function φ in the above definition any $\varphi \in C^2(D) \cap L^1(D)$ such that $\varphi = \nabla\varphi = 0$ at ∂D.

We want to prove now that the measure whose density is a nonnegative adjoint solution is doubling. To do that let us denote by $\phi(r, P)$ the cube centered at $P \in \mathbf{R}^n$ and radius r

$$\phi(r, P) = \{x \in \mathbf{R}^n : |x_i - p_i| < r \text{ for } i = 1, 2, \ldots, n\}.$$

Let us consider $\phi(r,P)\cap\Omega$ and smooth out its corners uniformly in the sense that its curvatures are bounded from above by $\frac{c}{r^2}$ (we can obtain it as a dilation by r of a fixed smooth domain).

We will denote by $\tilde{\phi}(r,P)$ the new domain obtained in this way.

LEMMA 2.3. *There exists a constant $c>0$ depending only on λ,η and D such that for any nonnegative weak solution v of L and for all cubes $\phi(r,P)$ we have*

$$\int_{\tilde{\phi}(r,P)} v(y)dy \le c \int_{\tilde{\phi}(r/2,P)} v(y)dy.$$

Proof. Using the dilation $x\to\frac{x}{r}$ we may assume $r=1$. Denote by $\phi(s)$ the image of $\tilde{\phi}(sr,P)$ under this dilation and consider in $\phi(1)$ the function

$$\delta(x)=\operatorname{dist}(x,\phi(1))$$

(or its regularized $\tilde{\delta}(x)$ obtained in Lemma 1.2).

Take $h(x)=\delta(x)^2$, which is a $C^2(\mathbb{R}^n)$ function such that $h=\nabla h=0$ on $\partial\phi(1)$.

We have

$$Lh(x)=2\sum_{i,j=1}^{n} a_{ij}(x)\left(\delta_{x_i}\delta_{x_j}+\delta\delta_{x_ix_j}\right)+2\sum_{i=1}^{n} b_i(x)\delta\delta_{x_i}+c(x)\delta^2$$

Using now the bounds for δ

$$|\delta_{x_ix_j}|\le c_1$$
$|\nabla\delta|\ge c_2$ for x close to the boundary $\partial\phi(1)$ and
$$|b_i(x)|\le\frac{\eta(\delta)}{\delta},\ |c(x)|\le\frac{\eta(\delta)}{\delta^2}$$

we get

$$Lh(x)\ge c_3,\ \text{for all } x\in\phi(1)\backslash\phi(1-\in)$$

for some $c_3>0$ and some small $\in>0$ and

$$|Lh(x)|\le c_4 \text{ for all } x\in\phi(1).$$

Therefore

$$\int_{\phi(1)\backslash\phi(1-\in)} v(y)dy\le\frac{1}{c_3}\int_{\phi(1)\backslash\phi(1-\in)} v(y)Lh(y)dy=$$
$$=\frac{1}{c_3}\int_{\phi(1)} v(y)Lh(y)dy-\frac{1}{c_3}\int_{\phi(1-\in)} v(y)Lh(y)dy$$

but

$$\int_{\phi(1)} v(y)Lh(y)dy=0$$

since $v(y)$ is a nonnegative adjoint solution.

Thus

$$\int_{\phi(1)\setminus\phi(1-\epsilon)} v(y)dy \le \frac{1}{c_3}\int_{\phi(1-\epsilon)} v(y)|Lh(y)|dy \le \frac{c_4}{c_3}\int_{\phi(1-\epsilon)} v(y)dy$$

and therefore

$$\int_{\phi(1)} v(y)dy \le c\int_{\phi(1-\epsilon)} v(y)dy$$

for some $c > 0$. An iteration gives now

$$\int_{\phi(1)} v(y)dy \le c\int_{\phi(1/2)} v(y)dy$$

for some larger $c > 0$, which finishes the proof of the Lemma. □

From now on we will use the local representation of D given in (9).

Fix $Q = (x_0', \varphi(x_0')) \in \partial D$ and consider the cylinder $\Omega(r, Q)$ for $r \le r_0$. We smooth out the corners of $\Omega(r, Q) \cap \Omega$ uniformly in r (that is, they are all dilation of a fixed smooth domain). Let us call $\widetilde{\Omega}(r, Q)$ the new domain.

Let ω_r^x be the L-harmonic measure at x for the domain $\widetilde{\Omega}(r, Q)$ and let

$$\alpha_r = \overline{D} \cap \widetilde{\Omega}(r, Q)$$
$$\beta_r = \{(x', x_n) : x_n = \varphi(x_0) + ar\} \cap \partial\widetilde{\Omega}(r, Q)$$

and as before

$$A_r(Q) = \left(x_0', \varphi(x_0') + \frac{ar}{2}\right)$$

LEMMA 2.4. *Suppose u, v are positive solutions of $Lu = 0$ in $\widetilde{\Omega}(4r, Q) \cap D \cap \{(x', x_n) : x_n < \varphi(x_0) + 2ar\}$, $0 < 4r < r_0$, which vanish continuously on the bottom and sides, that is, on*

$$\partial\widetilde{\Omega}(4r, Q) \setminus \{(x', x_n) : x_n = \varphi(x_0) + 4ar\}$$

There exists a constant $c > 0$ depending only on λ, η and D such that

$$\frac{u(x)}{u(A_r(Q))} \le c\frac{v(x)}{v(A_r(Q))}$$

for every $x \in \widetilde{\Omega}(4r, Q) \cap \{(x', x_n) : x_n < \varphi(x_0) + 2ar\}$.

Proof. By the maximum principle it is enough to prove the above inequality for all

$$x \in \widetilde{\Omega}(4r, Q) \cap \{(x', x_n) : x_n = \varphi(x_0) + 2ar\} = \Sigma.$$

Fix $P_1 \in \widetilde{\Omega}(4r,Q) \cap \sum$ and consider the exterior ball $B_{r/4}(P)$ to the domain $\widetilde{\Omega}(4r,Q)$, where $P = P_1 - \frac{r}{4}N$ and where N the unit inward normal to $\widetilde{\Omega}(4r,Q)$ at P_1.

If $\rho = |x-P|$ then the function

$$w(x) = e^{-\alpha} - e^{-\alpha\frac{16\rho^2}{r^2}}$$

satisfies

$$Lw(x) = \frac{32\alpha}{r^2}e^{-\frac{16\rho^2}{r^2}}\left[\sum_{i=1}^n a_{ii} - \frac{32\alpha}{r^2}\sum_{i,j=1}^n a_{ij}(x)(x_i-p_i)(x_j-p_j)+ \right.$$
$$\left. + \sum_{i=1}^n b_i(x)(x_i-p_i)\right] + c(x)w(x) \leq$$
$$\leq \frac{32\alpha}{r^2}e^{-\frac{16\rho^2}{r^2}}\left[n\lambda - \frac{32\alpha}{\lambda r^2}|x-P|^2 + \frac{n\eta(x)}{\operatorname{dist}(x,\partial D)}|x-P|\right] + c(x)w(x)$$

but since on $B_{r/2}(P) \cap \widetilde{\Omega}(4r,Q)$ is $|x-P| \leq r, \operatorname{dist}(x,\partial D) \geq r$ and $c(x)w(x) \leq 0$ we get

$$Lw(x) \leq 0$$

for all $x \in B_{r/2}(P) \cap \widetilde{\Omega}(4r,Q)$ if we take α large enough.

Also, by Lemma 2.2 we have

$$u(x) \leq cu(A_r) \text{ for every } x \in B_{r/2}(P_1) \cap \widetilde{\Omega}(4r,Q).$$

Since $w(x) \geq c_1$ for some constant $c_1 > 0$ on $\partial B_{r/2}(P_1) \cap \widetilde{\Omega}(4r,Q)$ we obtain

$$u(x) \leq \frac{c}{c_1}u(A_r)w(x)$$

for $x \in \partial(B_{r/2}(P) \cap \widetilde{\Omega}(4r,Q))$ and since $Lw(x) \leq 0$ the maximum principle extends the above inequality to all $B_{r/2}(P) \cap \widetilde{\Omega}(4r,Q)$.

A calculation shows that the normal derivative $\frac{\partial w}{\partial N}$ is uniformly bounded as

$$\frac{\partial w}{\partial N}(P_1 + tN) \leq \frac{c}{r}$$

for $0 \leq t \leq \frac{r}{4}$, and since $w(P_1) = 0$ we finally get

$$u(P_1 + tN) \leq \frac{c}{r}u(A_r)t \tag{13}$$

for $0 \leq t \leq \frac{r}{4}$.

Let $P_2 = P_1 + \frac{r}{2}N$ and consider the function

$$h(x) = e^{-\alpha\frac{4\rho^2}{r^2}} - e^{-\alpha}$$

where $\rho = |x - P_2|$.

A calculation as before shows

$$Lh(x) \geq 0$$

for every x on the set $E = B_{\frac{r}{2}}(P_2) \cap \left\{x \in \widetilde{\Omega}(4r, Q) : \operatorname{dist}(x, \partial\widetilde{\Omega}(4r, Q)) \leq \frac{r}{2}\right\}$ and since by Harnack's inequality

$$u(x) \geq cu(A_r)$$

on $\partial E \backslash \partial B_{r/2}(P_2)$ and $h(x) \leq c_1$ on E we get

$$u(x) \geq \frac{c}{c_1} u(A_r) h(x)$$

for every $x \in \partial E \backslash \partial B_{r/2}(P_2)$.

The same inequality holds on $\partial B_{r/2}(P_2) \cap E$ since $h(x) = 0$ there and thus the inequality holds on ∂E.

Maximum principle extends the inequality to all E and since

$$\inf\left\{\frac{\partial h}{\partial N}(P_1 + tN) : 0 \leq t \leq \frac{r}{4}\right\} \geq \frac{c}{r}$$

and $h(P_1) = 0$ we get

$$u(P_1 + tN) \geq \frac{c}{r} u(A_r) t \tag{14}$$

whenever $0 \leq t \leq \frac{r}{4}$.

Inequalities (13) and (14) hold also for v, therefore

$$\frac{u(x)}{u(A_r(Q))} \leq c \frac{v(x)}{v(A_r(Q))}$$

for $x \in \sum \cap \left\{x : \operatorname{dist}(x, \partial\widetilde{\Omega}(4r, Q)) \leq r/4\right\}$. By Harnack we can extend the same inequality to all $\sum$ with some larger constant c, which finishes the proof of the Lemma. □

LEMMA 2.5. *There exists a constant $c > 0$ depending only on λ, η and D such that*

$$\omega_r^x(\alpha_r) \leq c\ \omega_r^x(\beta_r) \tag{15}$$

for every $x \in \widetilde{\Omega}(r, Q), 0 < r < r_0$.

Proof. If $L = \sum_{i,j=1}^n a_{ij}(x) D_{ij} + \sum_{i=1}^n b_i(x) D_i + c(x)$ let $b_i^{(m)}(x)$ and $c^{(m)}(x)$ be the truncated coefficients

$$b_i^{(m)}(x) = \begin{cases} b_i(x) & \text{if } \operatorname{dist}(x, \partial D) \geq \frac{ar}{2m} \\ 0 & \text{if } \operatorname{dist}(x, \partial D) \leq \frac{ar}{2m} \end{cases}$$

$$i = 1, 2, \ldots, n$$

$$c^{(m)}(x) = \begin{cases} c_m(x) & \text{if } \operatorname{dist}(x, \partial D) \geq \frac{ar}{2m} \\ 0 & \text{if } \operatorname{dist}(x, \partial D) \leq \frac{ar}{2m} \end{cases}$$

Let L_m be the operator

$$L_m = \sum_{i,j=1}^{n} a_{ij}(x)D_{ij} + \sum_{i=1}^{n} b_i^{(m)}(x)D_i + c^{(m)}(x)$$

and ω_m^x its corresponding harmonic measure for the domain $\widetilde{\Omega}(4r, Q)$.

We will show first that

$$\omega_m^x(\alpha_r) \leq c\omega_m^x(\beta_r) \tag{15'}$$

with c a positive constant independent of m, (15) will follow then by approximation.

Choose $h \in C^\infty(\mathbf{R}^n)$ such that $0 \leq h \leq 1$ in $\mathbf{R}^n$, $h = 0$ on the cube $\phi_{r/8}(A_r(Q))$, $h = 1$ on $\mathbf{R}^n \backslash \widetilde{\Omega}(r, Q)$ and

$$\|h\|_{C^2(\mathbf{R}^n)} \leq \frac{c}{r^2}$$

where c only depends on D.

We have for $x \in \phi_{r/8}(A_r(Q))$

$$\begin{aligned} \omega_m^x(\alpha_r) &\leq \int_{\partial\widetilde{\Omega}(4r,Q)} h(P)d\omega_m^x(P) = \int_{\partial\widetilde{\Omega}(4r,Q)} h(P)d\omega_m^x(P) - h(x) = \\ &= \int_{\widetilde{\Omega}(4r,Q)} g_r^m(x,y)L_m h(y)dy \leq \frac{c}{r^2} \int_{\widetilde{\Omega}(4r,Q)\backslash\phi_{r/8}(A_r(Q))} g_r^m(x,y)dy \end{aligned}$$

because of the definition of harmonic measure and where $g_r^m(x,y)$ is the Green's function associated to L_m and domain $\widetilde{\Omega}(4r, Q)$. Recall that in view of the Alessandrov-Bakelman-Pucci estimate

$$\|u\|_\infty \leq c\|Lu\|_n$$

which holds for uniformly elliptic L with bounded lower order terms and u vanishing at the boundary, the linear functional $Lu \to u(x)$ is continuous and then by the Riesz representation Theorem can be represented using an integral kernel, which is by definition the Green's function associated to L.

Since for $x \neq y$ $g_r^m(x,y)$ is a nonnegative adjoint solution of L_m, the doubling property given in Lemma 2.3 implies

$$\int_{\widetilde{\Omega}(4r,Q)\backslash\phi_{r/8}(A_r(Q))} g_r^m(x,y)dy \leq c \int_{\phi_{r/16}(A_{3/4r}(Q))} g_r^m(x,y)dy$$

uniformly for $x \in \phi_{r/16}(A_r(Q))$ and where c is a constant that depends only on ellipticity λ, η and D.

Therefore

$$\omega_m^x(\alpha_r) \leq \frac{c}{r^2} \int_{\phi_{\frac{r}{16}}\left(A_{\frac{3r}{4}}(Q)\right)} g_r^m(x,y)dy = f(x).$$

But now, the functions $\omega_m^x(\beta_r)$ and $f(x)$ satisfy the hypothesis of Lemma 2.4 in $\widetilde{\Omega}(4r,Q)$. Also, as we can see by a dilation argument, $\omega_m^{A_r(Q)}$ and $f(A_r(Q))$ are bounded from above and below by constants depending only on λ,η and D.

Thus

$$f(x) \le c\omega_m^x(\beta_r)$$

for every $x \in \widetilde{\Omega}(r,Q)$ with constant $c = c(\lambda,\eta,D)$. This concludes the proof of (15′). □

THEOREM 2.6 (COMPARISON THEOREM FOR SOLUTIONS). *Given a C^2 domain D in $\mathbf{R}^n$ and $L \in \mathcal{L}_\lambda$, there exists a constant c depending only on λ,η and D such that if u and v are two nonnegative solutions to $Lw = 0$ in $\widetilde{\Omega}(4r,Q)$ which vanish continuously on $\Delta(2r,Q)$ then*

$$\frac{1}{c}\frac{u(x)}{u(A_r(Q))} \le \frac{v(x)}{v(A_r(Q))} \le c\frac{u(x)}{u(A_r(Q))} \tag{16}$$

for every $x \in \widetilde{\Omega}(r,Q)$ and $0 < r \le r_0$.

Proof. We will closely follow the proof given in [11] for elliptic equations in divergence form, using the tools already developed in this section.

By Lemma 2.2 and the maximum principle

$$u(x) \le c_1 u(A_r(Q))\omega_r^x(\alpha_r)$$

for all $x \in \widetilde{\Omega}(4r,Q)$ with c_1 a positive constant that depends only on λ,η and D.

By Harnack's inequality

$$\omega_r^x(\beta_r)v(A_r(Q)) \le c_2 v(x)$$

for all $x \in \beta_r$ and then for all $x \in \widetilde{\Omega}(4r,Q)$ by the maximum principle, with c_2 a constant depending only on λ,η and D.

Therefore, since $\omega_r^x(\alpha_r) \le c\omega_r^x(\beta_r)$ for all $x \in \widetilde{\Omega}(r,Q)$ we get

$$\begin{aligned} u(x) &\le c_1 u(A_r(Q))\omega_r^x(\alpha_r) \le c_1 c u(A_r(Q))\omega_r^x(\beta_r) \\ &\le c_1 c c_2 \frac{u(A_r(Q))}{v(A_r(Q))} v(x) \end{aligned}$$

which finishes the proof of (16). □

3. In this section we will sketch the construction of an operator in $\mathcal{L}_\lambda$ with singular harmonic measure. For a more complete exposition we refer to [6].

In dimension $n = 2$ a K-quasiconformal map

$$w(t,s) = (f(t,s), g(t,s))$$

from a domain $D \subset \mathbf{R}^2$ onto another domain $D' \subset \mathbf{R}^2$ is a homeomorphism with continuous partial derivatives such that

$$f_t^2 + f_s^2 + g_t^2 + g_s^2 \leq K(f_t g_s - f_s g_t)$$

Geometrically this means that w maps small balls onto small domains which are almost ellipsoids with a uniform bounded excentricity.

In [13] Carlesson constructed a continuous strictly increasing function $h(t)$ on the real line, purely singular, such that $h(\mathbf{R}) = \mathbf{R}$ and satisfies the quasisymmetry condition

$$\frac{1}{\rho} \leq \frac{h(t+s) - h(t)}{h(t) - h(t-s)} \leq \rho \tag{17}$$

for every real t and s, $s > 0$ and that for small $s > 0$

$$\frac{h(t+s) - h(t)}{h(t) - h(t-s)} = 1 + 0(\eta(s)) \tag{18}$$

uniformly in compact sets in t.

In order to construct the counterexample we need a C^2 quasiconformal extension of $h(t)$

THEOREM 3.1. *Given a continuous increasing function $\eta(s)$ for $s \geq 0$ with $\eta(0) = 0$, $s^{-1/4}\eta(s)$ decreasing for $s > 0$ and*

$$\int_0^1 \frac{\eta(s)^2}{s} ds = \infty$$

then there exists a quasiconformal map

$$w(t,s) = (f(t,s), g(t,s))$$

with the following properties

1) $w \in C^2(\mathbf{R}^2_+) \cap C(\overline{\mathbf{R}^2_+})$
2) *$w(t,0) = h(t)$ is a purely singular increasing function and then the measure $dw(t,0)$ is completely singular with respect to Lebesgue measure on the real line.*
3) *For small s*

$$f_t = \left[\frac{h(t+s) - h(t)}{s}\right](1 + 0(\eta(s))) \qquad f_s = \left[\frac{h(t+s) - h(t)}{s}\right]0(\eta(s))$$

$$g_s = \left[\frac{h(t+s) - h(t)}{s}\right](1 + 0(\eta(s))) \qquad g_s = \left[\frac{h(t+s) - h(t)}{s}\right]0(\eta(s))$$

$$f_{tt} = \left[\frac{h(t+s) - h(t)}{s^2}\right]0(\eta(s)) \qquad f_{st} = \left[\frac{h(t+s) - h(t)}{s^2}\right]0(\eta(s))$$

$$f_{ss} = \left[\frac{h(t+s) - h(t)}{s^2}\right]0(\eta(s)) \qquad g_{tt} = \left[\frac{h(t+s) - h(t)}{s^2}\right]0(\eta(s))$$

$$g_{st} = \left[\frac{h(t+s) - h(t)}{s^2}\right]0(\eta(s)) \qquad g_{ss} = \left[\frac{h(t+s) - h(t)}{s^2}\right]0(\eta(s))$$

uniformly on compact sets in t.

Sketch of the proof. Let $\chi(u)$ be the tent function

$$\chi(u) = \begin{cases} u & \text{if } 0 \le u \le 1 \\ 2-u & \text{if } 1 \le u \le 2 \\ 0 & \text{otherwise} \end{cases}$$

and let us define

$$\alpha(t,s) = \int \chi(u)h(t+us)du$$
$$\beta(t,s) = \int \chi(u)h(t-us)du$$

Our quasiconformal map will then be

$$w(z) = w(t,s) = \frac{1}{2}(\alpha+\beta) + \frac{1}{2}(\alpha-\beta) = f + ig$$

Notice that $\alpha(t,s)$ can be considered as an "average" of the values of h over the interval $[t, t+2s]$ and $\beta(t,s)$ as an "average" of h over $[t-2s, t]$.

A computation of the partial derivatives of α, β and the use of the quasisymmetry conditions (17) and (18) show that $w(t,s)$ is a C^2 quasiconformal map and that the relations 3) in Theorem 3.1 are satisfied.

Also, it can be verified that $w(z)$ is a homeomorphism from the upper half-plane onto itself.

Finally, by construction $w(t,0) = h(t)$ and therefore $dw(t,0)$ is completely singular with respect to Lebesgue measure. □

THEOREM 3.2. *There exists an elliptic operator defined on the upper half-plane* $\mathbf{R}^2_+ = \mathbf{R} \times (0,\infty)$

$$\begin{aligned} L = a_{11}(t,s)\frac{\partial^2}{\partial t^2} + 2a_{12}(t,s)\frac{\partial^2}{\partial s \partial t} + a_{22}(t,s)\frac{\partial^2}{\partial s^2} + \\ + b_1(t,s)\frac{\partial}{\partial t} \qquad + b_2(t,s)\frac{\partial}{\partial t} + b_2(t,s)\frac{\partial}{\partial s} \end{aligned} \tag{19}$$

with continuous coefficients $a_{i,j}(t,s) i,j = 1,2$ *up to the boundary, whose associated harmonic measure is completely singular with respect to Lebesgue measure on* $\mathbf{R}$.

Proof. Let us consider the quasiconformal map

$$w(t,s) = (f(t,s), g(t,s)) = (x,y)$$

given by Theorem 3.1.

Let $F(x,y) = (\phi(x,y), \psi(x,y))$ be the inverse function of $w(t,s)$. If $u(x,y)$ satisfies Laplace's equation

$$\Delta u = u_{xx} + u_{yy} = 0$$

then $v(t,s) = u \circ w(t,s)$ satisfies the second order linear equation

$$(\phi_x^2 + \phi_y^2)v_{tt} + 2(\phi_x\psi_x + \phi_y\psi_y)v_{ts} + (\psi_x^2 + \psi_y^2)v_{ss} + \\ + (\phi_{xx} + \phi_{yy})v_t \qquad + (\psi_{xx} + \psi_{yy})v_s = 0$$

If $\rho = f_t g_s - f_s g_t$ is the determinant of the Jacobian of w a computation shows that if we multiply the above equation by ρ then we obtain an elliptic operator L in $\mathcal{L}_\lambda$ of the form (7).

Now remember that the harmonic measure associated to L, w_L^z, gives the representation

$$v(z) = \int_{\mathbf{R}} g(t)dw_L^z(t) \tag{20}$$

for the solution $v(t,s)$ of the Dirichlet problem

$$\begin{cases} Lv(t,s) = 0 & \text{in } \mathbf{R}_+^2 \\ v(t,0) = g(t) & \text{on } \mathbf{R} \end{cases}$$

but $v = u \circ w$ and $u(x,y)$ is harmonic, therefore

$$u(x,y) = \int_{\mathbf{R}} Q(x,y,\sigma)h(\sigma)d\sigma \tag{21}$$

where $Q(x,y,\sigma) = P(x-\sigma, y), P(x,y) = \frac{1}{\pi}\frac{y}{x^2+y^2}$ is the Poisson kernel and $h(\sigma) = u(\sigma, 0) = g \circ F(\sigma, 0)$.

After the change of variables $z = F(x,y)$ in (21) we finally get

$$v(z) = \int_{\mathbf{R}} Q(w(z), w(t,0))g(t)dw(t) \tag{22}$$

Therefore comparing (20) and (22)

$$dw_L^z(t) = Q(w(z), w(t,0))dw(t)$$

and thus dw_L^z is completely singular with respect to Lebesgue measure because $dw(t)$ is. □

THEOREM 3.3. *There exists an elliptic operator L defined on the unit ball B_1 in $\mathbf{R}^n$, $L \in \mathcal{L}_\lambda$, whose harmonic measure is completely singular with respect to the Lebesgue measure on the unit sphere $S^{n-1} = \partial B_1$.*

Notice that given any function $\eta(s)$ as in Theorem 1, this $\eta(s)$ is the modulus of continuity of the leading coefficients $a_{ij}(x)$ for that L. We omit the details of the proof of Theorem 3.3 which is similar to the one given in [12].

4. In this section we will give part of the proof of the Theorem 1 stated at the Introduction, which basically says that the Poisson kernel $K(x,Q)$ is in $L^2(d\sigma)$ if the modulus of continuity $\eta(s)$ of the main coefficients at the boundary satisfies the Dini condition

$$\int_0^{\varepsilon_0} \frac{\eta(s)^2}{s} ds < \infty$$

In §4.1 we introduce some preliminaries and notation. In §4.2 we consider the equation in $\mathbf{R}_+^{n+1}$

$$Lu(x,t) = \sum_{i,j=1}^{n+1} a_{ij} u_{x_i x_j} + \sum_{i,j=1}^{n+1} \varepsilon_{ij}(x) u_{x_i x_j} + \sum_{i=1}^{n+1} b_i(x,t) u_{x_i} \tag{23}$$

where we write $t = x_{n+1}, A = (a_{ij})$ is an elliptic matrix of constant complex coefficients, and the coefficients of the drift term $b = (b_1, b_2, \ldots, b_{n+1})$ satisfy

$$\sup_{\substack{0 \le s \le t \\ x \in \mathbf{R}^n}} |b_i(x,s)| \le \frac{\eta(t)}{t}$$

where $\eta(t)$ is an increasing function such that $\eta(0) = 0$, $\eta(t) = 0$ for $t \ge \varepsilon$ for some $\varepsilon > 0$ to be fixed later and

$$\int_0^{\infty} \frac{\eta(s)^2}{s} ds < \infty$$

Theorem 4.3 gives a priori estimates for the initial value problem

$$\begin{cases} Lu(x,t) & = \eta(t) f(x,t) \\ u(x,0) & = g(x) \end{cases}$$

In §4.3, Theorem 4.6, we first solve the Dirichlet problem for an operator that satisfies a Dini condition in t and is close to a constant coefficients in x. This implies that the Poisson kernel is locally in L^2 at the boundary $\mathbf{R}^n$. Then we can pass these estimates to a C^2 domain by change of variables. This will give the proof of Theorem 1.

Later in §4.4 we will solve the Dirichlet problem for $\mathcal{L}_\lambda$ and boundary data in L^2:

THEOREM 4.0. *Let D be a bounded C^2 domain in $\mathbf{R}^n$. If $g \in L^2(\partial D, d\sigma)$ then there exists a solution u to $Lu = 0$ in D such that u converges nontangentially* a.e. *$d\sigma$ to g and*

$$\|N_\alpha(u)\|_{L^2(d\sigma)} \le c_\alpha \|g\|_{L^2(d\sigma)}. \tag{24}$$

Here if $\Gamma_\alpha(Q)$ is the cone of aperture $\alpha > 0$

$$\Gamma_\alpha(Q) = \{x \in D \big| |x - Q| < (1+\alpha)\mathrm{dist}(x, \partial D)\}$$

$N_\alpha(u)$ is the nontangential maximal function

$$N_\alpha(u)(Q) = \sup_{x\in\Gamma_\alpha(Q)} |u(x)|$$

4.1. Preliminaries and Notation. In the following we will denote points in the upper half-space $\mathbf{R}^{n+1}_+ = \mathbf{R}^n \times (0,\infty)$ by (x,t) or (y,s) where $x,y \in \mathbf{R}^n$ and $t = x_{n+1}, s = y_{n+1}$ are, respectively, the spatial and time components of the corresponding points.

If $f(x,t)$ is either a scalar valued or $(n+1)$-vector valued function defined in $\mathbf{R}^{n+1}_+$, we will denote the $L^2(\mathbf{R}^n)$ norm in x for any fixed $t > 0$ by

$$\|f(t)\| = \left(\int_{\mathbf{R}^n} |f(x,t)|^2 dx\right)^{1/2}$$

and its $L^2(\mathbf{R}^{n+1}_+)$ norm by

$$|||f(t)||| = \left(\int_0^\infty \|f(t)\|^2 dt\right)^{1/2} = \left(\int_0^\infty\int_{\mathbf{R}^n} |f(x,t)|^2 dx dt\right)^{1/2}$$

in which we still display its dependence on t.

If P, Q etc. are functions in $\mathbf{R}^n$ that act as convolution operators, we will denote

$$P_t(x) = t^{-n} P\left(\frac{x}{t}\right)$$

and if $f(x,s)$ is a scalar or $(n+1)$-vector valued function the convolution of P_t and f at the level s is the new function on $x \in \mathbf{R}^n$

$$(P_t * f(s))(x) = \int_{\mathbf{R}^n} P_t(x-y) f(y,s) dy.$$

We will use sometimes the notation $P_t f$, instead of $P_t * f$, for the convolution if that is clear in the context.

Similarly, the Fourier transform of f with respect to the spatial variable x is given by

$$\hat{f}(t)(\xi) = \int_{\mathbf{R}^n} f(x,t) e^{-ix\cdot\xi} dx \qquad \xi \in \mathbf{R}^n$$

Any $(n+1)\times(n+1)$ complex valued matrix function $E(x)$ that depends only on x gives a multiplication operator on vector-valued functions in $L^2(\mathbf{R}^n)$ with the norm

$$\|E\|_\infty = \max_{i,j} \sup_x |\in_{ij}(x)|$$

if $E(x) = (\in_{ij}(x))$.

A matrix valued function $S(x) = (s^{i,j}(x))_{i,j=1}^{n+1}$ acts also as a convolution operator on vector-valued functions in $\mathbf{R}^n$ as

$$(S * f)_i = \sum_{j=1}^{n+1} s^{i,j} * f_j \qquad i = 1, 2, \ldots, n+1$$

being its norm on $L^2(\mathbf{R}^n)$

$$\|S\| = \max_{i,j} \|s^{i,j}\| = \max_{i,j} \sup_{\xi} |\hat{s}^{i,j}(\xi)|.$$

For a fixed integer d we consider the operator $R(t)$ defined on scalar $L^2(\mathbf{R}^n)$ functions by

$$R(t)f = a(\delta - P_t) * S * f \tag{25}$$

where δ is the unit mass at $x = 0$, that is, the identity operator under convolution; $a(x)$ is a function that acts as a multiplication and $\sup_x |a(x)| \leq 1$, $\hat{S}(\xi)$ is homogeneous of degree 0 smooth away from the origin and

$$\left| \left(\frac{\partial}{\partial \xi} \right)^{\alpha} \hat{S}(\xi) \right| \leq 1 \text{ for all } |\alpha| \leq d, |\xi| = 1$$

$\hat{P}(\xi)$ is smooth away from the origin, $\hat{P}(0) = 1$ and

$$\left| \left(\frac{\partial}{\partial \xi} \right)^{\alpha} \hat{P}(\xi) \right| \leq A|\xi|^{-\frac{1}{2}-|\alpha|} \text{ for all } |\alpha| \leq d, |\xi| \geq 1$$

and

$$\left| \left(\frac{\partial}{\partial \xi} \right)^{\alpha} (1 - \hat{P}(\xi) \right| \leq A|\xi|^{\frac{1}{2}-|\alpha|} \text{ for all } |\alpha| \leq d, |\xi| \leq 1.$$

Let $\hat{Q}(\xi)$ be smooth away from the origin and satisfy

$$\left| \left(\frac{\partial}{\partial \xi} \right)^{\alpha} \hat{Q}(\xi) \right| \leq \min\left(|\xi|^{-\frac{1}{2}}, |\xi|^{\frac{1}{2}} \right) |\xi|^{-\alpha} \text{ for all } |\alpha| \leq d \tag{26}$$

Any $P_t(x) = t^{-n} P\left(\frac{x}{t}\right)$ such that $\hat{P}(\xi)$ satisfies $\hat{P}(0) = 1$ and (5), (6) is called a P_t-type operator. Similarly, any $Q_t(x) = t^{-n} Q\left(\frac{x}{t}\right)$ such that $\hat{Q}(\xi)$ satisfies (26) is called a Q_t-type operator.

The main tool to be used in this section is the following multilinear Littlewood-Paley estimate which proof can be found in [19] or [20]:

THEOREM 4.1. *There exists an integer $d = d(n)$ and a constant $C = C(n, A)$ such that for all integers k and all $f \in L^2(\mathbf{R}^n)$ we have*

$$|||t^{-1/2} Q_t * (R_1(t) R_2(t) \ldots R_k(t) f)(t)||| \leq C^{k+1} \|f\|,$$

in which $R_1(t), R_2(t), \ldots, R_k(t)$ are operators of the form of $R(t)$ above in (25) corresponding respectively to functions $a_1, a_2, \ldots, a_k$ and Q satisfies condition (26).

We will use also many times

LEMMA 4.2 (HARDY'S INEQUALITY). *If $K(t,s)$ is a homogeneous function of degree -1 and $\int_0^\infty |K(t,s)|s^{-1/2}ds \leq M$, then*

$$\int_0^\infty \left| \int_0^\infty K(t,s)f(s)ds \right|^2 dt \leq M^2 \int_0^\infty |f(s)|^2 ds.$$

4.2. For L as in (23) and the initial value problem

$$\begin{cases} Lu(x,t) & = \eta(t)f(x,t) \\ u(x,0) & = g(x) \end{cases} \tag{27}$$

we have

THEOREM 4.3. *There exists an $\varepsilon > 0$ for which there is a solution $u(x,t)$ satisfying*

(28)

$$\sup_{t>0} \|u(t)\| + \sup_{t>0} \left\| \int_0^t sD^2u(s)ds \right\| + |||t^{1/2} \nabla u(t)||| +$$

$$+ |||t^{3/2}D^2u(t)||| \leq M \left\{ \|g\| + \left[\sup_{t>0} \eta(t) + \left(\int_0^\infty \frac{\eta(t)^2}{t} dt \right)^{1/2} \right] |||t^{3/2}f(t)||| \right\}$$

for every $|\varepsilon_{ij}(x)| \leq \varepsilon$ and every increasing function $\eta(t)$ such that $\eta(0) = 0, \eta(t) = 0$ for $t \geq \varepsilon$ and the right hand side of (28) is finite.

Proof. The proof is technically involved, so in order to give the idea we will consider the more simple equation

$$Lu = \sum_{i,j=1}^{n+1} a_{ij}u_{x_ix_j} + \sum_{i,j=1}^{n+1} \varepsilon_{ij}(x)u_{x_ix_j}$$

and give the bounds for $\|u(t)\|$ and $|||t^{1/2} \nabla u(t)|||$ for it. A complete proof of Theorem 1 is given in [5]. For this L, we can express the solution of (23) as

$$u(x,t) = P_t * g(x) + \int_0^\infty \int_{\mathbf{R}^n} G(x-y,t,s)\eta(s)f(y,s)dyds -$$

$$- \int_0^\infty \int_{\mathbf{R}^n} G(x-y,t,s) \sum_{i,j=1}^{n+1} \varepsilon_{ij}(y)u_{y_iy_j}(y,s)dyds$$

where $G(x,t)$ is the Green's function of L and P_t is its Poisson kernel.

Let us introduce the new notation

$$u(t) = P_t * g + \int_0^\infty G(t,s)\eta(s)f(s)ds + \int_0^\infty G(t,s)ED^2u(s)ds$$

where the dependence on x is understood and we denote

$$G(t,s)f(s) = \int_{\mathbf{R}^n} G(x-y,t,s)f(y,s)dy$$

$$ED^2u(s) = \sum_{i,j=1}^{n+1} \varepsilon_{ij}(y)u_{y_iy_j}(y,s).$$

According to [21] if $\ell(i\xi, i\xi_{n+1}) = -\sum_{j,k=1}^{n+1} a_{jk}\xi_j\xi_k$ for $(\xi,\xi_{n+1}) \in \mathbf{R}^n \times \mathbf{R}$ is the symbol of the constant coefficient operator $L_0u = \sum_{j,k=1}^{n+1} a_{jk}u_{x_jx_k}$, then the equation in τ

$$\ell(\xi,\tau) = 0$$

has the two imaginary roots

$$\tau_+(\xi) = \alpha(\xi) + i\beta(\xi) \text{ and } \tau_-(\xi) = \alpha(\xi) - i\beta(\xi)$$

homogeneous of degree 1, smooth functions on the unit sphere $|\xi| = 1$ and with bounds for each C^k norm depending only on λ, n and k. We also can find a constant $m > 0$ depending on λ, h and k such that

$$|\beta(\xi)| \geq m \text{ for } \xi \text{ on the unit sphere } |\xi| = 1.$$

The Fourier transform in x of the fundamental solution $\Gamma(x,t)$ of L_0 is

$$\hat{\Gamma}(\xi,t) = \frac{1}{2\pi}\int_{-\infty}^{+\infty} \frac{e^{it\tau}}{\ell(i\xi_1, i\xi_{n+1})}d\tau = \frac{-e^{it\tau_\pm(\xi)}}{i\gamma(\tau_+(\xi) - \tau_-(\xi))}$$

where $\gamma = a_{n+1,n+1}$ and by $\tau_\pm(\xi)$ we mean $\tau_+(\xi)$ when $t \geq 0$ and $\tau_-(\xi)$ when $t \leq 0$.

Using these facts the Green's function is

$$G(t,s) = \Gamma(t-s) - P_t * \Gamma(-s)$$

and the Poisson kernel P_t satisfies

$$\hat{P}_t(\xi) = e^{it\tau_+(\xi)} \text{ for } t \geq 0$$

1) Let us estimate $\|u(t)\|$ first. Remember that $u(t)$ is the sum of the three terms

$$u(t) = P_t * g + \int_0^\infty G(t,s)\eta(s)f(s)ds + \int_0^\infty G(t,s)ED^2u(s)ds.$$

We get

$$\|P_t * g\| \leq M\|g\|$$

for the first term by Plancherel's Theorem.

Now

$$\left\| \int_0^\infty G(t,s)\eta(s)f(s)ds \right\| \leq M \int_0^\infty s\eta(s)\|f(s)\|ds \leq$$

$$\leq M \left(\int_0^\infty \frac{\eta(s)^2}{s} ds \right)^{1/2} \||s^{3/2}f(s)\||$$

where we have used the fact that

$$\|G(t,s)\| \leq Ms$$

as we can see by computing the Fourier transform of $G(t,s)$.

For the third term

$$\int_0^t G(t,s)ED^2u(s)ds = \int_{t/2}^t G(t,s)ED^2u(s)ds + \int_0^{t/2} G(t,s)ED^2u(s)ds$$

and

$$\left\| \int_{t/2}^t G(t,s)ED^2u(s)ds \right\| \leq M\epsilon \int_{t/2}^t s\|D^2u(s)\|ds \leq$$

$$\leq M\epsilon \left(\int_{t/2}^t s^{-1}ds \right)^{1/2} \||s^{3/2}D^2u(s)\|| \leq M\epsilon\||s^{3/2}D^2u(s)\||.$$

To bound

$$\int_0^{t/2} G(t,s)ED^2u(s)ds$$

observe that for $s \leq t/2$ the Fourier transform of $G(t,s)$ is

$$G(t,s)\hat{}(\xi) = \frac{1}{i\gamma(\tau_+(\xi) - \tau_-(\xi))}(e^{-t\tau_+(\xi)}e^{-is\tau_-(\xi)} - e^{i(t-s)\tau_+(\xi)})$$

and if P_t is the P_t-type operator with symbol

$$\hat{P}_t(\xi) = \frac{1}{\gamma}e^{it\tau_+(\xi)}$$

by applying the mean value Theorem to their symbols we can write

$$G(t,s) = sP_t + (G(t,s) - sP_t)$$

where $G(t,s) - sP_t$ becomes an error term satisfying

$$\|G(t,s) - sP_t\| \le M\frac{s^2}{t}.$$

Therefore

$$\int_0^{t/2} G(t,s)ED^2u(s)ds = \int_0^{t/2} (G(t,s) - sP_t)ED^2u(s)ds + \tag{29}$$
$$+ P_t \int_0^{t/2} sED^2u(s)ds$$

and

$$\left\| \int_0^{t/2} (G(t,s) - sP_t)ED^2u(s)ds \right\| \le \frac{M\epsilon}{t} \int s^2 \|D^2u(s)\| ds \le \tag{30}$$
$$\le M\epsilon\frac{1}{t}\left(\int_0^{t/2} sds\right)^{1/2} |||s^{3/2}D^2u(s)||| \le M\epsilon|||s^{3/2}D^2u(s)|||.$$

For the second integral in the expression (29)

$$P_t \int_0^{t/2} sED^2u(s)ds = P_t \int_0^t sED^2u(s)ds - P_t \int_{t/2}^t sED^2u(s)ds$$

$$\left\|P_t \int_0^t sED^2u(s)ds\right\| \le M\left\| \int_0^t sED^2u(s)ds\right\| = M\left\|E\int_0^t sD^2u(s)ds\right\| \le$$
$$\le M\epsilon\left\| \int_0^t sD^2u(s)ds\right\|$$

and

$$\left\|P_t \int_{t/2}^t sED^2u(s)ds\right\| \le M\epsilon \int_{t/2}^t s\left\|D^2u(s)\right\| ds \le M\epsilon|||s^{3/2}D^2u(s)|||$$

Collecting all these estimates we finally obtain

$$\|u(t)\| \le M\left[\|g\| + \left(\int_0^t \frac{\eta(t)^2}{t}dt\right)^{1/2} |||t^{3/2}f(t)|||\right]$$
$$+ \quad M\epsilon\left[\|\int_0^t sD^2u(s)ds\| + |||s^{3/2}D^2u(s)|||\right]$$

2) Let us compute now $|||s^{3/2}D^2u(s)|||$. From

$$u(t) = P_t * g + \int_0^\infty G(t,s)\eta(s)f(s)ds + \int_0^\infty G(t,s)ED^2u(s)ds$$

we get

$$\begin{aligned} D^2u(t) &= D^2P_t * g + \\ &+ \int_0^\infty K(t,s)\eta(s)f(s)ds + \int_0^\infty K(t,s)ED^2u(s)ds \end{aligned}$$

where $K(t,s) = D^2G(t,s)$ is the $(n+1)\times(n+1)$ matrix with entries

$$\begin{cases} K^{j,k}(t,s) &= \frac{\partial^2}{\partial x_j \partial x_k} G(t,s) \quad j,k = 1,\dots,n \\ K^{j,n+1}(t,s) &= \frac{\partial^2}{\partial x_j \partial t} G(t,s) \quad j = 1,\dots,n+1 \end{cases}$$

Therefore we can write

$$D^2u(t) - \int_0^\infty K(t,s)ED^2u(s)ds = F(t)$$

where

$$F(t) = (D^2P_t * g) + \int_0^\infty K(t,s)\eta(s)f(s)ds$$

that is

$$(I-T)D^2u(t) = F(t)$$

where T is the operator on vector valued functions h defined by

$$T(h) = \int_0^\infty K(t,s)Eh(s)ds.$$

Inverting formally $(I-T)$ we get the Newman series

$$D^2u(t) = \sum_{n=0}^\infty h_n(t) \tag{31}$$

where $h_0(t) = F(t)$ and $h_{n+1}(t) = T(h_n)(t)$.

Our goal is to study the convergence of (31). Before that we need a previous Lemma:

Let K be the operator defined on vector valued functions h by

$$Kh(t) = \int_0^\infty K(t,s)h(s)ds$$

Lemma 4.4. *For some Q as in (2.6) we have*

$$|||t^{3/2}Kh(t)||| \leq M\left[|||t^{3/2}h(t)||| + \left|\left|\left|t^{-1/2}Q_t\int_0^t sh(s)ds\right|\right|\right|\right].$$

Proof. We split

$$\begin{aligned} t^{3/2}Kh(t) = &\int_0^\infty s^{3/2}K(t,s)h(s)ds \\ &+\int_{t/2}^\infty (t^{3/2}-s^{3/2})K(t,s)h(s)ds+ \\ &+\int_0^{t/2} -s^{3/2}K(t,s)h(s)ds \\ &+\int_0^{t/2} t^{3/2}K(t,s)h(s)ds. \end{aligned} \tag{32}$$

Since $K(t,s)$ is a singular integral kernel we have

$$\left|\left|\left|\int_0^\infty s^{3/2}K(t,s)h(s)ds\right|\right|\right| \leq M\left|\left|\left|s^{3/2}h(s)\right|\right|\right|.$$

Now, using the bound $\|k(t,s)\| \leq \frac{M}{|t-s|}$ for the kernel $K(t,s)$ obtained by estimating its Fourier transform in (30), we get

$$\left\|\int_{1/2}^\infty (t^{3/2}-s^{3/2})K(t,s)h(s)ds\right| \leq M\int_{t/2}^\infty \frac{t^{3/2}-s^{3/2}}{|t-s|s^{1/2}}\|s^{3/2}h(s)\|ds.$$

Applying Hardy's inequality to this last integral and since

$$\int_{t/2}^\infty \left|\frac{1-s^{3/2}}{1-s}\right|\frac{1}{s^{3/2}s^{3/2}}ds \leq M$$

we obtain

$$\left|\left|\left|\int_{t/2}^\infty (t^{3/2}-s^{3/2})K(t,s)h(s)ds\right|\right|\right| \leq M|||s^{3/2}h(s)|||.$$

For the third term in the sum (32)

$$\left\|\int_0^{t/2} -s^{3/2}K(t,s)h(s)ds\right\| \leq \int_0^{t/2}\frac{M}{|t-s|}\|s^{3/2}h(s)\|ds$$

and thus

$$\left|\left|\left| \int_0^{t/2} -s^{3/2} K(t,s)h(s)ds \right|\right|\right| \leq M|||s^{3/2}h(s)|||$$

by Hardy's inequality again.

The last term in (32) is more delicate. Observe first that for $s \leq t/2$ the symbol of the kernel $K^{j,k}(t,s)$ for $j \neq n+1, k \neq n+1$ is

$$K^{j,k}(t,s)\hat{}(\xi) = \frac{\xi_i \xi_k}{2\gamma\beta(\xi)} e^{it\tau_+(\xi)}(e^{-is\tau_-(\xi)} - e^{-is\tau_+(\xi)})$$

and since it vanishes at $s = 0$, using the mean value Theorem we obtain

$$\begin{aligned} K^{j,k}(t,s)(\xi) = is\tau_+(\xi)\frac{\xi_i\xi_k}{2\gamma\beta(\xi)} e^{it\tau_-(\xi)}e^{-is\theta\tau_+(\xi)}- \\ - is\tau_-(\xi)\frac{\xi_j\xi_k}{2\gamma\beta(\xi)} e^{it\tau_-(\xi)}e^{-is\theta\tau_-(\xi)} \end{aligned}$$

where $\theta = \theta(s)$ is $0 < \theta < 1$.

Observe now that since $s \leq t/2$ we can write the first term in this symbol as

$$\begin{aligned} is\tau_+(\xi)\frac{\xi_j\xi_k}{2\gamma\beta(\xi)} e^{it\tau_-(\xi)}e^{-is\theta\tau_+(\xi)} = \\ = is\tau_+(\xi)\frac{\xi_j\xi_k}{2\gamma\beta(\xi)} e^{-it\tau_-(\xi)} + E(t,s)(\xi) \end{aligned}$$

where the first term on the right hand side is the symbol of an operator of the form

$$\frac{s}{t^2}Q_t$$

and the second term $E(t,s)(\xi)$ produces an error operator $E(t,s)$ bounded as

$$\|E(t,s)\| \leq M\frac{s^2}{t^3}.$$

The same reasoning applies to the second term in the symbol of $K^{j,k}(t,s)$, therefore we can put

$$K^{j,k}(t,s) = \frac{s}{t^2}Q_t + E(t,s) \text{ where } \|E(t,s)\| \leq M\frac{s^2}{t^3}.$$

The other entries $K^{j,k}(t,s)$ where $j = n+1$ or $k = n+1$ of $K(t,s)$ produce the same result, thus

$$\begin{aligned} \int_0^{t/2} t^{3/2}K(t,s)h(s)ds = t^{-1/2}Q_t \int_0^t sh(s)ds- \\ - t^{-1/2}Q_t \int_{t/2}^t sh(s)ds+ \\ + \int_0^{t/2} t^{3/2}E(t,s)h(s)ds \end{aligned}$$

and

$$\left\| -t^{-1/2}Q_t \int_{t/2}^{t} sh(s)ds + \int_0^{t/2} t^{3/2}E(t,s)h(s)ds \right\| \leq$$

$$\leq Mt^{-1/2} \int_{t/2}^{t} \|s^{3/2}h(s)\|ds + Mt^{-3/2} \int_0^{t/2} s^{1/2}\|s^{3/2}h(s)\|ds.$$

By Hardy's inequality the triple norm of this last expression is bounded by $M|||t^{3/2}h(t)|||$ and this finishes the proof of Lemma 4.4. □

Let us study now

$$D^2u(t) = \sum_{j=0}^{\infty} h_j(t)$$

to this end denote

$$a = \|g\| + \left[\sup_{t>0} \eta(t) + \left(\int_0^{\infty} \frac{\eta(t)^2}{t} dt\right)^{1/2}\right] |||t^{3/2}f(t)|||$$

$$a_j = |||t^{3/2}h_j(t)|||$$

$$b_j = |||t^{-1/2}Q_t E \int_0^t sh_j(s)ds|||.$$

LEMMA 4.5. *There exists a constant M depending only on λ and n such that*

(i) $a_0 \leq Ma$

(ii) $a_{j+1} \leq M\epsilon a_j + Mb_j$

(iii) $b_0 \leq M\epsilon a$

(iv) $b_{j+1} \leq \sum_{k=1}^{j+1} \left[(M\epsilon)^k a_{j-k+1} + (M\epsilon)^k b_{j-k+1} + (M\epsilon)^{j+2} a\right]$.

Notice that if $\epsilon > 0$ is sufficiently small this Lemma implies

$$|||t^{3/2}D^2u(t)||| \leq Ma$$

which will finish part 2) of the proof of Theorem 1. The proof of Lemma 4.5 uses the multilinear Littlewood-Paley estimates given in Theorem 4.1. We omit the details.

4.3. Let us consider the uniformly elliptic equation in $\mathbb{R}_+^{n+1}$ $Lu(x,t) = \sum_{i,j=1}^{n+1} a_{ij}(x,t)\, u_{x_i,x_j} + \sum_{i=1}^{n} b_i(x,t)u_{x_i}$ where the leading coefficients form a matrix $A(x,t) = (a_{ij}(x,t))_{i,j=1}^{n+1}$ of complex valued functions that satisfy

(33) $$a_{ij}(x,t) = a_{ij}(x,0) \text{ for } t \geq 2R$$

for some fixed $R > 0$, and if

$$\sup_{\substack{i,j \\ x\in\mathbf{R}^n}} |a_{ij}(x,t) - a_{ij}(x,0)| \le \eta(t)$$

then the coefficients of the lower order terms satisfy

$$\sup_{\substack{0\le s\le t \\ x\in\mathbf{R}^n}} |b_i(x,s)| \le M\frac{\eta(t)}{t}$$

where now $\eta(t)$ is a continuous increasing function such that $\eta(0) = 0$ and satisfies

$$\int_0^\infty \frac{\eta(t)^2}{t}dt < \infty.$$

For the Dirichlet problem

$$(34) \qquad \begin{cases} Lu(x,t) = 0 & \text{in } \mathbf{R}_+^{n+1} \\ u(x,0) = g(x) & \text{in } \mathbf{R}^n \end{cases}$$

we obtain

THEOREM 4.6. *There exist two positive constants $\epsilon > 0$ and $c > 0$ depending only on ellipticity λ and dimension n such that if the coefficient matrix of L, $A(x,t)$, satisfies (33) and moreover $\|A(x,0) - A_0\| \le \epsilon$ for some constant matrix A_0, then there is a solution u to the Dirichlet problem (34) satisfying*

$$(35) \qquad \begin{aligned} \sup_{t>0} \|u(t)\| + \sup_{t>0} \left\| \int_0^t sD^2u(s)ds \right\| + |||t^{1/2} \nabla u(t)|||+ \\ + |||t^{3/2}D^2u(t)||| \le C\|g\| \end{aligned}$$

Proof. If ϵ and M are the constants given in Theorem 2.6, take $t_0 > 0$ such that

$$\eta(t_0) < \frac{1}{4M} \text{ and } \int_0^{t_0} \frac{\eta(t)^2}{t}dt < \frac{1}{16M^2}$$

and let $\epsilon = \min\{\epsilon, t_0\}$ and

$$f(x,t) = \begin{cases} \frac{1}{\eta(t)} \sum_{i,j=1}^{n+1}(a_{ij}(x,0) - a_{ij}(x,t))u_{x_ix_j}(x,t) \text{ if } t \le \delta \\ 0 \quad \text{for} \quad t > \delta \end{cases}$$

Then, the solution of the Dirichlet problem (33) in $T_0 = \mathbf{R}^n \times (0,\delta)$ is the solution of

$$\begin{cases} L_0u(x,t) &= \eta(t)f(x,t) \\ u(x,0) &= g(x) \end{cases}$$

in $T_\delta = \mathbf{R}^n \times (0,\delta)$ where L_0 is the operator

$$L_0 u(x,t) = \sum_{i,j=1}^{n+1} a_{ij}(x,0) u_{x_i x_j} + \sum_{i=1}^{n+1} b_i(x,t) u_{x_i}.$$

By Theorem 4.3

$$sup_{t>0}\|u(t)\| + \sup_{t>0}\left\| \int_0^t s D^2 u(s) ds \right\| + \left\|\left| t^{1/2} \nabla u(t) \right|\right\| + \left\|\left| t^{3/2} D^2 u(t) \right|\right\| \leq$$

$$\leq M\|g\| + M\left[\eta(\delta) + \left(\int_0^\delta \frac{\eta(t)^2}{t} dt\right)^{1/2}\right] \left\|\left| t^{3/2} D^2 u(t) \right|\right\| \leq$$

$$\leq M\|g\| + \frac{1}{2}\left\|\left| t^{3/2} D^2 u(t) \right|\right\|$$

which gives (35) in T_δ after absorbing $\frac{1}{2}|||t^{3/2} D^2 u(t)|||$ in the left hand side.

Theorem 4.6 follows after repeating this argument a finite number of times till we reach $T_R = \mathbf{R}^n \times (0, 2R)$. □

Proof of Theorem 1. Part (a) follows from (b) using some very well known real variable techniques. To prove (b) let us consider the domain

$$\Omega = B(0,R) \cap \mathbf{R}_+^{n+1}$$

and let us denote by ω^P the harmonic measure at $P \in \Omega$ for the operator L and domain Ω.

Thus, if g is continuous and compactly supported in $\Delta(0,R)$, the function

$$v(P) = \int_{\Delta(0,R)} g(y) d\omega^P(y)$$

gives the solution to the Dirichlet problem $Lu = 0$ in $D, u = g$ on $\Delta(0,R)$ and $u = 0$ on $\partial\Omega \backslash \Delta(0,R)$.

Let us assume $g \geq 0$ and let $u(x,t)$ be the solution to $Lu(x,t) = 0$ in $\mathbf{R}_+^{n+1}$, $u = g$ on $\mathbf{R}^n$ given by Theorem 4.6 which we can apply by assuming R small enough.

By the maximum principle $v \leq u$, and if $P = (x,t)$, by Harnack's inequality

$$u(x,t) \leq M\left(\frac{1}{t^{n+1}} \int\int_{B(P,t/2)} |u(y,s)|^2 dy ds\right)^{1/2}.$$

Since this last expression is dominated by $t^{-n/2} \sup_{s>0} \|u(s)\|$ we get by Theorem 3.1

$$\int_{\Delta(0,R)} g(y) d\omega^P(Q) = v(P) \leq u(P) \leq M t^{-n/2} \|g\|$$

This implies that the harmonic measure ω^P is absolutely continuous with respect to Lebesgue measure dy on Δ and if $d\omega^P(y) = k(P,y)dy$ then by duality

$$\left(\int_{\Delta(0,R)} K(P,y)^2 dy\right)^{1/2} \leq Mt^{-n/2} \text{ for } P = (x,t) \in \Omega.$$

Let us fix now $P_0 = (0, R/2)$. By the Comparison Theorem (see [2] Corollary 12) it follows that

$$\frac{c_1}{\omega^{P_0}(\Delta)} \leq \frac{\omega(\Delta')}{\omega^{P_0}(\Delta')} \leq \frac{1}{\omega^{P_0}(\Delta)} \tag{36}$$

for every $\Delta' \subset \Delta(x, t/2)$ if $\Delta = \Delta(x,t)$ for $t < R/4$, every $P = (x,t) \in B(0, R/4)$ and where c_1 and c_2 are two positive constants independent of P.

Therefore, the Radon-Nikodym derivative

$$\frac{K(P,y)}{K(P_0,y)} = \frac{d\omega^P}{d\omega^{P_0}}(y) \simeq \frac{1}{\omega^{P_0}(\Delta)}$$

for every $y \in \Delta(0, R/4)$.

Since

$$\omega^{P_0}(\Delta(x,t)) = \int_{\Delta(x,t)} K(P_0, y)dy$$

we finally get

$$\left(\frac{1}{\sigma(\Delta(x,t))} \int_{\Delta(x,t)} K(P_0,y)^2 dy\right)^{1/2} \leq \frac{1}{\sigma(\Delta(x,t))} \int_{\Delta(x,t)} K(P_0,y)dy$$

for every $(x,t) \in B(0, R/4)$.

Given now the domain D of class C^2 in $\mathbf{R}^{n+1}$, there exist a finite collection of diffeomorphisms φ_j of class C^2 for $j = 1, 2, \ldots, N$ from the ball $B(0,R)$ in $\mathbf{R}^{n+1}$ into $\mathbf{R}^{n+1}$, such that $\varphi_j(\Delta(0,R)) \subset \partial D$, $D_j = \varphi_j(\Omega) \subset D$, $\Delta_j = \varphi_j(\Delta(0, R/4)) \subset \partial D$ and $\partial D = \cup_{j=1}^N \Delta_j$. Moreover we can choose φ_j in such a way that the inward normals to ∂D come from the normal vectors to $\mathbf{R}^n$ in $\mathbf{R}^{n+1}_+$. In this way, the solutions u to $Lu = 0$ in D_j come from the solutions $v = u \circ \varphi_j$ of an elliptic operator $L_j = \sum a_{ij} D_{ij} + \sum b_i D_i + cu$ of the same type as L, that satisfies the hypothesis of Theorem 3.1.

Therefore the harmonic measures $\omega_j^{P_j}$ for D_j at $P_j = \varphi_j(P_0)$ are absolutely continuous with respect to surface measure on Δ_j and if $K_j(y)$ is its Poisson kernel, $K_j(y) = \frac{d\omega_j}{d\sigma}(y)$. we get

$$\left(\frac{1}{\sigma(\Delta)} \int_\Delta K_j(Q)^2 d\sigma(Q)\right)^{1/2} \leq M \int_\Delta K_j(Q) d\sigma(Q)$$

for any spherical cap $\Delta \subset \varphi_j(\Delta(0, R/2))$.

Again, by the Comparison Theorem, as in (36).

$$\frac{c_1}{\omega^{x_0}(\Delta)} \leq \frac{d\omega^j}{d\omega^{x_0}}(Q) \leq \frac{c_2}{\omega^{x_0}(\Delta)}$$

where x_0 is in D and Δ is a large spherical cap. Since $\frac{dw^j}{d\omega^{x_0}}(Q) = \frac{k_j}{k}(Q)$ and $c_3 \leq w^{x_0}(\Delta) \leq 1$, this finishes the proof of part (b). □

4.4. We will give now the proof of Theorem 4.10 stated at the introduction of §4. We will use first a few Lemmas:

LEMMA 4.7. *Let B_r be a ball such that its concentric double $B_{2r} \subset D$. If $g(x,y)$ denotes the Green's function for L and D we have for x outside B_r*

$$\left(\frac{1}{|B_r|}\int_{B_r} g(x,y)^2 dy\right)^{1/2} \leq c\frac{1}{|B_r|}\int_{B_r} g(x,y)dy$$

with c a constant independent of $x \notin B_r$ and r.

Proof. We may assume B_r is centered at the origin. Notice also that by a dilation by r, the function $g(x,y) = r^{n-2}g(r\tilde{x}, r\tilde{y})$ is the Green's function corresponding to the operator

$$L_r u(\tilde{x}) = \sum a_{ij}(r\tilde{x})u_{\tilde{x}_i\tilde{x}_j} + \sum rb_i(r\tilde{x}) + r^2 c(r\tilde{x})u.$$

L_r is elliptic with the same ellipticity constant λ as in L and because of the assumptions on the coefficients for the lower order terms of L in (1), L_r has bounded lower order terms with bounds independent of r.

Therefore, changing notation, it is enough to prove the inequality

$$\left(\int_{B_1} g(x,y)^2 dy\right)^{1/2} \leq c\int_{B_1} g(x,y)dy$$

where $g(x,y)$ is the Green's function corresponding to an elliptic operator L with bounded lower order coefficients and B_1 is a fixed ball such that its concentric double B_2 lies also inside the domain.

We have

$$\left(\int_{B_1} g(x,y)^2 dy\right)^{1/2} = \sup \int_{B_1} g(x,y)f(y)dy \tag{37}$$

where the sup is taken over all $f \in C_0^\infty(B_1)$ functions such that $\|f\|_2 \leq 1$. Let u solve $Lu = f$ on $B_1, u = 0$ at ∂B_1 and pick $\phi \in C_0^\infty(B_2)$ with bounded derivatives satisfying $\phi \equiv 1$ on B_1. We get

$$\int_{B_1} g(x,y)f(y)dy \leq \int_{B_2} g(x,y)Lu(y)\phi(y)dy$$

and since

$$L(u\phi) = (Lu)\phi + u(L\phi - c\phi) + 2a_{ij}u_{x_i}\phi_{x_j}$$

and

$$\int_{B_2} g(x,y)L(u\phi)(y)dy = 0$$

because $g(x,y)$ is an adjoint solution and $u\phi$ is compactly supported on B_2, we obtain

$$\int_{B_2} g(x,y)Lu(y)\phi(y)dy \leq c\int_{B_2} g(x,y)|u(y)|dy + \\ + c\int_{B_2} g(x,y)|Lu(y)|dy + c\int_{B_2} g(x,y)|\nabla u(y)|dy$$

On B_2 Schauder's estimates give $\|u\|_{W^{2,2}} \leq c\|f\|_2$ and Sobolev's inequalities give $\|u\|_r + \|\nabla u\|_s \leq c\|u\|_{W^{2,2}}$ where $r = \frac{2n}{n-4}$, $s = \frac{2n}{n-2}$.

We conclude that

$$\left(\int_{B_1} g(x,y)^2 dy\right)^{1/2} \leq c\left(\int_{B_2} g(x,y)^{\frac{2n}{n-2}} dy\right)^{\frac{n-2}{2n}}$$

For a fixed n we can iterate this process a finite number of times till we reach an inequality like

$$\left(\int_{B_1} g(x,y)^2 dy\right)^{1/2} \leq c\left(\int_{\widetilde{B}} g(x,y)^q dy\right)^{1/q}$$

where $q \leq \frac{n}{n-1}$ and $\widetilde{B}$ is some larger ball $\widetilde{B} \subset D$.

Now we can use the estimate for the Green's function

$$\left(\int_{\widetilde{B}} g(x,y)^{\frac{n}{n-1}} dy\right)^{\frac{n-1}{n}} \leq c\int_{\widetilde{B}} g(x,y)dy$$

obtained by Fabes and Stroock in [14] and the doubling property for $g(x,y)$ to finally get (37). □

LEMMA 4.8. *Let $B_\epsilon \subset D$ be a ball such that $B_{2\epsilon} \cap \partial D \neq \phi$. If $Q \in \partial D$ is the nearest point to the center of B_ϵ and $\Delta_\epsilon = \{x \in \partial D|\ |x - Q| \leq \epsilon\}$ then*

$$\frac{1}{\epsilon^2}\int_{B_\epsilon} g(x,y)dy \leq c\omega^x(\Delta_\epsilon)$$

where $g(x,y)$ is the Green's function associated to L and D and ω^x is its harmonic measure. The constant c is independent of ϵ and x.

Proof. Let

$$u(x) = \frac{1}{\epsilon^2} \int_{B_\epsilon} g(x,y)dy$$

Notice that $u \in W^{2,P}_{\mathrm{loc}}(D) \cap C(\overline{D})$ for all $p \in (1,\infty)$, $u = 0$ on ∂D and

$$Lu(x) = \begin{cases} 0 & \text{if } x \notin B_\epsilon \\ -\frac{1}{\epsilon^2} & \text{if } x \in B_\epsilon \end{cases}$$

As a consequence of the Comparison Theorem for non-negative solutions that vanish at a portion of the boundary, Theorem 2.6, we have

$$u(x) \leq cu(x_\epsilon) w^x(\Delta_\epsilon)$$

where x_ϵ is a point at distance ϵ to ∂D.

We only have to prove that $u(x_\epsilon)$ is bounded by above by a constant independent of ϵ. To this end consider the barrier $w \in C^2(A_{\epsilon_0}) \cap C(\overline{A}_{\epsilon_0})$ associated to L, where

$$A_{\epsilon_0} = \{x \in D | \delta(x) = \mathrm{dist}(x, \partial D) \leq \epsilon_0\}$$

for some $\epsilon_0 > 0$ and such that $w(x) \simeq \delta(x)^\alpha$ for some $0 < \alpha < 1$ and $Lw(x) \leq -c\delta(x)^{\alpha-2}$ for some constant $c > 0$ (Lemma 1.2).

Take

$$h(x) = w(x) - M\epsilon^\alpha u(x)$$

where $M > 0$ is a constant to be determined. Observe that $Lh(x) \leq 0$ on A_{ϵ_0} if M is small enough and since $u(x)$ is bounded on ∂A_{ϵ_0} we can get also $h(x) \geq 0$ on ∂A_{ϵ_0} by taking M sufficiently small. By the maximum principle $h(x) \geq 0$ for all $x \in A_{\epsilon_0}$ and thus

$$u(x_\epsilon) \leq \frac{1}{M\epsilon^\alpha} w(x_\epsilon) \leq \text{constant}.$$

□

LEMMA 4.9. *Let $B_r \subset D$ be a ball such that its concentric double $B_{2r} \subset D$. If $u \in W^{2,p}_{\mathrm{loc}}$ is a solution to $Lu = 0$ in D we have*

$$\int_{B_r} \left| \nabla u(y) \right|^2 dy \leq \frac{c}{r^2} \int_{B_r} \left| u(y) \right|^2 dy \tag{38}$$

where the constant c is independent of r.

Proof. By a dilation by r as we did in the proof of Lemma 3.3 we may assume $r = 1$ and L an elliptic operator with bounded lower order terms. Inequality (38) is then a part of Schauder's estimates.

The constant in the Schauder's estimates depends on ellipticity, modulus of continuity of the coefficients and the relative position of the balls inside the domain. Since by the dilation the ellipticity λ remains the same, the moduli of continuity are bounded by the same constants and if B_1 is the dilated of B_r we still have that its concentric double B_2 lies inside the dilated of the domain $D, \widetilde{D}$, we conclude that the constant c is independent of r. $\square$

Proof of Theorem 4.0. Existence: If g is continuous at ∂D then there exists a solution u to the Dirichlet problem $Lu = 0$ in $D, u = g$ at ∂D and such that

$$N_\alpha(u) \le C_\alpha \mathcal{M}_\omega(g) \tag{39}$$

where

$$\mathcal{M}_\omega(g)(Q) = \sup_{r>0} \frac{1}{\omega(\Delta(r,Q))} \int_{\Delta(r,Q)} |g(y)| d\omega(y)$$

is the Hardy-Littlewood maximal function of g with respect to the harmonic measure ω, see [2].

Since property (b) in Theorem 1 implies

$$\left(\frac{1}{\sigma(\Delta)} \int_\Delta k^{2+\epsilon} d\sigma \right)^{\frac{1}{2+\epsilon}} \le c \frac{1}{\alpha(\Delta)} \int_\Delta k d\sigma$$

for some $\epsilon > 0$, by Hölder's inequality we get

$$\frac{1}{\omega(\Delta)} \int_\Delta |g| d\omega \le \frac{1}{\omega(\Delta)} \left(\int_\Delta |g|^{\frac{2+\epsilon}{1+\epsilon}} \right)^{\frac{1+\epsilon}{2+\epsilon}} \left(\int_\Delta k^{2+\epsilon} d\sigma \right)^{\frac{1}{2+\epsilon}} \le$$
$$\le c \left(\frac{1}{\sigma(\Delta)} \int_\Delta |g|^{\frac{2+\epsilon}{1+\epsilon}} \right)^{\frac{1+\epsilon}{2+\epsilon}} \le c\mathcal{M}_\sigma \left(|g|^{\frac{2+\epsilon}{1+\epsilon}} \right)^{\frac{1+\epsilon}{2+\epsilon}}$$

and since

$$\|\mathcal{M}_\sigma(g)\|_q \le c\|g\|_q$$

for $q > 1$ we finally get

$$\|\mathcal{M}_\omega(g)\|_{L^2(d\sigma)} \le c\|g\|_{L^2(d\sigma)}$$

which combined with (39) gives the conclusion (2.4) of Theorem 4.0.

The existence of the solution for general boundary data in L^2 follows now by a limiting argument.

Uniqueness: We have to prove that given a solution u to $Lu = 0$ in D such that $N_\alpha(u) \in L^2(d\sigma)$, if $u(x)$ converges nontangentially to 0 at a.e. $d\sigma$ boundary point then $u \equiv 0$ in D.

To do that consider a $C_0^\infty(D)$ function φ_ϵ such that $0 \le \varphi_\epsilon \le 1$, $\varphi_\epsilon = 1$ on $\{x \in D | \mathrm{dist}(x, \partial D) \ge 2\epsilon\}$, $\varphi_\epsilon = 0$ on $\{x \in D | \mathrm{dist}(x, \partial D) \le \epsilon\}$ and

$$\left| \frac{\partial^\alpha \varphi_\epsilon}{\partial x^\alpha} \right| \le c_\alpha \epsilon^{-|\alpha|} \text{ for } |\alpha| \le 2.$$

Given a fixed $x \in D$, if $\epsilon > 0$ is small enough we have

$$u(x) = u(x)\varphi_\epsilon(x) = -\int_D g(x,y) L(u\varphi_\epsilon)(y)dy$$

and since

$$L(u\varphi_\epsilon) = (Lu)\varphi_\epsilon + 2\sum a_{ij} u_{x_i} (\varphi_\epsilon)_{x_j} + u(L\varphi_\epsilon - c\varphi_\epsilon)$$

and $Lu = 0$ we get

$$|u(x)| \le \frac{c}{\epsilon} \int_{T_\epsilon} g(x,y) | \nabla u(y)| dy + \frac{c}{\epsilon^2} \int_{T_\epsilon} g(x,y) |u(y)| dy$$

where the integrals are taken over $T_\epsilon = \{x \in D | \epsilon \le \mathrm{dist}(x, \partial D) \le 2\epsilon\}$.

The uniqueness will follow if we prove that both terms go to 0 as $\epsilon \to 0$.

We will treat the first term, the second being similar

Let us consider a family of balls of radius ϵ, B_ϵ^j, with finite overlapping such that T_ϵ is contained in $U_j B_\epsilon^j$.

If Δ_ϵ^j is the portion of the boundary lying in front of B_ϵ^j we have by Lemma 4.7, 4.8 and Theorem 1

$$\int_{T_\epsilon} g(x,y)^2 dy \le c \sum_j \int_{B_\epsilon^j} g(x,y)^2 dy \le c \sum_j \left[\epsilon^{\frac{n}{2}-n} \int_{B_\epsilon^j} g(x,y) dy \right]^2 \le$$

$$\le c\epsilon^{-n+4} \sum_j \omega^x(\Delta_\epsilon^j)^2 = c\epsilon^{-n+4} \sum_j \left(\int_{\Delta_\epsilon^j} k d\sigma \right)^2 \le \tag{40}$$

$$\le c\epsilon^{-n+4} \sum_j \left(\int_{\Delta_\epsilon^j} k^2 d\sigma \right) \epsilon^{n-1} \le c\epsilon^3.$$

By Lemma 4.9

$$\int_{T_\epsilon} |\nabla u|^2 dy \le c \sum_j \int_{B_\epsilon^j} |\nabla u|^2 dy \le \frac{c}{\epsilon^2} \sum_j \int_{B_\epsilon^j} u^2 dy \le \frac{c}{\epsilon^2} \int_{T_\epsilon} u^2 dy \tag{41}$$

Using (40) and (41), we finally obtain

$$\frac{1}{\epsilon}\int_{T_\epsilon} g(x,y)|\nabla u(y)|dy \leq \frac{1}{\epsilon}\left(\int_{T_\epsilon} g(x,y)^2 dy\right)^{1/2}\left(\int_{T_\epsilon} |\nabla u|^2 dy\right)^{1/2} \leq$$

$$\leq \frac{c}{\epsilon^{1/2}}\left(\int_{T_\epsilon} u^2 dy\right)^{1/2}$$

which goes to 0 as $\epsilon \to 0$ as we can see by using the dominated convergence theorem and the assumption $N_\alpha(u) \in L^2(d\sigma)$.

REFERENCES

[1] S. Agmon, A. Douglis and L. Niremberg, *Estimates near the boundary for solutions of elliptic PDE satisfying general boundary conditions*, Comm. Pure Appl. Math. 12 (1959), pp. 623–727.

[2] A. Ancona, *Principe de Harnack à la frontier et théorème de Fatou pour un operateur elliptique dans un domain Lipschitzien*, Ann. Inst. Fourier (Grenoble) 28 (1972), pp. 169–213.

[3] A. Ancona, *Théorème de Fatou et frontiere de Martin pour une class d' operateurs elliptiques dans un demi-espace*, C.R. Acad. Sc. Paris, t. 290, Serie A (1980), pp. 401–404.

[4] O. Arena, *Bounds for the nonhomogeneous GASPT Equation*, Annali di Matematica Pura ed applicata (IV), vol. CXXXVI (1984), pp. 153–182.

[5] T. Barcelo, *On the harmonic measure for nondivergence elliptic equations*, To appear in Communications in PDE.

[6] T. Barcelo, *A completely singular harmonic measure for a nondivergence form elliptic operator with a drift*, Communications in PDE, vol 13, no. 7 (1989), pp. 931–958.

[7] T. Barcelo, *A comparison and Fatou theorem for a class of nondivergence elliptic equations with singular lower order terms*, To appear in Indiana Univ. Math. J.

[8] P. Bauman, *Positive solutions of elliptic equations in nondivergence form and their adjoints*, Arkiv for Matematic, Vol. 22, no. 2 (1984), pp. 153–173.

[9] A. Beurling, L. Ahlfors, *The boundary correspondence under quasiconformal mappings*, Acta Math., 96 (1956), pp. 125–142.

[10] J. Bony, *Principe du maximum dans les espaces de Sobolev*, C.R. Acad. Sci. Paris Ser. A 265 (1967), pp. 333–336.

[11] L. Caffarelli, E. Fabes, S. Mortola, S. Salsa, *Boundary behaviour of nonnegative solutions of elliptic operators in divergence form*, Indiana J. Math. 30 (4) (1981), pp. 621–640.

[12] L. Caffarelli, E. Fabes, C. Kenig, *Completely singular elliptic-harmonic measures*, Indiana Univ. Math. J. Vol. 30, no. 6 (1981), pp. 917–924.

[13] L. Carlesson, *On mappings conformal at the boundary*, J. Analyse Math. XIX (1967), pp. 1–13.

[14] R. Coifman, D. Deng, Y. Meyer, *Domaine de la racine cantée de certains opérateurs differentiels accrétifs*, Annals Inst. Fourier , 33, 2 (1983), pp. 123–134.

[15] B. Dahlberg, *On estimates of harmonic measure*, Arch. Rational Mech. Anal., 65 (1977), pp. 272–288.

[16] B. Dahlberg, *On the absolute continuity of elliptic measures*, Amer. J. of Math. 108 (1986), pp. 1119–1138.

[17] B. Dahlberg, *On estimates of harmonic measure*, Arch. Rational Mech. Anal. 65 (1977), pp. 272–288.

[18] B. Dahlberg, *Weighted norm inequalities for the Lusin area integral and the nontangential maximal function for functions harmonic in a Lipschitz domain*, Studia Math. 67 (1980), pp. 297–314.

[19] G. David, J. Journé, *A boundedness interior for Calderon Zygmund operators*, Annls of Math 120 (1984), pp. 371–397.

[20] E. Fabes, D. Jerison, C. Kenig, *Multilinear Littlewood-Paley estimates with applications to partial differential equations*, Proc. NAS 79 91982),, pp. 5748–5750.

[21] E. Fabes, O. Jerison, C. Kenig, *Necessary and sufficient conditions for absolute continuity of elliptic-harmonic measure*, Annals of Math 119 (1984), pp. 121–141.

[22] E. Fabes, D. Jerison, C. Kenig, *Multilinear square functions and partial differential equations*, Amer. J. of Math. 107 (1985), pp. 1325–1367.

[23] E. Fabes, D. Stroock, *The L^p-integrability of Green's functions and fundamental solutions for elliptic and parabolic equations*, Duke Mathematical J. Vol. 51, no. 4 (1984), pp. 997–1016.

[24] R. Fefferman, *A criterion for the absolute continuity of the harmonic measure associated with an elliptic operator*, Journal of the AMS, vol. 1, Number 1 (1989), pp. 127–136.

[25] R. Fefferman, C. Kenig, J. Pipher, *The theory of weights and the Dirichlet problem for elliptic equations*, Preprint.

[26] D. Gilbarg, N. Trudinger, *Elliptic partial differential equations of secodn order*, Second Edition Springer-Verlag, 1983.

[27] R. Hunt, R. Wheeden, *On the boundary values of harmonic functions*, Trans. Amer. Math. Soc. 132 (1968), pp. 307–322.

[28] R. Hunt, R. Wheeden, *Positive harmonic functions on Lipschitz domains*, Trans. Amer. Math. Soc. 147 (1970), pp. 507–527.

[29] D. Jerison, C. Kenig, *Boundary behaviour of harmonic functions in non-tangentially accessible domains*, Adv. in Math. vol. 46, no. 1 (1982), pp. 80–147.

[30] N. Krylov, M. Safanov, *Anf estimate of the probability that a diffusion process hits a set of positive measure*, Dokl. Akad. Nauk SSSR 245 (1979), pp. 253–255; English translation, Soviet Math. Dokl. 20 (1979), pp. 253–255.

[31] M. Lehtinen, *A real-analytic quasiconformal extension of a quasisymmetric function*, Ann. Acad. Sci. Fenn. Ser. A, 1,Math. Vol 3 (1977), pp. 207–213.

[32] G. Lieberman, *Regularized distance and its applications*, Pacific journal of Mathematics, vol. 117 no. 2 (1985), pp. 329–352.

[33] K. Miller, *Barriers on cones for uniformly elliptic operators*, Ann. Mat. Pura Appl. 76 (1967), pp. 93–105.

[34] L. Modica, S. Mortola, S. Salsa, *A nonvariational secodn order elliptic operator with singular elliptic measure*, Proceedings of AMS, vol. 84, no 2 (1982), pp. 225–230.

[35] E.M. Stein, *Singular integrals and differentiability properties of functions*, Princeton University Press, 1970.

[36] A. Weinstein, *Generalized axially symmetric potential theory*, Bulletin Ann. Math. Ser. 69 (1953), p. 20.

SOME QUESTIONS CONCERNING HARMONIC MEASURE*

CHRISTOPHER J. BISHOP**

Key words. Harmonic measure, twist points, Hausdorff dimension, eigenvalues of Laplacian

AMS(MOS) subject classifications. Primary 30C35, Secondary 30C55

The purpose of this note is to discuss some conjectures concerning harmonic measure. We will start by considering harmonic measure on a simply connected plane domain Ω, but eventually we will also consider some multiply connected and higher dimensional domains. Most of these questions are trivial if $\partial\Omega$ has tangents a.e. and many are easy if Ω is only a quasicircle. Thus they are really questions about very non-smooth domains.

Let Ω be a simply connected plane domain and let ω denote harmonic measure on Ω. For our purposes the choice of base point is unimportant. Fix $x \in \partial\Omega$ and define a continuous branch of $\arg(z-x)$ on Ω. We say x is a *twist point* of Ω if both

$$\liminf_{z \to x, z \in \Omega} \arg(z-x) = -\infty$$

and

$$\limsup_{z \to x, z \in \Omega} \arg(z-x) = +\infty.$$

On the other hand, we say Ω has an *inner tangent* at x if there is a unique $\theta_0 \in [0, 2\pi)$ such that for every $0 < \epsilon < \pi/2$ there is a $\delta > 0$ such that :

$$\{x + re^{i\theta} : 0 < r < \delta, |\theta - \theta_0| < \pi/2 - \epsilon\} \subset \Omega$$

Up to a set of Λ_1 (one dimensional Hausdorff) measure zero, this is the same as simply requiring x to be the vertex of a cone in Ω. McMillan's twist point theorem [22] states that (with respect to harmonic measure) almost every boundary point of Ω is of one of these two types. We will let $\text{Tan}(\Omega)$ and $\text{Tw}(\Omega)$ denote the set of inner tangents and twist points respectively.

On the set of inner tangents, harmonic measure is mutually absolutely continuous with Λ_1 [22]. ¿From the point of view of this note, this is the "trivial" case. What is more interesting to us is what harmonic measure looks like on the twist points. Only a few things are known. First, there is a subset $A \subset \text{Tw}(\Omega)$ such that $\omega(A) = \omega(\text{Tw}(\Omega))$ but such that $\Lambda_1(A) = 0$ (see [20], [24]). Precise estimates concerning the Hausdorff dimension of A have been given by Makarov (see our remarks below). Also, $\Gamma \cap \text{Tw}(\Omega)$ has zero harmonic measure whenever Γ is rectifiable ([10]). This says that even though harmonic measure on the twist points is concentrated

*The author is partially supported by an NSF Postdoctoral Fellowship
**Dept. of Mathematics, UCLA

on a set of zero length, this set is so dispersed in the plane that it cannot lie on any rectifiable curve.

In [24] it is shown that

$$\limsup_{r\to 0} \frac{\omega(D(x,r))}{r} = \infty$$

at almost every (ω) twist point. (At almost every tangent point this ratio has a finite, nonzero limit.) Our first conjecture is

CONJECTURE 1. *At almost every (ω) twist point*

$$\liminf_{r\to 0} \frac{\omega(D(x,r))}{r} = 0.$$

Thus at twist points the ratio must oscillate between 0 and ∞ as $r \to 0$. This is sometimes called the "lower density" conjecture. This is true if Ω is a quasidisk. We should also point out that this conjecture implies the result of [10] about twist points and rectifiable curves. For suppose $E \subset \mathrm{Tw}(\Omega)$ has positive harmonic measure and $E \subset \Gamma$. Choose ϵ very small and note that a.e. $x \in E$ is contained in arbitrarily small disks D such that $\omega(D)/\mathrm{rad}(D) < \epsilon$. We can choose a covering of a.e. point of E by a collection of such disks which has bounded overlap, say each point in at most M disks. Then

$$\Lambda_1(\Gamma) \geq M^{-1} \sum \mathrm{rad}(D_j) \geq M^{-1}\epsilon^{-1} \sum \omega(D_j) \geq M^{-1}\epsilon^{-1}\omega(E).$$

Taking $\epsilon \to 0$ shows $\Lambda_1(\Gamma) = \infty$.

An equivalent formulation of Conjecture 1 is:

CONJECTURE 2. *Suppose $E \subset \partial\Omega$ has the following property. For any disjoint collection of disks $\{D_j\}$ with centers in E and radii bounded by diam(E), $\sum$ rad(D_j) $\leq M(E) < \infty$. Then $\Lambda_1(E) = 0$ implies $\omega(E) = 0$.*

This condition is a generalization of being a subset of a rectifiable curve (if $E \subset \Gamma$, the sum of radii in the conjecture is bounded above by the length of Γ), but also includes other sets, such as the "1/4" square Cantor set. To see that it implies Conjecture 1, suppose $E \subset \mathrm{Tw}(\Omega)$ is a set where the liminf is greater than some $\epsilon > 0$. Then E satisfies the hypothesis of Conjecture 2 since

$$\sum \mathrm{rad}(D_j) \leq \epsilon^{-1} \sum \omega(D_j) \leq \epsilon^{-1} \sum \omega(E).$$

Since $E \subset \mathrm{Tw}(\Omega)$ it can be chosen to have zero Λ_1 measure so Conjecture 2 implies it has zero harmonic measure. To prove (1) $\Rightarrow$ (2) use (1) to cover E by a disjoint collection of disks $\{D_j\}$ with $\omega(D_j) \leq \epsilon r_j$. Then $\omega(E) \leq \epsilon M$ is as small as we wish.

Our next few questions require some notation. We will now consider a fixed closed Jordan curve Γ in the plane and its two complementary components Ω_1 and Ω_2. A point $x \in \Gamma$ is a tangent of Γ if it is an inner tangent of both sides and is a twist point of Γ if it is a twist point of either side. Harmonic measures for these domains are denoted ω_1 and ω_2. For $r < t < 1$ let $\theta_i(t)$ denote the angle measure of the largest arc in $\partial D(x,t) \cap \Omega_i$ for $i = 1, 2$. Then the Ahlfors distortion theorem implies

$$\omega_i(\Gamma \cap D(x,r)) \leq C \exp\{-\pi \int_r^1 \frac{dt}{t\theta_i(t)}\},$$

Since $\theta_1(t) + \theta_2(t) \leq 2\pi$ this easily implies

$$\omega_1(\Gamma \cap D(x,r))\omega_2(\Gamma \cap D(x,r)) \leq Cr^2.$$

Now define

$$\epsilon(x,t) = \max_{i=1,2}\{|\pi - \theta_i(t)|\}.$$

A simple calculation shows

$$\frac{1}{\theta_1(t)} + \frac{1}{\theta_2(t)} \geq \frac{2}{\pi} + \frac{2}{\pi}(\frac{\epsilon(x,t)}{\pi})^2,$$

so

$$\frac{\omega_1(D(x,r))\omega_2(D(x,r))}{r^2} \leq C_1 \exp(-C_2 \int_r^1 \epsilon^2(x,t)\frac{dt}{t}).$$

On the tangent points of Γ the left hand side is bounded away from zero so the integral on the right must converge. Our next conjecture (due to L. Carleson) is that the converse holds.

CONJECTURE 3. *Except for a set of zero Λ_1 measure, x is a tangent point of Γ iff*

$$\int_0^1 \epsilon^2(x,t)\frac{dt}{t} < \infty.$$

This is some times called the "epsilon squared" conjecture. Note that it does not mention harmonic measure at all, although it has equivalent reformulations that do. For example: if $E \subset \Gamma$ then $\omega_1 \perp \omega_2$ on E iff

$$\int_0^1 \epsilon^2(x,t)\frac{dt}{t} = \infty,$$

at a.e. (Λ_1) point of E. A weaker version of this conjecture has been proved in [11]. It should also be compared with the Stein-Zygmund theorem of L^2 differentiability of functions [26, Theorem VIII.6].

A related question for simply connected domains has been asked by McMillan.

CONJECTURE 4. *For a.e. $x \in \partial\Omega$ (ω), $\liminf_{r\to 0} \theta(x,r) \leq \pi$.*

Conjecture 4 is easy if Ω is a quasidisk. By definition this holds at inner tangent points, so this is really only a question about the twist points.

The estimate on the product of harmonic measures given above for the plane is also true in higher dimensions, i.e., if Ω_1 and Ω_2 are disjoint domains in $\mathbf{R}^n$ then (with appropriate normalizations)

$$\omega_1(D(x,r)\cap\partial\Omega_1)\omega_2(D(x,r)\cap\partial\Omega_2)\le Cr^{2(n-1)}.$$

For $n>2$ this follows from an estimate of Friedland and Hayman [15] concerning eigenvalues of the Laplacian. To describe this result suppose u is positive and subharmonic on $\mathbf{R}^n$ and vanishes on $\partial\Omega$, and for $r>0$ let $S(x,r)=\partial D(x,r)$ and define

$$m_r(u)=(\int_{S(x,r)}u^2d\sigma)^{1/2}$$

where σ is surface measure on the sphere normalized to have mass 1. They show

$$m_r(u)\le Cm_1(u)\exp(-\int_r^{1/2}\alpha(t)\frac{dt}{t})$$

where $\alpha(t)$ is the characteristic constant of the $n-1$ dimensional set $\Omega(t)$ which is the radial projection of $\Omega\cap S(x,t)$ onto the unit sphere. This can be defined by $\alpha(\alpha+n-2)=\lambda$ where

$$\lambda(\Omega(t))=\inf\frac{\int|\nabla_S f|^2d\sigma}{\int|f|^2d\sigma},$$

where the "inf" is over all Lipschitz, nonnegative functions vanishing off $\Omega(t)$ and $\nabla_S f$ denotes the spherical gradient of f. The constant $-\lambda$ is also the first eigenvalue for the Dirichlet problem with vanishing boundary conditions, at least if $\Omega(t)$ is smooth enough. If f is the eigenfunction corresponding to λ then $u(x)=|x|^\alpha f(x/|x|)$ is harmonic in the cone defined by $\Omega(t)$ iff $\alpha(\alpha+n-2)=\lambda$. See [15] for details.

Now suppose $\Omega_1,\dots,\Omega_m$ are disjoint domains in $\mathbf{R}^n$ and $x\in\cap_j\partial\Omega_j$. Then

$$\prod_{j=i}^m\omega_j(D(x,r))\le Cr^{m(n-2)}\exp(-\int_r^1\sum_{j=1}^m\alpha_j(t)\frac{dt}{t})$$

where α_i are the characteristic constants for the Dirichlet problem on the domains $\Omega_j(t)$. It follows from results in [15] that if Ω_1,Ω_2 are disjoint domains on the sphere then $\alpha_1+\alpha_2\ge 2$, which proves the product formula mentioned above. For $n=2$ and $m>2$ it easy to see that the sum of the α_j's is minimized when each Ω_j is a arc of length $2\pi/m$, but in higher dimensions the extremal configuration is not obvious. For example, when $n=m=3$ I suspect the worst case cases should consist of three spherical domains constructed by connecting two antipodal points on the sphere by three geodesics meeting at 120 degrees. In this case $\alpha=3/2$ and $\lambda=15/4$ so we expect:

Conjecture 5. *If Ω_1, Ω_2 and Ω_3 are disjoint domains in $\mathbf{R}^3$, $x \in \partial\Omega_1 \cap \partial\Omega_2 \cap \partial\Omega_3$ and $D = D(x,r)$ then*

$$\omega_1(D)\omega_2(D)\omega_3(D) \leq Cr^{15/2}.$$

Conjecture 6. *If Ω_1, Ω_2 and Ω_3 are disjoint domains in $\mathbf{S}^2$, the unit sphere of $\mathbf{R}^3$, then*

$$\lambda_1 + \lambda_2 + \lambda_3 \geq \frac{45}{4},$$

where λ_i is the first eigenvalue of the Laplacian for Ω_i with Dirichlet boundary values. (The minimum should be attained by joining antipodal points by three geodesics meeting at 120 degrees.)

Given a domain Ω we define

$$\dim(\omega) = \inf\{\alpha : \exists E \subset \partial\Omega \text{ such that } \omega(E) = 1 \text{ and } \dim(E) = \alpha\}.$$

Here $\dim(E)$ refers to the Hausdorff dimension of E. Makarov has proven that $\dim(\omega) = 1$ for every simply connected domain Ω in the plane. Peter Jones and Tom Wolff [16] have shown that $\dim(\omega) \leq 1$ for any planar domain, verifying a conjecture of Øksendal. Øksendal had also conjectured that $\dim(\omega) \leq n-1$ for any domain $\Omega \subset \mathbf{R}^n$, but Wolff [27] has shown this is false. In the other direction, Jean Bourgain [12] has shown that there exists an $\epsilon > 0$ such that $\dim(\omega) \leq n - \epsilon$ for every $\Omega \subset \mathbf{R}^n$ (and his ϵ depends on n). The obvious problem is to determine what the best value of ϵ is.

Conjecture 7. *For any domain $\Omega \subset \mathbf{R}^n$, $\dim(\omega) \leq d_n \equiv n-1+(n-2)/(n-1)$.*

There is no strong reason for choosing this value. It has been suggested only because of the fact that if f is harmonic on $\mathbf{R}^n$ then $|\nabla f|^q$ is subharmonic for $q \geq (n-2)/(n-1)$. Suggestions for a better value are welcome.

If Ω_1 and Ω_2 are disjoint (not necessarily simply connected) domains in $\mathbf{R}^2$, then it is known exactly when their harmonic measures will be mutually singular (see below), but this is not understood in higher dimensions. In particular, if the harmonic measures for $\Omega_1, \Omega_2 \subset \mathbf{R}^2$ are not mutually singular then one can show $\partial\Omega_1 \cap \partial\Omega_2$ must intersect a Lipschitz graph in positive length and that the 2 harmonic measures be mutually absolutely continuous to Λ_1 on this intersection.

Conjecture 8. *If $\Omega_1, \Omega_2 \subset \mathbf{R}^n$ are disjoint domains with harmonic measures ω_1, ω_2 that are mutually absolutely continuous on a set $E \subset \partial\Omega_1 \cap \partial\Omega_2$, of positive measure, then there exists $F \subset E$ of positive measure such that ω_1 and ω_2 are mutually absolutely continuous with Λ_{n-1} on F*

This may be false, but it would be interesting even to prove Conjecture 7 under the additional hypothesis of mutually continuity, i.e., if $\omega_1 \ll \omega_2 \ll \omega_1$ on E then there exists a set $F \subset E$ of positive measure with $\dim(F) \leq d_n$.

There is another refinement of the Jones-Wolff result suggested by the results in [8]. The theorem on mutually singularity of harmonic measures mentioned above is stated in terms of a Wiener type condition involving the logarithmic capacity of $\partial\Omega$ as follows. For $x \in \mathbf{R}^2$, $\delta > 0$, $\epsilon > 0$ and $\theta \in [0, 2\pi)$ we define the cone and wedge

$$C(x, \delta, \epsilon, \theta) = \{x + re^{i\psi} : 0 < r < \delta, |\psi - \theta| < \epsilon\}$$

$$W(x, \delta, \epsilon, \theta) = C(x, \delta, \epsilon, \theta) \cap \{z : \delta/2 \le |z - x|\}.$$

We also let $\operatorname{cap}(E)$ denote the logarithmic capacity of E. For a fixed x, ϵ and θ let

$$\gamma(k) = \operatorname{cap}(2^{k-2}(W(x, 2^{-k}, \epsilon, \theta) \backslash \Omega)),$$

i.e., $\gamma(k)$ is the capacity of $\Omega^c \cap W(x, 2^{-k}, \epsilon, \theta)$ after we have dilated it to have diameter about $1/2$. We say a point $x \in \partial\Omega$ satisfies a weak cone condition (WCC) if there exists ϵ and θ such that

$$\sum_{k=1}^{\infty} \gamma(k) < \infty.$$

We refer to this as a "weak" condition because it generalizes the cone condition stated in [6], [9] which requires that

$$C(x, \delta, \epsilon, \theta) \subset \Omega.$$

It is clear that this condition implies the WCC since all but finitely many of the terms in the sum will be zero.

THEOREM. *Suppose Ω_1 and Ω_2 are disjoint subdomains in $\mathbf{R}^2$ and let ω_1 and ω_2 be their harmonic measures. Then $\omega_1 \perp \omega_2$ iff the set of points in $\partial\Omega_1 \cap \partial\Omega_2$ satisfying a weak cone condition with respect to both Ω_1, Ω_2 has zero 1 dimensional measure, Λ_1. Moreover, if ω_1 and ω_2 are mutually absolutely continuous on a set E then there is Besicovitch regular $F \subset E$ with $\omega_i(F) = \omega_i(E)$ and ω_i mutually absolutely continuous with Λ_1 on F for $i = 1, 2$.*

Wolff has proven that

$$F = \{x \in \partial\Omega : \limsup_{r \to 0} \frac{\omega(D(x, r) \cap \partial\Omega)}{r} > 0\}$$

has sigma finite length and full harmonic measure for any planar domain (unpublished). It is shown in [8] that if $E \subset \partial\Omega$ and every point of E is the vertex of a cone with convergent capacity series (as above) then

$$\lim_{r \to 0} \frac{\omega(D(x, r) \cap \partial\Omega)}{r} < \infty$$

Λ_1 a.e. on E. It seems possible that the converse is also true, i.e.,

CONJECTURE 9. *If $E \subset \partial\Omega$, $0 < \Lambda_1(E) < \infty$ and no point of E satisfies the weak cone condition then for ω a.e. $x \in E$*

$$\limsup_{r \to 0} \frac{\omega(D(x,r) \cap \partial\Omega)}{r} = \infty.$$

In particular, there exists $F \subset E$ with $\omega(F) = \omega(E)$, but $\Lambda_1(F) = 0$.

Wolff's theorem tells us that we can split the boundary of any planar domain into two pieces E, $\partial\Omega \backslash E$ such that harmonic measure is mutually continuous with Λ_1 on E and singular to Λ_1 on $\partial\Omega\backslash E$. If Conjecture 9 is true it gives a geometric characterization of this set. The conjecture is consistent with what is known in the simply connected case. It also has the following consequence which is of interest in its own right:

CONJECTURE 10. *If $\Omega \subset \mathbf{R}^2$ is a domain and $E \subset \partial\Omega$ is Besicovitch irregular then there exists $F \subset E$ with $\omega(F) = \omega(E)$ and $\Lambda_1(F) = 0$.*

Peter Jones has pointed out that this is true in the case when $E = \partial\Omega$ satisfies a capacitary "thickness" condition: there exists $\epsilon > 0$ such that for every $x \in \partial\Omega$ and $0 < r < r_0$, $\text{cap}(r^{-1}(D(x,r/4) \cap \partial\Omega)) \geq \epsilon$.

Our next question is not about harmonic measure, but rather about Brownian motion. We will say a Brownian path γ in $\Omega = \mathbf{R}^2\backslash E$ separates E if there are $0 \leq s < t < \tau$ (τ is the time the particle hits E) such that $\gamma(s) = \gamma(t)$ and so that the closed curve $\Gamma = \gamma([s,t])$ separates E (i.e., has points of E in more than one of its complementary components). For example, if $E \subset \mathbf{R}$ and $|E| > 0$ then it is clear that a Brownian path (say starting at $(0,1)$) has a positive probability of hitting E the first time it hits $\mathbf{R}$ (so it does not separate E). If E is the middle third Cantor set then almost every Brownian path separates E. One might think that the probability of a path separating E is 1 whenever $|E| = 0$. However, this in not the case. There is an example of a compact set $E \subset \mathbf{R}$ with $|E| = 0$ but such that a Brownian particle in $\Omega = \mathbf{R}^2\backslash E$ has a positive probability of hitting E without separating E [7].

The idea of a Brownian path separating the boundary is related to the problem of characterizing the compact sets E in $\mathbf{R}^2$ such that $E \cap \partial\Omega$ has zero harmonic measure in Ω for any simply connected Ω. We shall call such a set a SC-null set. If $\Omega = \mathbf{R}^2\backslash E$ and Brownian paths separate E a.s. then E has this property, since Brownian paths will have to hit $\partial\Omega\backslash E$ a.s. before hitting E. The example mentioned above was originally constructed to show the converse is false. This is because a theorem of Øksendal says that a subset of $\mathbf{R}$ is SC-null iff it has zero length. Thus the set constructed here must have zero harmonic measure in every simply connected domain even though it is not a.s. separated by Brownian paths. As mentioned earlier, a subset of a rectifiable curve is SC-null iff it has zero length [10]. Makarov [20] has proven that if $\Lambda_h(E) = 0$ where Λ_h is the Hausdorff measure associated to the function

$$h(t) = t\exp(C\sqrt{\log(1/t)\log\log\log(1/t)})$$

then E is SC-null, and that this is sharp except for the choice of $C > 0$. Neither of these results is easy, and a characterization of SC-null sets is probably quite difficult.

K. Burdzy has pointed out to me that if the set E has small enough Hausdorff dimension, then Brownian motion in the complement of E necessarily separates E. Consider a Brownian path starting at distance ϵ from the origin and let it run until the first time it hits the unit circle. Let $P(\epsilon)$ denote the probability that the origin and and infinity are in the same connected component of the path's complement, and suppose that it satisfies $P(\epsilon) \leq C\epsilon^{\alpha}$ for some fixed $\alpha > 0$. If E has finite α-dimensional Hausdorff measure, then an application of the Borel-Cantelli lemma shows that E is necessarily separated by Brownian paths hitting it. In [13] Burdzy and Lawler have shown that $P(\epsilon) \leq \epsilon^{\pi^{-2}}$. In light of Makarov's theorem we expect

CONJECTURE 11. *If* $E \subset \mathbb{R}^2$ *is compact and* $\dim(E) < 1$ *then Brownian paths separate it almost surely.*

This has also been conjectured by Terry Lyons. The conjecture is true if $E \subset \mathbb{R}$ or if we assume $\dim(E) < 1/2$ [7]. Another set of interesting questions concern the analogs of various theorems on harmonic measure in the discrete case, i.e., theorems about the hitting probabilities of random walks on lattices. Not very much is known about harmonic measure for random walks. One exception is a result of Harry Keston [19] which is a random walk analogy of a classical estimate of Beurling. Keston uses his result to estimate the rate of growth of Diffusion Limited Aggregation (DLA) [18].

DLA has a continuous and a discrete version. The continuous version can be described as follows. Fix a unit disk at the origin. Start another disk near infinity and move it along a Brownian path until the first time it hits the first disk and then stop it. Successively add new disks in the same way and try to describe what the resulting collection looks like. In the discrete version (as in Kesten's work), we move along a lattice by a random walk until we reach a vertex which is adjacent to a previously occupied vertex. We stop here and call this new vertex occupied. Among the questions we can ask, one of the simplest is what the average rate of growth of the diameter, D, is as a function of the number of disks, N. Trivially, $C\sqrt{N} \leq D \leq N$ and Kesten proved that $D \leq CN^{2/3}$ almost surely [18].

CONJECTURE 12. *There exists* $\epsilon > 0$ *such that* $D \geq CN^{1/2+\epsilon}$ *almost surely.*

On the the basis of computer pictures this seems obvious. Proving it, however, seems very difficult. One should also attempt to establish the exact order of growth. One interesting aspect of the computer simulations is that when DLA is grown on a lattice and a large enough number of disks are considered, the resulting object is not the same in all directions (as had been expected). Instead the growth is fastest along the axis directions, so the DLA resembles a cross. Proving this actuals happens would be another interesting problem. See [25] for background on this and related physical processes.

REFERENCES

[1] L.V. AHLFORS, *Conformal Invariants*, McGraw-Hill, New York, 1973.

[2] ——————, *Lectures on quasiconformal mappings*, Wadsworth and Brooks/Cole, Monterey, California, 1987.

[3] A. ANCONA, *Une propiété de la compactification de Martin d'un domaine euclidien*, Ann. Inst. Fourier (Grenoble), 29 (1979), pp. 71–90.

[4] M. BENEDICKS, *Positive harmonic functions vanishing on the boundary of certain domains in* $\mathbf{R}^n$, Ark. Mat., 18 (1980), pp. 53–72.

[5] A. BEURLING, *Études sur un Problem de Majoration*, thesis, Upsala 1933.

[6] C.J. BISHOP, *Harmonic measures supported on curves*, thesis, University of Chicago, 1987.

[7] ——————, *Brownian motion in Denjoy domains*, preprint, 1990.

[8] ——————, *A characterization of Poissonian domains*, to appear in Arkiv Mat..

[9] C.J. BISHOP, L. CARLESON, J.B. GARNETT AND P.W. JONES, *Harmonic measures supported on curves*, Pacific J. Math., 138 (1989), pp. 233–236.

[10] C.J. BISHOP AND P.W. JONES, *Harmonic measure and arclength*, to appear in Ann. Math..

[11] ——————, *Harmonic measure, L^2 estimates and the Schwarzian derivative* , preprint 1990.

[12] J. BOURGAIN, *On the Hausdorff dimension of harmonic measure in higher dimensions*, Inv. Math., 87 (1987), pp. 477–483.

[13] K BURDZY AND G.F. LAWLER, *Non-intersection exponents for Browian paths. Part II. Estimates and applications to a random fractal*, to appear in Ann. Probab..

[14] L. CARLESON, *On the support of harmonic measure for sets of Cantor type*, Ann. Acad. Sci. Fenn. Ser. A.I Math., 10 (1985), pp. 113–123.

[15] S. FRIEDLAND AND H.K. HAYMAN, *Eigenvalue inequalities for the Dirichlet problem on spheres and the growth of subharmonic functions*, Comment. Math. Helv., 51 (1976), pp. 133-161.

[16] P.W. JONES AND T.H. WOLFF, *Hausdorff dimension of harmonic measures in the plane*, Acta. Math., 161 (1988), pp. 131–144.

[17] D.S. JERISON AND C.E. KENIG, *Boundary behavior of harmonic functions in non-tangentially accessible domains*, Adv. in Math., 46 (1982), pp. 80-147.

[18] H. KESTEN, *How long are the arms in DLA?*, J. Phy. A, 20 (1987), pp. L29-L33.

[19] ——————, *Hitting probabilities of random walks on* $\mathbb{Z}^d$, Stochastic Process. Appl., 25 (1987), pp. 165–184.

[20] N.G. MAKAROV, *On distortion of boundary sets under conformal mappings*, Proc. London Math. Soc., 51 (1985), pp. 369-384.

[21] ——————, *Metric properties of harmonic measure*, in *Proceedings of the International Congress of Mathematicians, Berkeley 1986*, Amer. Math. Soc., 1987, pp. 766–776.

[22] J.E. MCMILLAN, *Boundary behavior of a conformal mapping*, Acta. Math., 123 (1969), pp. 43–67.

[23] CH. POMMERENKE, *Univalent functions*, Vanderhoeck and Ruprecht, Göttingen, 1975.

[24] ——————, *On conformal mapping and linear measure*, J. Analyse Math., 46 (1986), pp. 231–238.

[25] H.E. STANLEY AND N. OSTROWSKY, *On Gowth and form: fractal and non-fractal patterns in physics*, Martinus Nijhoff Publishers, Boston, 1986.

[26] E.M. STEIN, *Singular integrals and differentiability properties of functions*, Princeton University Press, Princeton, New Jersey, 1971.

[27] T.H. WOLFF, *Counterexamples with harmonic gradients*, preprint, 1987.

THE TRACE OF THE HEAT KERNEL IN DOMAINS WITH NONSMOOTH BOUNDARIES

RUSSELL M. BROWN*

In this note, I would like to describe some recent work that considers the relationship between the smoothness of the boundary of a domain in $\mathbf{R}^n$ and the spectral properties of the Laplacian in the domain.

To fix notation, we will let D denote a bounded open subset of $\mathbf{R}^n$ and we use $\Delta = \sum_{i=1}^{n} \frac{\partial^2}{\partial x_i^2}$ to indicate the Laplacian on D. It is well-known that there exists a sequence of eigenvalues $0 < \lambda_1 < \lambda_2 \leq \lambda_3 \ldots$ going to infinity and eigenfunctions ϕ_j which satisfy

$$\begin{cases} -\Delta\phi_j = \lambda_j \phi_j, & \text{in } D \\ \phi_j(Q) = 0, & Q \in \partial D. \end{cases}$$

These Dirichlet eigenfunctions form a complete set in $L^2(D)$ which we require to be orthonormal. A classical question asks the asymptotic behavior of the function

$$N(\lambda) = \#\{j : \lambda_j \leq \lambda\}$$

which counts the number of eigenvalues less than or equal to λ.

H. Weyl obtained that

$$N(\lambda) = \frac{|D|\lambda^{n/2}}{(4\pi)^{n/2}\Gamma(\frac{n}{2}+1)} + o(\lambda^{n/2}), \tag{1}$$

and this result is valid without any assumption on the boundary. The expansion (1) may be obtained by studying the Laplace transform of the measure

$$\mu(t) = \sum_j \delta_{\lambda_j}(t)$$

which places unit mass on each of the eigenvalues counted according to multiplicity. The Laplace transform of μ also arises as the trace of the heat kernel on D:

$$\text{tr}\,(g)(t) = \sum_j e^{-\lambda_j t} = \int_D g(t,x,x)\,dx \tag{2}$$

where $g(t,x,y)$ satisfies

$$\begin{cases} \Delta_x g(t,x,y) - \partial_t g(t,x,y) = 0, & \text{in } (0,\infty)\times D \\ g(0,\cdot,y) = \delta_y(\cdot), & \text{on } D \\ g(s,Q) = 0 & \text{on } (0,\infty)\times\partial D. \end{cases}$$

---

*Department of Mathematics, University of Kentucky, Lexington, KY 40506-0027. The author gratefully acknowledges the support of the National Science Foundation.

The identity (2) follows easily from the eigenfunction expansion

$$g(t,x,y)=\sum_j e^{-\lambda_j t}\phi_j(y)\phi_j(x).$$

Now to obtain Weyl's formula (1) for $N(\lambda)$ one needs to show that

$$\text{tr}\,(g)(t)=(4\pi t)^{-n/2}(|D|+o(1)) \tag{3}$$

and apply a Tauberian theorem.

This approach may be found in M. Kac's paper [9]. Asymptotic expansions for tr (g) may be found [12] where the relationship between tr (g) and geometrical properties of D are developed in the setting of smooth Riemannian manifolds. The best results for asymptotics of $N(\lambda)$ may be found in Ivrii's work [8]. We return now to domains which are minimally smooth and see what improvement we may make in the error terms in (1) and (3).

One approach to this question has been developed by J. Brossard and R. Carmona [2]. They consider the trace of the heat kernel in domains where the boundary is of fractional dimension. They observe that the correct way to measure the boundary is with the Minkowski content. We will define the d-dimensional Minkowski content of a set $E\subset\partial D$ (with respect to D) by

$$\mathcal{M}^d(E)=\lim_{\epsilon\to 0^+}\frac{|\{X\in D:\ \delta(X;E)<\epsilon\}|}{\omega_{n-d}\epsilon^{n-d}} \tag{4}$$

if the limit exists. The normalizing constant $\omega_m=\pi^{m/2}/(2\Gamma(1+m/2)$ is half the volume of the unit m-ball, when m is an integer. Similarly, we may define the upper and lower d-dimensional Minkowski contents by replacing the limit in (4) by lim sup or lim inf, respectively. These will be denoted by $\overline{\mathcal{M}}^d$ and $\underline{\mathcal{M}}^d$.

We make several remarks about the Minkowski content. First, it is not a measure. This is easy to see since $\mathcal{M}(E)=\mathcal{M}(\bar{E})$. Even for closed sets, the Hausdorff dimension and Minkowski dimension may disagree, see [2] for example. However, when k is an integer and E is nice, then $\mathcal{M}$ and $\mathcal{H}$ agree. In particular if E is closed and $(n-1)$-rectifiable (see [7], Theorem 3.2.39).

With these notations we may state

THEOREM 1. (Brossard and Carmona) *Let D be a bounded open set in $\mathbf{R}^n$ with ∂D of finite d-dimensional Minkowski content, then*

$$\text{tr}\,(g)(t)\geq((4\pi t)^{-n/2}|D|-c_1t^{-d/2}\overline{\mathcal{M}}^d(\partial D)+o(t^{-d/2})). \tag{5}$$

In addition, if ∂D satisfies a capacitary density condition, then

$$\text{tr}\,(g)(t)\leq((4\pi t)^{-n/2}|D|-c_2t^{-d/2}\underline{\mathcal{M}}^d(\partial D)+o(t^{-d/2})). \tag{6}$$

We remark that Brossard and Carmona give a probabilistic statement of these results which may provide additional information about the constants c_1 and c_2 in (5) and (6). We will give a sketch of an analytic restatement of their argument.

Proof of Theorem 1. To establish the first estimate, we write the Green's function

$$g(t,x,y)=p(t,x,y)-v(t,x,y)$$

where p is the usual fundamental solution of the heat equation:

$$p(t,x,y) = (4\pi t)^{-n/2} \exp\left(\frac{-|x-y|^2}{4t}\right).$$

And, for y fixed, $v(\cdot,\cdot,y)$ is the solution of the boundary value problem

$$\begin{cases} \partial_t v(t,x,y) - \Delta_x v(t,x,y) = 0, & \text{in } (0,\infty)\times D \\ v(s,Q,y) = p(s,Q,y), & \text{on } (0,\infty)\times \partial D \\ v(0,x,y) = 0, & x \in D. \end{cases}$$

Using the maximum principle for the heat equation we have

$$\begin{aligned} v(t,x,y) &\le \sup_{\substack{0<s<t \\ Q\in\partial D}} p(s,Q,y) \\ &\le Ct^{-n/2} \exp\left(\tfrac{-\delta(y)^2}{Ct}\right) \end{aligned} \tag{7}$$

for some constant C depending only on the dimension. The lower bound (5) in Theorem 1 now follows from the estimate (7) and the following Lemma:

LEMMA 1. *Let D be an open set and suppose that f is decreasing on $(0,\infty)$, $\lim_{t\to\infty} f(t) = 0$ and that*

$$\int_0^\infty s^{n-d}|f'(s)|\,ds < \infty,$$

then

$$\omega_{n-d}(n-d)\underline{\mathcal{M}}^d(\partial D)\int_0^\infty s^{n-d-1} f(s)ds \le \liminf_{\epsilon\to 0+} \tfrac{1}{\epsilon^{n-d}} \int_D f\left(\tfrac{\delta(x)}{\epsilon}\right) dx$$

$$\limsup_{\epsilon\to 0+} \tfrac{1}{\epsilon^{n-d}} \int_D f\left(\tfrac{\delta(x)}{\epsilon}\right) dx \le \omega_{n-d}(n-d)\overline{\mathcal{M}}^d(\partial D)\int_0^\infty s^{n-d-1} f(s)ds.$$

Sketch of proof. Let $\psi_\epsilon(r) = \epsilon^{d-n}|\{x\in D : \delta(x) < \epsilon r\}|$. It is fairly easy to see that

$$\epsilon^{d-n}\int_D f(\delta(x)/\epsilon)\,dx = \int_0^\infty f(r)\,d\psi_\epsilon(r)$$

where the integral on the right is an improper Riemann-Stieltjes integral. Now, we may integrate by parts, use our assumption on f' and the definition of Minkowski content to compare the limit of the last integral with the integral $\int_0^\infty f(s)\,ds^{n-d}$. See [3] where a special case of this Lemma is established. □

Finally, the upper bound in Theorem 1 follows from a lower bound for v. The maximum principle only gives the trivial lower bound of zero for v. Using the capacitary density hypothesis we can show that $v(t,x,x) > c_0 t^{-n/2}$ for $\delta(x) < \sqrt{t}$ which quickly leads to (6). We refer the reader to [2] for precise definitions and details. □

A similar result may be established for the counting function, $N(\lambda)$. This was announced by M. Lapidus and J. Fleckinger-Pelle in [11] and a proof is available in [10].

THEOREM 2. (Fleckinger-Pelle and Lapidus) *Let D be a bounded open set with ∂D of finite d-dimensional Minkowski content, $d > n-1$, then*

$$N(\lambda) = \frac{\lambda^{n/2}|D|}{(4\pi)^{n/2}\Gamma\left(\frac{n}{2}+1\right)} + O(\lambda^{d/2}).$$

The proof of Lapidus [10] depends on the classical min-max characterization of eigenvalues. We sketch another proof of Theorem 2.

Proof of Theorem 2. Let

$$e(\lambda, x, y) = \sum_{\{j:\ \lambda_j \leq \lambda\}} \phi_j(x)\phi_j(y)$$

denote the spectral function for D. This is the kernel of the projection onto an initial segment of the spectrum. We will need two facts about e:

$$\int_D e(\lambda, x, x)\,dx = N(\lambda), \tag{8}$$

$$|e(\lambda, x, x) - c_n\lambda^{n/2}| \leq C \begin{cases} \frac{\lambda^{(n-1)/2}}{\delta(x)}, & \sqrt{\lambda} > \frac{1}{\delta(x)} \\ \lambda^{n/2}, & \sqrt{\lambda} \leq \frac{1}{\delta(x)}. \end{cases} \tag{9}$$

In (9), $c_n = \frac{1}{\Gamma\left(\frac{n}{2}+1\right)(4\pi)^{n/2}}$.

The first fact is obvious, the second is established by R. Seeley in [13] where the proof relies on the finite propagation speed of the support of solutions of the wave equation and a Tauberian argument. Now the conclusion of Theorem 2 follows by integrating (9) over D and applying Lemma 1. □

These results give a fairly complete picture for the roughest domains where we use the rather coarse criterion of the dimension of the boundary to measure smoothness. On the other hand, smooth domains can be studied using parametrix methods to construct the heat kernel as a Levi sum (see [1,12]).

Thus we turn to the question of establishing asymptotics for domains whose boundary is of dimension $n-1$, but not smooth. Here, the best result for the counting function $N(\lambda)$ is due to Courant who showed that

$$N(\lambda) = \frac{|D|\lambda^{n/2}}{(4\pi)^{n/2}\Gamma\left(\frac{n}{2}+1\right)} + O(\lambda^{(n-1)/2}\log(\lambda)), \quad \text{as } \lambda \to \infty.$$

This may also be proven using the techniques of Seeley discussed above.

However, for the heat kernel, we may do better. We will say that D is Lipschitz if locally the boundary of D may be represented as the graph of a Lipschitz function.

THEOREM 3. *If D is Lipschitz, then*

$$\operatorname{tr}(g)(t) = (4\pi t)^{-n/2}(|D| - \frac{\sqrt{\pi t}}{2}\mathcal{H}^{n-1}(\partial D) + o(t^{1/2})). \tag{10}$$

Several remarks about this theorem are in order. 1) If ∂D is C^1, then this result is obtained by Brossard and Carmona in [2]. 2) As observed earlier, for closed subsets of Lipschitz graphs, the Hausdorff measure and the Minkowski content agree; hence this result may be viewed as a sharpening of Theorem 1. 3) This theorem continues to hold for slightly more general domains than Lipschitz (see [3] for details).

We briefly sketch the proof of Theorem 3 and refer the reader to [3] for details.

Sketch of Proof. In order to obtain the second term in the expansion (10) we need to deal more carefully with the boundary. We may divide this argument into three steps. 1) Construct a set of points where the boundary is almost flat. 2) Show that near this set, the heat kernel g can be approximated by the heat kernel for a half-space defined by some tangent plane of ∂D. This leads to the approximation

$$g(t,x,x) \approx (4\pi t)^{-n/2}\left(1 - \exp\left(\frac{-\delta(x)^2}{t}\right) + \text{error}\right),$$

which via Lemma 1 integrates to give the desired expansion. 3) Show that the bad set, where ∂D has no tangent plane may be ignored for the purposes of computing the trace. The first two steps are carried out by Brossard and Carmona who obtain Theorem 3 under the hypothesis that ∂D is C^1 and hence step 3 is unnecessary. In [3], the extension to Lipschitz domains is established. While the set of points where ∂D has no tangent plane is of measure zero in ∂D, it may be dense. Hence, carrying out step 3 is not completely trivial. □

Finally, we remark that an analogue of Theorem 3 has recently been established for Neumann boundary conditions ([5]). The outline of the argument for Neumann boundary conditions is the same as for Dirichlet boundary conditions. The main difference comes when we try to estimate the error that arises when we compare the heat kernel for D with the heat kernel for the half space bounded by a tangent plane to ∂D. We no longer have the maximum principle at our disposal; as a substitute we have the following estimate for solutions of the Neumann problem in D. For data f on $(0,\infty) \times \partial D$, the initial-Neumann problem seeks a function u satisfying

$$\begin{cases} \partial_t u - \Delta u = 0, & \text{in}(0,\infty) \times D \\ \partial_\nu u(s,Q) = f(s,Q), & \text{on } (0,\infty) \times \partial D \\ u(0,x) = 0, & \text{in } D. \end{cases} \tag{11}$$

We are using ∂_ν to denote differentiation in the direction of the unit inner normal.

LEMMA 2. *Suppose that ∂D is Lipschitz and that $f \in L^p(D)$, $p > n+1$. Then the solution of the Neumann problem with data f satisfies*

$$\|u\|_{L^\infty((0,1)\times\partial D)} \le c_p \|f\|_{L^p((0,1)\times\partial D)}.$$

The constant c_p also depends on the diameter of D.

This estimate should be viewed as asserting that the solution operator behaves like a parabolic fractional integral of order one. This explains the condition on p. A proof may be found in [5]. The proof depends on techniques used by Dahlberg and Kenig [6] and the author [4] to study the Neumann problem with L^p-data for Laplace's equation and the heat equation, respectively. This lemma is useful since control of solutions to (11) in the L^∞-norm allows us to estimate $g(t,x,y)$ at the point $y = x$.

With this estimate, we may prove

THEOREM 3′. *Let D be a Lipschitz domain and let g be the heat kernel which has zero Neumann data on the lateral boundary, then*

$$\mathrm{tr}\,(g)(t) = (4\pi t)^{-n/2}(|D| + \frac{\sqrt{\pi t}}{2}H^{n-1}(\partial D) + o(t^{1/2})).$$

We remark that, unlike Theorem 3, we are unable to relax the hypothesis that ∂D is Lipschitz in Theorem 3′. Lemma 2 uses this hypothesis in a substantial way.

REFERENCES

[1] P.B. BAILEY AND F.H. BROWNELL, *Removal of the log factor in the asymptotic estimates of polyhedral membrane eigenvalues,* J. Math. Anal. and Appl., 4 (1962), pp. 212–239.

[2] JEAN BROSSARD AND RENÉ CARMONA, *Can one hear the dimension of a fractal?,* Comm. Math. Phys., 104 (1986), pp. 103–122.

[3] RUSSELL M. BROWN, *The trace of the heat kernel in nonsmooth domains,* Preprint, 1989.

[4] RUSSELL M. BROWN, *The initial-Neumann problem for the heat equation in Lipschitz cylinders,* Trans. Amer. Math. Soc., 302 (1990), pp. 1–52.

[5] RUSSELL M. BROWN, *The trace of heat kernel in nonsmooth domains: Neumann boundary conditions,* In preparation, 1990.

[6] B.E.J. DAHLBERG AND C.E. KENIG, *Hardy space and the Neumann problem in L^p for Laplace's equation in Lipschitz domains,* Ann. of Math., 125 (1987), pp. 437–466.

[7] H. FEDERER, *Geometric Measure Theory,* Springer Verlag, 1969.

[8] V. JA. IVRII, *Second term of the spectral asymptotic expansion of the Laplace-Beltrami operator on manifolds with boundary,* Funct. Anal. Appl., 14 (1980), pp. 98–106.

[9] M. KAC, *Can one hear the shape of a drum?,* Amer. Math. Monthly, 73 (1966), pp. 1–23. Part II of H.E. Slaught Memorial Papers.

[10] M. LAPIDUS, *Fractal drum, inverse spectral problems for elliptic operators and a partial resolution of the Weyl-Berry conjecture,* Preprint, 1989.

[11] M.L. LAPIDUS AND J. FLECKINGER-PELLE, *Tambour fractale: vers une resolution de la conjecture de Weyl-Berry pour les valeurs propres du Laplacien,* C. R. Acad. Sci. Paris Ser. I, 306 (1988) pp. 171–175.

[12] H.P. MCKEAN AND I.M. SINGER, *Curvature and eigenvalues of the Laplacian,* J. Diff. Geom., 1 (1967) pp. 43–69.

[13] R. SEELEY, *A sharp remainder estimate for the eigenvalues of the Laplacian in a domain in* $\mathbf{R}^3$, Adv. in Math., 29 (1978) pp. 244–269.

A NOTE ON L^p ESTIMATES FOR PARABOLIC SYSTEMS IN LIPSCHITZ CYLINDERS

RUSSELL M. BROWN* AND ZHONGWEI SHEN**

Let Ω be a bounded Lipschitz domain in $\mathbf{R}^n$. Consider the parabolic system

$$\frac{\partial \vec{u}}{\partial t} = \mu \Delta \vec{u} + (\lambda + \mu)\nabla(\mathrm{div}\vec{u}) \quad \text{in} \quad \Omega_T = \Omega \times (0,T) \tag{1}$$

where $0 < T < \infty$, $\mu > 0$ and $\lambda > -2\mu/n$ are constants. We are interested in the solvability of the initial–Dirichlet problem:

$$\text{(iDP)} \qquad \begin{cases} \vec{u} \mid_{\Sigma_T} = \vec{g} \in L^p(\Sigma_T) \\ \vec{u} \mid_{t=0} = \vec{0} \end{cases}$$

and the initial traction problem:

$$\text{(iTP)} \qquad \begin{cases} \dfrac{\partial \vec{u}}{\partial \nu} = \lambda(\mathrm{div}\vec{u})N + \mu((\nabla\vec{u}) + (\nabla\vec{u})^{\mathrm{tr}})N \mid_{\Sigma_T} = \vec{g} \in L^p(\Sigma_T) \\ \vec{u} \mid_{t=0} = \vec{0} \end{cases}$$

where $\Sigma_T = \partial\Omega \times (0,T)$, N denotes the outward unit normal to $\partial\Omega$ and $(\ )^{\mathrm{tr}}$ denotes the transpose of a matrix. In **[S]**, the solvability of (iDP) and (iTP) was obtained for data in $L^2(\Sigma_T)$ and $n \geq 3$. In this note, we establish the L^p estimates for the sharp ranges of p when $n = 3$.

THEOREM 1. *If $n = 3$, there exists $\varepsilon > 0$, so that, given any $\vec{g} \in L^p(\Sigma_T)$, $2 - \varepsilon < p < \infty$, there exists a unique $\vec{u}$ in Ω_T satisfying (1) and (iDP) and such that $(\vec{u})^*$ (the nontangential maximal function) is in $L^p(\Sigma_T)$. Moreover,*

$$\|(\vec{u})^*\|_{L^p(\Sigma_T)} \leq C\, \|\vec{g}\|_{L^p(\Sigma_T)}.$$

In the next theorem, $\Lambda^\alpha(\Sigma_T)$ denotes the space of functions on Σ_T which are Hölder continuous (in the parabolic sense) of exponent α. We let $\Lambda_0^\alpha(\Sigma_T)$ denote the subspace of functions which vanish when $t = 0$.

THEOREM 2. *If $n = 3$, then, given any $\vec{g} \in L^\infty(\Sigma_T)$, there exists a unique $\vec{u}$ in Ω_T satisfying (1), (iDP) and $\vec{u} \in L^\infty(\Omega_T)$. In fact, $\|\vec{u}\|_{L^\infty(\Omega_T)} \leq C\|\vec{u}\|_{L^\infty(\Sigma_T)}$. Moreover, there exists $\alpha_0 > 0$, so that, if $\vec{g} \in \Lambda_0^\alpha(\Sigma_T)$, $0 < \alpha < \alpha_0$, then $\vec{u} \in \Lambda_0^\alpha(\Omega_T)$ and $\|\vec{u}\|_{\Lambda_0^\alpha(\Omega_T)} \leq c\|\vec{g}\|_{\Lambda_0^\alpha(\Sigma_T)}$.*

THEOREM 3. *If $n = 3$, there exists $\varepsilon > 0$, so that given any $\vec{g} \in L^p(\Sigma_T)$, $1 < p < 2 + \varepsilon$, there exists a unique $\vec{u}$ satisfying (1), (iTP) and $(\nabla\vec{u})^* \in L^p(\Sigma_T)$. In fact, $\|(\nabla\vec{u})^*\|_{L^p(\Sigma_T)} + \|(\partial_t^{1/2}\vec{u})^*\|_{L^p(\Sigma_T)} \leq C\, \|\vec{g}\|_{L^p(\Sigma_T)}$.*

In Theorems 1, 2 and 3, C is a constant depending at most on Ω, T,λ, μ and p.

*Department of Mathematics, University of Kentucky, Lexington, KY 40506.
**Department of Mathematics, Princeton University, Princeton, NJ 08544.

For the system of elastostatics, $\mu\Delta\vec{u}+(\lambda+\mu)\nabla(\text{div}\vec{u}) = \vec{0}$ in Lipschitz domains, the corresponding results were obtained in [**DK2**].

Following the ideas of B. Dahlberg and C. Kenig [**DK1**] and J. Pipher and G. Verchota [**PV2**], we begin the proofs of our theorems by considering the initial-traction problem with atomic data.

MAIN LEMMA. *Let Ω be a bounded Lipschitz domain in $\mathbf{R}^3$. Suppose $\vec{u}$ is a solution of (1) in Ω_T with $\|(\nabla\vec{u})^*\|_{L^2(\Sigma_T)} < \infty$. Assume that, $\vec{u}\mid_{t=0}= \vec{0}$ and $\frac{\partial \vec{u}}{\partial \nu} = \vec{a}$ where $\vec{a}$ is a parabolic atom, i.e., supp $\vec{a} \subset \{(P,t) \in \Sigma_T \mid |P-P_0| < r, |t-t_0| < r^2\}$ for some $r > 0$, $(P_0,t_0) \in \Sigma_T$, $\|\vec{a}\|_{L^2(\Sigma_T)} \leq \frac{1}{r^2}$ and $\iint_{\Sigma_T} \vec{a} = \vec{0}$. Then*

$$\iint_{\Sigma_T} (\nabla\vec{u})^* \leq C.$$

In the parabolic case, the dimension of the time variable t should be counted twice, thus we are working in a space of (homogeneous) dimension five. This is where the difficulty lies when we try to apply the argument from [**DK2**] for the elliptic system of elastostatics in three dimensions. However, since our domains are cylindrical, we are able to overcome this difficulty. This is done by using Fourier analysis in the time variable to show that we may solve the initial boundary value problem with data in the mixed L^p–spaces

$$L^{p,q}(\Sigma_T) = L^q((0,T), L^p(\partial\Omega))$$

for $p \in (2-\varepsilon, 2]$ and $q \in (1,2]$ where $\varepsilon > 0$. Since the time index may be arbitrarily close to one, the extra dimensions do not introduce any additional difficulties.

We start with the boundedness of the potential operator on $L^{p,q}$. Let

$$\vec{v}(X,t) = \int_0^t \int_{\partial\Omega} \left(\frac{\partial\Gamma}{\partial\nu_Q}(X-Q,t-s)\right)^{\text{tr}} \vec{f}(Q,s)dQds$$

be the double layer potential with density $\vec{f}$, where $\Gamma(X,t)$ is a matrix of fundamental solutions for (1). Then

$$\vec{v}\mid_{\Sigma_T}= (-\frac{1}{2}I + K)\vec{f}$$

where K is a singular integral operator (see [**S**]).

LEMMA 1. *Let $1 < p,q < \infty$. We have*
(i) K is bounded on $L^{p,q}(\Sigma_T)$
(ii) $\|(\vec{v})^*\|_{L^{p,q}(\Sigma_T)} \leq C\|\vec{f}\|_{L^{p,q}(\Sigma_T)}$.

The next lemma is essential to the application of Dahlberg and Kenig's argument [**DK2**]. This lemma follows from the L^2–estimates in [**S**], Calderón-David-Semmes perturbation theorem (see [**DKV1**]) and the Calderón-Zygmund theory of singular integral operators applied to Banach space valued functions.

LEMMA 2. *If $n \geq 3$, there exists $\varepsilon > 0$, such that $\pm\frac{1}{2}I + K$ is invertible on $L^{p,q}(\Sigma_T)$ for $p \in (2-\varepsilon, 2]$ and $q \in (1,2]$. Consequently, the initial–Dirichlet problem is uniquely solvable for $\vec{g} \in L^{p,q}(\Sigma_T)$, $p \in (2-\varepsilon, 2]$ and $q \in (1,2]$.*

We need some notations for Lemma 3.

For $(P,t) \in \Sigma_T$, $R > 0$ small, let

$$\begin{aligned} J_R(P) &= \{Q \in \partial\Omega \mid |Q-P| < R\} \\ \Delta_R(P,t) &= J_R(P) \times (t - R^2, t + R^2) \\ D_R(p) &= \{X \in \Omega \mid |X - P| < R\} \\ Z_R(P,t) &= D_R(P) \times (t - R^2, t + R^2). \end{aligned}$$

For a function $\vec{u}$ on Ω_T, we define

$$(\vec{u})^*_R(P,t) = \sup_{\substack{(Y,s)\in\gamma(P,t) \\ |Y-P|\leq R}} |\vec{u}(Y,s)|$$

and

$$(\vec{u})^{*R}(P,t) = \sup_{\substack{(Y,s)\in\gamma(P,t) \\ |Y-P|\geq R}} |\vec{u}(Y,s)|$$

where $\gamma(P,t) = \{(Y,s) : |Y-P| + |t-s|^{1/2} < 2\,\mathrm{dist}(Y,\partial\Omega)\} \cap \Omega_T$.

LEMMA 3. *Suppose $\vec{u}$ is a solution of (1) in Ω_T. Assume $\frac{\partial\vec{u}}{\partial\nu} = \vec{0}$ on Δ_{2R} and $\|(\nabla\vec{u})^*\|_{L^2(\Sigma_T)} < \infty$. Then*

(i) $\iint_{Z_R} |\nabla\vec{u}|^2 \leq \frac{C}{R^2} \iint_{Z_{2R}} |\vec{u}|^2$;

(ii) $\sup\limits_{|t-t_0|<R^2} (\int_{D_R} |\vec{u}(X,t)|^{5/2} dX)^{2/5} \leq \frac{C}{R^{13/10}} (\iint_{Z_{2R}} |\vec{u}|^2)^{1/2}$;

(iii) $\iint_{\Delta_R} (\nabla\vec{u})^*_R \leq C\,R^{1/2} (\iint_{Z_{2R}} |\vec{u}|^2)^{1/2}$.

We now sketch the proof of the Main Lemma.

Proof of Main Lemma: By the L^2–theory for the initial-traction problem (see **[S]**), we have

$$\begin{aligned} \iint_{\Delta_{100r}(P_0,t_0)} (\nabla\vec{u})^* &\leq C\,r^2 \left(\iint_{\Delta_{100r}(P_0,t_0)} (\nabla\vec{u})^{*2} \right)^{1/2} \\ &\leq C\,r^2 \|\vec{a}\|_{L^2(\Sigma_T)} \leq C. \end{aligned}$$

Thus, it is easy to see that the estimate for $(\nabla u)^*$ will follow if we can prove that

$$\iint_{\Lambda(R)} (\nabla\vec{u})^* \leq C\left(\frac{r}{R}\right)^{\alpha_0} \quad \text{for some } \alpha_0 > 0$$

where $\Lambda(R) = \Delta_R(P_1,t_1)$ for some $(P_1,t_1) \in \Sigma_T$, $R \geq 5r$ and $\frac{\partial\vec{u}}{\partial\nu} = \vec{0}$ on $\Lambda(10R)$.

First, we estimate $(\nabla\vec{u})^{*R}$ on $\Lambda(R)$. By interior estimates, for $(P,t)\in\Lambda(R)$

$$(\nabla\vec{u})^{*R}(P,t)\le\frac{C}{R^5}\iint_{\Lambda(2R)}(\vec{u})^*.$$

Consequently,

$$\begin{aligned}\iint_{\Lambda(R)}(\nabla\vec{u})^{*R}&\le\frac{C}{R}\iint_{\Lambda(2R)}(\vec{u})^*\\&\le C\,R^{3-\frac{2}{q}-\frac{2}{p}}\|(\vec{u})^*\|_{L^{p,q}(\Sigma_T)}.\end{aligned}$$

Choose $p\in(2-\varepsilon,2)$, ε as in Lemma 2 and q close to 1 such that $3-\frac{2}{p}-\frac{2}{q}=-\alpha_1<0$. Note that, by the divergence theorem, we have

$$(\frac{1}{2}I+K)(\vec{u}\mid_{\Sigma_T})=\mathcal{S}(\frac{\partial\vec{u}}{\partial\nu})\Big|_{\Sigma_T}=\mathcal{S}(\vec{a})\mid_{\Sigma_T}$$

where $\mathcal{S}$ denotes the single layer potential defined by

$$\mathcal{S}(\vec{g})(X,t)=\int_0^t\int_{\partial\Omega}\Gamma(X-Q,t-s)\vec{g}(Q,s)\,dQds.$$

A computation shows that $\|\mathcal{S}(\vec{a})\mid_{\Sigma_T}\|_{L^{p,q}(\Sigma_T)}\le C\,r^{\frac{2}{p}+\frac{2}{q}-3}$. Thus, by Lemma 2,

$$\begin{aligned}\|(\vec{u})^*\|_{L^{p,q}(\Sigma_T)}&\le C\|\vec{u}\|_{L^{p,q}(\Sigma_T)}\le C\|(\frac{1}{2}I+K)(\vec{u}\mid_{\Sigma_T})\|_{L^{p,q}(\Sigma_T)}\\&=C\|\mathcal{S}(\vec{a})\mid_{\Sigma_T}\|_{L^{p,q}(\Sigma_T)}\le C\,r^{\frac{2}{p}+\frac{2}{q}-3}=C\,r^{\alpha_1}.\end{aligned}$$

Hence,

$$\iint_{\Lambda(R)}(\nabla\vec{u})^{*R}\le C\left(\frac{r}{R}\right)^{\alpha_1}$$

Next, we estimate $(\nabla\vec{u})^*_R$ on $\Lambda(R)$. By Lemma 3,

$$\iint_{\Lambda(R)}(\nabla\vec{u})^*_R\le C\,R^{1/2}\left(\iint_{Z_{2R}(P_1,t_1)}|\vec{u}|^2\right)^{1/2}$$

and

$$\begin{aligned}&\iint_{Z_{2R}(P_1,t_1)}|\vec{u}|^2\\&\le C\,R^3\int_{|t-t_1|<4R^2}\left(\frac{1}{R^3}\int_{D_{2R}(P_1)}|\vec{u}|^p\right)^{1/p}\left(\frac{1}{R^3}\int_{D_{2R}(P_1)}|\vec{u}|^{5/2}\right)^{2/5}\\&\le C\,R^{\frac{1}{2}-\frac{2}{p}}\int_{|t-t_1|<4R^2}\left(\int_{J_{3R}(P_1)}(\vec{u})^{*p}\right)^{1/p}\cdot\left(\iint_{Z_{3R}(P_1,t_1)}|\vec{u}|^2\right)^{1/2}\\&\le C\,R^{\frac{5}{2}-\frac{2}{p}-\frac{2}{q}}\|(\vec{u})^*\|_{L^{p,q}(\Sigma_T)}\left(\iint_{Z_{3R}(P_1,t_1)}|\vec{u}|^2\right)^{1/2}.\end{aligned}$$

As before, choose $p \in (2-\varepsilon, 2)$, q close to 1 such that $\alpha_1 = \frac{2}{p} + \frac{2}{q} - 3 > 0$. We get

$$a_R^2 \le C\left(\frac{r}{R}\right)^{\alpha_1} a_{2R} \tag{2}$$

where

$$a_R = R^{1/2}\left(\iint_{Z_{2R}(p_1,t_1)} |\vec{u}|^2\right)^{1/2}.$$

Note that $a_R \le CR\|(\vec{u})^*\|_{L^2(\Sigma_T)} \le CR\|S(\vec{a})\|_{L^2(\Sigma_T)} \le C(\frac{r}{R})^{-1}$. It then follows from (2) that

$$a_R \le C\left(\frac{r}{R}\right)^{\frac{\alpha_1 - 1}{2}}.$$

Repeating this argument many times, we obtain

$$a_R \le C\left(\frac{r}{R}\right)^{\alpha_1/4}.$$

Let $\alpha_0 = \frac{\alpha_1}{4}$. The proof is complete.

With the Main Lemma in hand, Theorems 1, 2 and 3 follow in a manner similar to the elliptic case.

REFERENCES

[B] R. Brown, *The initial-Neumann problem for the heat-equation in Lipschitz cylinders*, Trans. Amer. Math. Soc., 320 (1990), pp. 1–52.

[C] A. Calderón, *Boundary value problems in Lipschitzian domains*, in "Recent Progress in Fourier Analysis," 33-48. 1985.

[CMM] R. Coifman, A. McIntosh and Y. Meyer, *L'intégrale de Cauchy définit un opérateur borné sur L^2 pour les courbes Lipschitziennes*, Annals of Math., 116 (1982), pp. 361–387.

[DK1] B. Dahlberg and C. Kenig, *Hardy spaces and the Neumann problem in L^p for Laplace's equation in Lipschitz domains*, Annals of Math., 125 (1987), pp. 437–465.

[DK2] B. Dahlberg and C. Kenig, *L^p estimates for the three-dimension system of elastostatics on Lipschitz domains*, preprint, 1989.

[DKV1] B. Dahlberg, C. Kenig and G. Verchota, *The Dirichlet problem for the biharmonic equation in a Lipschitz domain*, Annales de l'Institute Fourier Grenoble 36 (1986), pp. 109-135.

[DKV2] B. Dahlberg, C. Kenig and G. Verchota, *Boundary value problems for the systems of elastostatics in Lipschitz domains*, Duke Math. J., 57 (1988), pp. 795–818.

[FR] E. Fabes and N. Riviere, Dirichlet and Neumann problems for the heat equation in C^1-cylinders, Proc. Symp. Pure Math. v. 35, 179-196, 1979.

[FSt] C. Fefferman and E. Stein, *Some maximal inequalities*, Amer J. Math., 93 (1971), pp. 107–115.

[NSt] A. Nagel and E. Stein, *Lectures on Pseudo-Differential Operators*, Princeton University Press, 1979.

[PV1] J. Pipher and G. Verchota, *The Dirichlet problem in L^p for biharmonic functions on Lipschitz domains*, preprint 1989.

[PV2] J. Pipher and G. Verchota, *The maximum principle for biharmonic functions*, in preparation.

[S] Z. Shen, *Boundary value problems for parabolic Lamé systems and a nonstationary linearized systems of Navier–Stokes equations in Lipschitz cylinders*, to appear in Amer. J. Math.

[St] E. Stein, *Singular integrals and differentiability properties of functions*, Princeton University Pess, 1970.

[V] G. Verchota, *Layer potentials and regularity for the Dirichlet problem for Laplace's equation in Lipschitz domains*, J. of Functional Analysis, 59 (1984), pp. 572–611.

INTRINSIC ULTRACONTRACTIVITY AND PROBABILITY

BURGESS DAVIS*

Let $p_t^D(x,y) = p_t(x,y)$ be the Dirichlet heat kernel for $\frac{1}{2}\Delta$ in a domain $D \subset \mathbf{R}^n$, $n \geq 2$. In [4] E.B. Davies and B. Simon define the semigroup connected with the Dirichlet Laplacian to be intrinsically ultracontractive if there is a positive (in D) eigenfunction ϕ_0 for $\frac{1}{2}\Delta$ in D and if for each $t > 0$ there are positive constants c_t, C_t depending only on D and t such that

$$c_t\phi_0(x)\phi_0(y) < p_t(x,y) < C_t\phi_0(x)\phi_0(y), \quad x, y \in D. \tag{1}$$

(To be precise, they show (1) is equivalent to their definition.) Here we will say D is IU when (1) holds. Among many results in their very interesting paper Davies and Simon prove that bounded Lipschitz domains are IU. Recently the investigation of IU has been taken up by a number of probabilists who approached the area via their study of distributions of the lifetimes of Doob's conditioned h-processes, a study which in its modern version started with Cranston and McConnell's solution ([3]) of a conjecture of Chung. Here we are going to give an intuitive sketch of the connection between lifetimes and IU, and describe the results in our paper [6].

The reader will note that Rodrigo Bañuelos's paper in this volume deals with the same topic as this paper. Bañuelos was more diligent than we were and wrote his paper first. Sometimes sloth is rewarded (besides being its own reward), and this is one of those times, for, since Bañuelos has surveyed the area, we do not have to. Accordingly only those papers with the most immediate connection to [6] will be mentioned.

Let $\mathbf{X} = \{X_t, t \geq 0\}$, be standard n dimensional Brownian motion and let P_x stand for probability associated with this motion given $X_0 = x$. The kernel $p_t(x,\cdot)$ has an immediate probabilistic interpretation as the density at time t of $\mathbf{X}$ killed when it leaves D, that is, if A is a Borel subset of D and $\tau_D = \inf\{t > 0 : X_t \notin D\}$,

$$P_x(X_t \in A,\ t < \tau_D) = \int_A p_t(x,y)dy.$$

Thus $p_t(x,y)/p_s(x,y)$ gives the ratio of the probabilities of killed Brownian motion being infinitessimally close to y (we will just say "hitting y") at times t and s, respectively.

Now it can be shown that for $\delta > 0$ fixed,

$$\lim_{\varepsilon \to 0} \frac{P_x(\exists s,\ \ t < \tau_D : |s-t| > \delta \text{ and } |B_s - y| < \varepsilon,\ |B_t - y| < \varepsilon)}{P_x(\exists t < \tau_D : |B_t - y| < \varepsilon)} = 0, \tag{2}$$

*Department of Mathematics and Statistics, Purdue University, West Lafayette, IN 47907.
Supported in part by NSF.

and roughly what this implies is that given y is hit by $\mathbf{X}$ before τ_D, it is hit at only one time, although some work must be done to make this rigorous, since the probability $\mathbf{X}$ ever hits y is zero. Thus intuitively, $p_t(x,y)/p_s(x,y)$ gives the ratio of the probabilities of $\mathbf{X}$ *first* hitting y before τ_D at times t and s, respectively, if $X_0 = x$, given $\mathbf{X}$ does hit y before time τ_D. Equivalently,

$$L_{x,y}(t) = \frac{p_t(x,y)}{\int\limits_0^\infty p_t(x,y)dt}$$

is the density of the time $\mathbf{X}$ (started at x) first hits y, given that y is hit before τ_D. The integral in the denominator is finite whenever $x \neq y$ and D has a Green function, assumptions always in force here. We use P_x^y and E_x^y to denote probability and expectation associated with $\mathbf{X}$, started at x, conditioned to hit y before τ_D, and use T to denote the time of hitting y, so that the density of T under P_x^y is $L_{x,y}(\cdot)$. Note $E_x^y T = \int\limits_0^\infty t L_{x,y}(t)dt$.

Cranston and McConnel's theorem is equivalent to the following.

THEOREM. *Let $n = 2$. There is an absolute constant C such that*

$$\int\limits_0^\infty t L_{x,y}(t)dt < C \text{ area } D, \; x \neq y, \; x,y \in D.$$

Thus the Cranston–McConnell theorem can be interpreted as as theorem about the shape of the normalized (in t) heat kernel which holds uniformly over all points x, y. We will call domains satisfying $\sup\limits_{\substack{x,y \in D \\ x \neq y}} E_x^y T < \infty$ Cranston–McConnell domains. Let $\hat{p}_t(x,\cdot) = p_t(x,\cdot)/\int\limits_D pt(x,y)dy$. It is shown in [6] that D is IU if and only if there is for each $t > 0$ a positive constant α_t such that $\hat{p}_t(x,y)/\hat{p}_t(x,y) < \alpha_t, \; x,y,z \in D$. The *only if* part is immediate, and the *if* part is easy. Thus IU is equivalent to a different kind of uniformity of the heat kernel, uniformity in the first variable under normalization in the second variable. Since $\hat{p}_t(x,y)$ is the density of killed Brownian motion, started at x, conditioned to be alive at time t, IU says all such densities are comparable, so it is in some sense a mixing condition for this motion.

Now it is immediate that if λ is the eigenvalue associated with the eigenfunction ϕ_0, both $c_t e^{-\lambda t}$ and $C_t e^{-\lambda t}$ may be chosen to be respectively nondecreasing and nonincreasing in t, since

$$\int\limits_D \phi_0(y) p_s(y,z)dy = e^{-\lambda s}\phi_0(z) \, ,$$

and

$$\int\limits_D p_t(x,y)p_s(y,z)dy = p_{s+t}(x,z).$$

Thus we immediately get from (1) that

$$c_{t_0}/C_{t_0} < L_{x,y}(t)/[e^{\lambda(s-t)}L_{x,y}(s)] < C_{t_0}/c_{t_0},\ t_0 < t < s,$$

a property that will call uniformly (in x, y) close to exponential decay at rate λ. Especially we see that this uniform decay immediately implies

$$\sup_{\substack{x,y\in D\\ x\neq y}} \int tL_{x,y}(t)dt < C,$$

that is, IU domains are Cranston–McConnell domains. We note there are Cranston–McConnell domains, in fact planar domains of finite area, which are not IU domains, and that both c_t and C_t in (1) can be chosen to converge to 1. See [4] or Bañuelos' paper in this volume.

In [5], Davies and Simon note without proof that D is IU if there exists a compact subset F of D and a constant $k_t > 0$ such that

(i) $$\int_F p_t(x,y)dy \geq k_t \int_D p_t(x,y)dy, \quad x \in D\,.$$

This idea was not exploited by Davies and Simon, rather, this was done in [6], where a weaker variant of (i) was used, and subsequently, in [1], where (i) is proved. A result proved in [6] is the following.

THEOREM 1. *Let f be a real valued upper semicontinuous function defined as $\{(x_1,\ldots,x_{n-1}) : \sum_{i=1}^{n-1} x_i^2 < 1\}$ such that $-M < f \leq 0$ for some constant $M < \infty$. Then the domain $D_f = \{(x_1,\ldots,x_n) \in \mathbb{R}^n : \sum_{i=1}^{n-1} x_i^2 < 1,\ f(x_1,\ldots,x_{n-1}) < x_n < 1\}$ is* IU.

We note that Bass and Burdzy independently a few months before our work proved, under the conditions of this theorem, that the domains D_f are Cranston–McConnell domains. Bass and Burdzy in [1] extend Theorem 1 to functions f in L^p, $p < n$. Both [1] and [2] show intrinsic ultracontractivity for other classes of domains, and for other semigroups than that connected with the Dirichlet Laplacian.

We also prove in [6]

THEOREM 2. *Let $D = (0,1)\times(0,1) \cup \bigcup_{i=1}^{\infty} R_i$, where $R_n = I_n \times (-a_n, 0]$ and $I_1, I_2, \ldots$ are disjoint open intervals contained in $(1/4,\ 3/4)$ and $a_i \geq 0$. Then D is* IU *if and only if $\lim_{n\to\infty}$ area $R_n = 0$.*

Our proof of the if part of this theorem resembles that of the last theorem but is a little harder. The proof of the only if part uses results of Xu [7] and the author about the mean and variance of T under P_x^y to show that if $x = (1/2, 1/2)$, and if the

y_i are close to the bottom of R_i, then if lim area $R_n \neq 0$, the lifetime densities L_{x,y_i} do not have a uniformly close to exponential shape which, as has been mentioned, is implied by IU. The if part of Theorem 2 answered affirmatively the question, raised in [4], whether there are domains of infinite area which are IU. Subsequently very different examples (although closely related to each other) have been given in [1], [2]. Previously Xu [7] had given an example of a domain of infinite area which is a Cranston–McConnell domain. Theorem 2 gives a view of part of the edge of IU. We hope that a geometric characterization of IU for the class of domains above the graph of a function, in two dimensions, will be proved, a goal that does not seem out of reach. There is a conjecture in [6]. At the least Theorem 2 will provide a test case for investigations into this, at most our method will extend. A more distant goal which may or may not be feasible is a nice geometric characterization of IU for all simply connected planar domains.

REFERENCES

[1] R. Bass and K. Burdzy, *Lifetimes of conditioned diffusions*, preprint.

[2] R. Bañuelos, *Intrinsic ultracontractivity and eigenfunction estimates for Schrödinger operators*, to appear J. Func. Anal.

[3] M. Cranston and T. McConnell, *The lifetime of conditioned Brownian motion*, Z. Wahrsch. Verw. Geb **65** (1983), pp. 1–11.

[4] E.B. Davies and B. Simon, *Ultracontractivity and the heat kernel for Schrödinger operators and the Dirichlet Laplacians*, J. Func. Anal. (1984), pp. 335–395.

[5] E.B. Davies and B. Simon, *L^1 properties of intrinsic Schrödinger semigroups*, J. Func. Analysis (1986), pp. 126–146.

[6] B. Davis, *Intrinsic Ultracontracivity and the Dirichlet Laplacians*, to appear, J. Func. Analysis.

[7] J. Xu, *The Lifetime of Conditioned Brownian Motion in Planar Domains of Infinite Area*, Prob. Theory. Rel. Fields, 87 (1991), pp. 469–487.

UNIQUENESS IN THE DIRICHLET PROBLEM FOR TIME INDEPENDENT ELLIPTIC OPERATORS

LUIS ESCAURIAZA*

Abstract. Uniqueness is proved for the Dirichlet problem associated to a second order non-divergence form elliptic operator whose coefficients are independent of one direction and have discontinuities along a countable set of lines which are parallel to the given direction and accumulate around a single line.

AMS(MOS) subject classifications. 35

0. Introduction. Let L denote a time independent elliptic operator of the form

$$(0.1)\qquad Lu(x,t)u = c(x)D_{tt}u(x,t) + \sum_{i,j=1}^{n} a_{ij}(x)D_{ij}u(x,t) + \sum_{i=1}^{n} b_i(x)D_{it}u(x,t),$$

where $x \in \mathbf{R}^n$, $t \in \mathbf{R}$ and satisfying for some $\lambda > 0$

$$(0.2)\quad \lambda^{-1}(|\eta|^2 + |\xi|^2) \geq c(x)|\eta|^2 + \sum_{i,j=1}^{n} a_{ij}(x)\xi_i\xi_j + \sum_{i=1}^{n} b_i(x)\xi_i\eta \leq \lambda(|\eta|^2 + |\xi|^2)$$

for all $x, \xi \in \mathbf{R}^n$ and $\eta \in \mathbf{R}$. Observe that when (0.2) is satisfied the operator on $\mathbf{R}^n$ given by

$$(0.3)\qquad Su = \sum_{i,j=1}^{n} a_{ij}(x)D_{ij}u(x)$$

defines an elliptic operator with coefficient matrix $A(x) = (a_{ij}(x))$ satisfying the ellipticity condition

$$(0.4)\qquad \lambda^{-1}|\xi|^2 \geq \sum_{i,j=1}^{n} a_{ij}(x)\xi_i\xi_j \geq \lambda|\xi|^2 \text{ for all } x, \xi \in \mathbf{R}^n.$$

In [3] it is shown that for all $\varphi \in C(\partial D)$, $f \in L^n(D)$ and $D \subset \mathbf{R}^n$ bounded Lipschitz domain, the Dirichlet problem

$$\begin{cases} Su = -f & \text{on } D \\ u = \varphi & \text{on } \partial D \end{cases}$$

has a unique good solution when the set of discontinuities E of the coefficient matrix $A(x)$ reduces to a countable set with at most one cluster point, and where by a good

*The University of Chicago, Department of Mathematics, 5734 University Avenue, Chicago, IL 60637

solution is meant a function $u \in C(\overline{D})$ which is the uniform limit on $\overline{D}$ of a sequence $\{u^k\}$ formed by the solutions to the Dirichlet problems

$$\begin{cases} S^k u^k = -f & \text{on } D \\ u^k = \varphi & \text{on } \partial D, \end{cases}$$

where S^k is an elliptic operator whose coefficient matrix $A^k(x)$ is smooth on $\mathbf{R}^n$ satisfying (0.4), and so that $A^k(x)$ converges pointwise to $A(x)$ almost everywhere, and where this convergence is assumed to be uniform on compact subsets of $\mathbf{R}^n \setminus E$.

In this paper we will consider the Dirichlet problem associated with an elliptic operator L whose coefficients are independent of one direction, as in (0.1) and where the definition of a good solution is the following.

DEFINITION 0.1. *Let D denote a Lipschitz domain in $\mathbf{R}^{n+1}$, $\varphi \in C(\partial D)$, $f \in L^{n+1}(D)$ and L be an elliptic operator as in (0.1) satisfying (0.2) and assume that the coefficient matrix $A(x)$ of the operator S in (0.3) is continuous on $\mathbf{R}^n$ except for a possibly nonempty set E. A function $u \in C(\overline{D})$ will be said to be a good solution to the problem*

$$\begin{cases} Lu = -f & \text{on } D, \\ u = \varphi & \text{on } \partial D \end{cases} \tag{0.5}$$

when there exists a sequence of operators $\{L^k\}$ with continuous time independent coefficients satisfying (1.2) so that the coefficients of L^k converge pointwise almost everywhere to those for L on $\mathbf{R}^n$, and if A^k denotes the coefficient matrix associated to the operator S^k, then $\{A^k\}$ converges uniformly to A on compact subsets of $\mathbf{R}^n \setminus \overline{E}$ and u is the uniform limit on $\overline{D}$ of the sequence $\{u^k\}$ formed by the strong solutions to the problem (0.5) associated to L^k

Analogously, if Ω is an open set in $\mathbf{R}^{n+1}$ we will say that a function $u \in C(\overline{\Omega})$ is a good solution to $Lu = -f$ on Ω when there exists a sequence of operators $\{L^k\}$ as before and functions $u^k \in C(\overline{\Omega}) \cap W^{2,n+1}(\Omega)$ so that $\{u^k\}$ converges uniformly to u on $\overline{\Omega}$ and $L^k u^k = -f$ on Ω.

As it is remarked in [3], the existence of good solutions follows from well known arguments.

In this work we will prove the following uniqueness theorem.

THEOREM 0.1. *Let L be a time independent elliptic operator as in (0.1) satisfying (0.2) and assume that the matrix $A(x)$ is continuous on $\mathbf{R}^n$ except possibly for a countable set of points E with at most one cluster point $x_0 \in \mathbf{R}^n$. Then, the Dirichlet problem (0.5) for L has a unique good solution on any bounded Lipschitz domain D in $\mathbf{R}^{n+1}$.*

In the first section of this paper we will show some regularity properties and representation formulas of good solutions to $Lu = 0$ which are needed to conclude the above theorem and in the second section we will prove the theorem.

1. Regularity properties of good solutions. In the following lemmas we will assume that the operator L has C^∞ coefficients on $\mathbf{R}^n$. $B_r(x_0)$ will denote an open ball of radius r centered at a point x_0 in $\mathbf{R}^n$, $R_r(x_0, a)$ will denote the set $B_r(x_0) \times (a-r, a+r)$ for $x_0 \in \mathbf{R}^n$, $a \in \mathbf{R}$, and R_r will stand for $B_r \times (-4, r)$, where $B_r = B_r(0)$.

LEMMA 1.1. *Let $u \in C(\overline{R}_r(x_0, a))$ be a smooth solution to $Lu = 0$ on $R_r(x_0, a)$. Then, the following estimate holds*

$$\|D_t u\|_{L^\infty(\overline{R}_{r/2}(x_0,a))} \leq C r^{-1} \|u\|_{L^\infty(\overline{R}_{r/2}(x_0,a))}, \tag{1.1}$$

where C depends only on λ and n.

Proof. We may reduce by translation and dilation to the case $r = 1$, $x_0 = 0$ and $a = 0$. Let φ be a smooth function with $\varphi = 1$ on $B_{1/2} \times [-1/2, 1/2]$ and $\varphi = 0$ on the complement of $R_{3/4}$. From the representation formula of smooth functions in terms of the Green's function $g(x,t,y,s)$ for L on R_1 we have

$$D_t u(x,t)\varphi(x,t) = -\int g(x,t,y,s) L[(D_t u)\varphi](y,s)\, dyds \quad \text{for all } (x,t) \in R_1.$$

Observe that $D_t u$ is again a solution to $Lu = 0$ on R_1. Developing the integrand above and recalling that from the Pucci-Aleksandrov-Bakelman inequality (see [6], [1] and [2]) there is a constant C depending only on λ and n so that

$$\left\{\int\!\!\int_{R_1} g(x,t,y,s)^{\frac{n+1}{n}}\, dyds\right\}^{\frac{n}{n+1}} \leq C \quad \text{for all } (x,t) \in R_1, \tag{1.2}$$

we have

$$|D_t u(x,t)\varphi(x,t)| \leq C\|u\|_{L^\infty(R_1)} + C\int g(x,t,y,s)|\nabla_{y,s} D_s u(y,s)||\nabla_{y,s}\varphi(y,s)|\, dyds.$$

Since $\nabla\varphi$ has its support contained in $R_{3/4}$ and the Green's function for L on R_1 is a super adjoint solution for L on R_1 we get from theorem 12 in [3] that the integral above is controlled by

$$C\left\{\int\!\!\int_{R_{7/8}} g(x,t,y,s)|\nabla_{y,s} u(y,s)|^2\, dyds\right\}^{1/2},$$

and the same theorem together with (1.2) imply that the above integral can be estimated in terms of the L^∞ norm of u on R_1 times a constant C depending only on λ and n.

LEMMA 1.2. *There exists a constant C depending on $p \in (1, \infty)$ such that for all $\varphi \in C_0^2(\mathbf{R}^2)$ and for all $\varepsilon > 0$ the following estimate holds*

$$\|D_{xy}\varphi\|_{L^p(\mathbf{R}^2)} \leq C\left\{\varepsilon\|D_{xx}\varphi\|_{L^p(\mathbf{R}^2)} + \varepsilon^{-1}\|D_{yy}\varphi\|_{L^p(\mathbf{R}^2)}\right\}.$$

Proof. Set $\phi(x,y) = \varphi(\varepsilon^1 2x, \varepsilon^{1/2} y)$. It is well known that

$$\|D_{xy}\phi\|_{L^p(\mathbf{R}^2)} \leq C\|\Delta\phi\|_{L^p(\mathbf{R}^2)} \leq C\left[\|D_{xx}\phi\|_{L^p(\mathbf{R}^2)} + \|D_{yy}\phi\|_{L^p(\mathbf{R}^2)}\right]$$

Writing the second partial derivatives of ϕ and undoing the change of variables the lemma follows.

The following two lemmas are very well known and we refer the reader for a proof to [5, chapter 9].

LEMMA 1.3. *Let S be an elliptic operator on $\mathbf{R}^n$ as in (0.3) satisfying (0.4) with coefficient matrix $A(x) = (a_{ij}(x))$. Then there exists a constant C depending on the modulus of continuity of the matrix A on $B_{2r}(x_0)$, λ, n and $p \in (1, \infty)$ such that for all $u \in C^\infty(B_{2r}(x_0))$ the following estimate holds*

$$\|D^2 u\|_{L^p(B_r(x_0))} \leq C \left[\|Su\|_{L^p(B_{2r}(x_0))} + r^{-2}\|u\|_{L^p(B_{2r}(x_0))}\right].$$

LEMMA 1.4. *There exists a constant C depending only on n and $p \in (1, +\infty)$ such that for all functions u belonging to $C^\infty(\overline{R}_1(x_0, a))$ and all $\varepsilon > 0$ the following estimate holds*

$$\sup_{\sigma\in(0,1)} (1-\sigma)\|\nabla_{x,t} u\|_{L^p(R_0(x_0,a))}$$
$$\leq C \left[\varepsilon \sup_{\sigma\in(0,1)} (1-\sigma)^2 \|D^2_{x,t} u\|_{L^p(R_0(x_0,a))} + \varepsilon^{-1} \sup_{\sigma\in(0.1)} \|u\|_{L^p(R_0(x_0,a))}\right]$$

LEMMA 1.5. *Let $u \in C(\overline{R}_r(x_0, .a))$ be a smooth solution to $Lu = 0$ on $R_r(x_0, a)$. Then for each $p \in (1, \infty)$ there is a constant C depending on λ, n, p and the modulus of continuity of the matrix $A(x) = (a_{ij}(x))$ on $\overline{B}_{2r}(x_0)$ such that the following estimate holds*

$$\|D^2_{x,t} u\|_{L^p(R_{r/2}(x_0,a))} \leq Cr^{-2}\|u\|_{L^\infty(R_r(x_0,a))}. \tag{1.3}$$

Proof. We may assume by dilation and translation that $x_0 = 0$, $a = 0$ and $r = 1$. Let $\varphi \in C_0^\infty(\mathbf{R}^{n+1})$ be equal to 1 on R_σ and identically zero outside $R_{(1+\sigma)/2}$ where $\sigma \in (0, 1)$. Observing that $L(u\varphi) = uL\varphi + 2a_{ij}D_iuD_j\varphi + b_i(D_iuD_t\varphi + D_i\varphi D_tu)$ and recalling the definition of the operator S in (0.3) we get

$$S(u\varphi) = -cD_{tt}(u\varphi) - b_iD_{it}(u\varphi) + L(u\varphi) \quad \text{for each } t \in [-1, 1].$$

From Lemma 1.3 we have for each $t \in [-1, 1]$

$$\begin{aligned}\|D^2(u\varphi)\|_{L^p(\mathbf{R})} \leq C\chi_\sigma \Big[&\|D_{tt}u\|_{L^p(B_{(1+\sigma)/2})} + (1-\sigma)^{-1}\|D_tu\|_{L^p(B_{(1+\sigma)/2})} \\ &+ (1-\sigma)^{-1}\|\nabla u\|_{L^p(B_{(1+\sigma)/2})} + \|D_{it}(u\varphi)\|_{L^p(\mathbf{R}^n)} \\ &+ (1-\sigma)^{-2}\|u\|_{L^p(B_{(1+\sigma)/2})}\Big],\end{aligned}$$

where C depends on the modulus of continuity of the matrix A on B_2, λ and n, and χ_σ denotes the characteristic function of the interval $[-(1+\sigma)/2, (1+\sigma)/2]$. Taking the p power of both sides of the above inequality and integrating in the variable t from -1 to 1 we obtain,

$$\begin{aligned}\|D^2_{x,t}(u\varphi)\|_{L^p(\mathbf{R}^{n+1})} \leq C\Big[&\|D_{tt}u\|_{L^p(R_{(1+\sigma)/2})} + (1-\sigma)^{-1}\|D_tu\|_{L^p(R_{(1+\sigma)/2})} \\ &+ (1-\sigma)^{-1}\|\nabla u\|_{L^p(R_{(1+\sigma)/2})} + \|D_{it}(u\varphi)\|_{L^p(\mathbf{R}^{n+1})} \\ &+ (1-\sigma)^{-2}\|u\|_{L^p(R_{(1+\sigma)/2})}\Big],\end{aligned}$$

and applying Lemma 1.2 to the fourth term on the right-hand side of the above inequality we obtain

$$\begin{aligned}\|D^2_{x,t}u\|_{L^p(\mathbf{R}_\sigma)} \le C\Big[&\|D_{tt}u\|_{L^p(R_{(1+\sigma)/2})} + (1-\sigma)^{-1}\|D_t u\|_{L^p(R_{(1+\sigma)/2})}\\ &+(1-\sigma)^{-1}\|\nabla u\|_{L^p(R_{(1+\sigma)/2})}\\ &+(1-\sigma)^{-2}\|u\|_{L^p(R_{(1+\sigma)/2})}\Big] \text{ for all } \sigma\in(0,1).\end{aligned}$$

From the above inequality and the Lemmas 1.1 and 1.4 the estimate (1.3) follows.

Remark 1.1. Let $u \in C(\overline{R}_r(x_0,a))$ be a good solution to $Lu = 0$ on $R_r(x_0,a)$, where L is as in (0.1). From our definition of good solution there is a sequence of operators L^k with time independent coefficients satisfying (0.2) and nice functions $\{u^k\}$ satisfying $L^k u^k = 0$ on $R_r(x_0,a)$ so that $\{u^k\}$ converges uniformly to u on $\overline{R}_r(x_0,a)$. From Lemma 1.1 it follows that the sequence $\{D_t u^k\}$ is uniformly bounded on any compact subset of $R_r(x_0,a)$ and at the same time $L^k(D_t u^k) = 0$ on $R_r(x_0,a)$. Hence, from well known arguments, some subsequence of $\{D_t u^k\}$ is going to converge uniformly on compact subsets of $R_r(x_0,a)$. This implies that u has a time derivative which satisfies the estimate (1.1) and is a good solution to $Lu = 0$ on any open set Ω with $\overline{\Omega}$ contained in $R_r(x_0,a)$. By repeating the same argument we see that any order partial derivative of u with respect to the t variable satisfies the same property.

In addition, if the matrix A is continuous on $B_{4r}(x_0)$ we have from the definition of good solution that the modulus of continuity of the matrices A^k on $B_{2r}(x_0)$ is uniformly controlled by the modulus of continuity of A on the same set. Thus, the estimate (1.3) holds uniformly on k with u^k replaced by u and therefore u is actually a strong solution to $Lu = 0$ on $R_{r/2}(x_0,a)$.

In the following lemmas we will consider the Green's function $g(x,y)$ and harmonic measure $d\omega^x$ associated to the operator S on a ball $B_s(x_0)$ centered at some point x_0 in $\mathbf{R}^n$. From the arguments in [3] it is clear that these objects can be uniquely defined when uniqueness in the Dirichlet problem associated to S in the class of good solutions is previously known or is assumed to hold. I will omit these details, leaving them as easy exercises for the interested reader.

LEMMA 1.6. *Let L be a time-independent elliptic operator as in (0.1) satisfying (0.2) and $u \in C(\overline{R}_r(x_0,a))$ a good solution to $Lu = 0$ on $R_r(x_0,a)$. Let us assume that the set of discontinuities E of the matrix A reduces to a countable set of points with at most one cluster point in $\mathbf{R}^n$. Then, there exists a measurable function f defined on $R_{r/2}(x_0,a)$ such that for all $s < r/2$ and $t \in (a - r/2, a + r/2)$ the following representation formula holds*

(1.4)

$$u(x,t) = \int_{B_s(x_0)} g_s(x,y)f(y,t)\,dy + \int_{\partial B_s(x_0)} u(Q,t)\,d\omega^x_s(Q) \quad \text{for all } x \in B-s(x_0),$$

where $g_s(x,y)$ and $d\omega^x_s$ are respectively the Green's function and harmonic measure for the operator S evaluated at x on $B_s(x_0)$. Moreover, the function f satisfies for each $t \in (a - r/2, a + r/2)$

$$\sup_{x \in B_{r/2}(x_0)} \int_{B_{r/2}(x_0)} g_r(x,y) f^2(y,t)\, dy < \infty. \tag{1.5}$$

Proof. Without loss of generality we may assume that $r = 1$, $x_0 = 0$ and $a = 0$. Let $\{L^k\}$, $\{u^k\}$ and $\{S^k\}$ be as in Definition 1.1, so that $\{u^k\}$ converges uniformly to u on $\overline{R}_1$ and

$$L^k w(x,t) = c^k(x) D_{tt} w(x,t) + \sum_{i,j=1}^{n} a_{ij}^k(x) D_{ij} w(x,t) + \sum_{i=1}^{n} b_i^k(x) D_{it} w(x,t).$$

Since $g_1^k(x,y)$ is a super adjoint solution for the operator S^k on B_1 it is easy to check that also $g_1^k(x,y)$ is a super adjoint solution for the operator L^k on F_1. Since $L^k(D_t^m u^k) = 0$ on R_1 for all $m \geq 0$ it follows from Theorem 12 in [3] that

$$\int_{R_{1/2}} g_1^k(x,y) |D_i D_t^m u^k(y,s)|^2\, dyds \leq C \int_{R_{3/4}} g_1^k(x,y) |D_t^m u^k(y,s)|^2\, dyds,$$

where C depends on λ and n. From Lemma 1.1 and the uniform estimate (see [6], [1], and [2])

$$\left\{ \int_{B_1} g_1^k(x,y)^{\frac{n}{n-1}}\, dy \right\}^{\frac{n-1}{n}} \leq C \quad \text{for all } x \in B_1, \tag{1.6}$$

where C depends only on λ and n, it follows that

$$\int_{R_{1/2}} g_1^k(x,y) |D_i D_t^m u^k(y,s)|^2\, dyds \leq C \|u^k\|_{L^\infty(R_1)}^2 \text{ for } m = 1, 2. \tag{1.7}$$

Hence, for each $k \geq 0$ and $x \in B_{1/2}$ there exists $s_{k,x} \in [-1/2, 1/2]$ so that

$$\int_{B_{1/2}} g_1^k(x,y) |D_i D_t u^k(y, s_{k,x})|^2\, dy C \|u^k\|_{L^\infty(R_1)}^2. \tag{1.8}$$

The fundamental theorem of calculus implies that for each $t \in [-1/2, 1/2]$ and $y \in B_{1/2}$

$$|D_{it} u^k(y,t)|^2 \leq |D_{it} u^k(y, s_{k,x})|^2 + \int_{-1/2}^{1/2} |D_i D_{tt} u^k(y,s)|^2\, ds.$$

Hence, multiplying both sides of the above Inequality by $g_1^k(x,y)$ and integrating in the y variable on $B_{1/2}$ we obtain from (1.7) and (1.8) that for some constant C independent of $k \geq 0$

$$\begin{aligned} \sup_{x \in B_{1/2}} \int_{B_{1/2}} g_1^k(x,y) |D_{it} u^k(y,t)|^2\, dy &\leq C \|u^k\|_{L^\infty(R_1)}^2 dy \\ &\leq C \|u^k\|_{L^\infty(R_1)}^2 \text{ for all } t \in (-1/2, 1/2). \end{aligned} \tag{1.9}$$

Now, from the representation formula for strong solutions in terms of the Green's function $g_s^k(x,y)$ and the harmonic measure $d\omega_{k,s}^x$ for the operator S^k on B_s, we have for each $s \in (0,1/2)$ and $t \in (-1/2,1/2)$

$$(1.10) \quad u^k(x,t) = \int_{B_s} g_s^k(x,y) f^k(y,t)\,dy + \int_{\partial B_s} u^k(Q,t)\,d\omega_{k,s}^x(Q) \text{ for all } x \in B_s,$$

where $f^k(x,t) = c^k(x) D_{tt} u^k(x,t) + b_i^k(x) D_{it} u^k(x,t)$.

If V^ε denotes an open set in $\mathbf{R}^n$ containing the set E of discontinuities of the matrix A with $|\overline{V}^\varepsilon| \leq \varepsilon$, it follows from definition 1.1 and lemma 1.5 that the sequence of functions $\{f^k(.,t)\}$ converges in $L^n(B_s \backslash \overline{V}^\varepsilon)$ to the measurable function $f(.,.)$ given by

$$f(x,t) = c(x) D_{tt} u(x,t) + b_i(x) D_{it} u(x,t).$$

Hence, if we rewrite the left-hand side of (1.10) as

$$u^k(x,t) = \int_{B_s \backslash \overline{V}^\varepsilon} g_s^k(x,y) f^k(y,t)\,dy + \int_{\overline{V}^\varepsilon} g_s^k(x,y) f^k(y,t)\,dy + \int_{\partial B_s} u^k(Q,t) d\omega_{k,s}^x(Q)$$

and let k tend to infinity we obtain

$$u(x,t) = \int_{B_s \backslash \overline{V}^\varepsilon} g_s(x,y) f(y,t)\,dy + \int_{\partial B_s} u(Q,t) d\omega_s^x(Q) + \lim_{k \to \infty} \int_{\overline{V}^\varepsilon} g_s^k(x,y) f^k(y,t)\,dy$$

for all $\varepsilon > 0, s \in (0,1/2)$ and $t \in (-1/2,1/2)$. But from (1.9), Lemma 1.1, the well-known fact that $0 \leq g_s^k(x,y) \leq g_1^k(x,y)$ for all $s \in (0,1)$, and (1.6) we have the third term of the right-hand side above controlled by

$$\limsup_{k \to \infty} \left[\int_{\overline{V}^\varepsilon} g_1^k(x,y) |f^k(y,t)|^2\,dy \right]^{1/2} \left[\int_{\overline{V}^\varepsilon} g_1^k(x,y)\,dy \right]^{1/2} \leq C \|u\|_{L^\infty(R_1)} |\overline{V}^\varepsilon|^{1/n}.$$

From (1.9) and Fatou's lemma it follows that

$$\sup_{x \in B_{1/2}} \int_{B_{1/2}} g_1(x,y) |f(y,t)|^2\,dy \leq C \|u\|_{L^\infty(R_1)}^2 \quad \text{for all } t \in (-1/2,1/2).$$

The dominated convergence theorem implies that the first term of the above equality converges as ε tends to zero to

$$\int_{B_s} g_s(x,y) f(y,t)\,dy$$

and therefore, we conclude that

$$u(x,t) = \int_{B_s} g_s(x,y) f(y,t)\,dy + \int_{\partial B_s} u(Q,t) d\omega_s^x(Q) \quad \text{for all } x \in B_{1/2} \text{ and } t \in (-1/2,1/2)$$

which finishes the proof of the lemma.

The following lemma is a very well known result but for the sake of completeness and since we need it in a more general setting, we will include its proof here.

LEMMA 1.7. (Hopf's lemma) *Let S be an elliptic operator on $\mathbf{R}^n$ as in (0.3) satisfying (0.4). Let $u \in C(\overline{B}_r(x_0))$ Let $u \in C(\overline{B}_r(x_0))$ be a good solution to $Su = 0$ on $B_r(x_0)$ satisfying $u(x) > u(y)$ for all x in $B_r(x_0)$ and where y is a point lying on $\partial B_r(x_0)$. Then, there is a constant C depending on λ and n, so that if $\nu = \frac{x-x_0}{|x-x_0|}$ and $0 < s < r/2$ the following holds*

$$u(y - s\nu) - u(y) \geq C\frac{s}{r} \inf_{\partial B_{r/2}(x_0)} [u(\cdot) - u(y)].$$

Proof. Without loss of generality we may assume that $y = 0$ and $u(y) = 0$. We introduce an auxiliary function v defined by

$$v(x) = e^{-\alpha r^{-2}|x-x_0|^2} - e^{-\alpha}$$

where α is a positive constant to be chosen. If $\{u^k\}$ denotes a sequence of nice functions converging uniformly to u on $\overline{B}_r(x_0)$ with corresponding elliptic operators S^k as in definition 1 in [3], there is a constant C as before so that

$$w^k(x) = u^k(x) - C \inf_{\partial B_{r/2}(x_0)} [u^k(\cdot)]v(x) \geq 0$$

for all x lying on $\partial B_{r/2}(x_0)$ and all k sufficiently large. Direct calculation gives $S^k w^k(x) \geq 0$ on $B_r(x_0)\backslash B_{r/2}(x_0)$ for some α depending on λ and n. From the Strong Maximum principle for strong solutions we conclude that $w^k(x) \geq \min\{0, \inf_{\partial B_{r/2}(x_0)}[u(\cdot)]\}$. for all x in $B_r(x_0)\backslash B_{r/2}(x_0)$ and letting k tend to infinity we get $u(x) \geq C \inf_{\partial B_{r/2}(x_0)}[u(\cdot)]v(x)$ on the annulus, which implies the lemma.

The following lemma is usually called the weak Harnack's Inequality and the reader can find its proof in [5, chapter 9].

LEMMA 1.8. *Let S be an elliptic operator on $\mathbf{R}^n$ as in (0.3) satisfying (0.4) and $u \in C(\overline{B}_r) \cap W^{2,n}_{\text{loc}}(B_r)$ satisfy $u \geq 0$ and $Su \leq 0$ on B_{2r}. Then for each $\sigma \in (0,1)$ there exist positive constants C and p depending on λ, n and σ so that*

$$\left[r^{-n} \int_{B_{\sigma r}} u(y)^p \, dy\right]^{1/p} \leq C \inf_{B_{\sigma r}} u$$

LEMMA 1.9. *Let S be an elliptic operator as in the above lemma and assume that it is known that the Dirichlet problem associated to S has a unique good solution on any bounded Lipschitz domain D in $\mathbf{R}^n$. Let u be a continuous function on $\overline{B}_1$ satisfying the following properties:*

(i) *u is nonnegative on $\overline{B}_1$ and u(0)=0.*

(ii) *There exists a measurable function h on $\overline{B}_1$ so that*

$$T^2 = \sup_{x \in B_{1/2}} \int_{B_{1/2}} g_1(x,y)h^2(y) \, dy < \infty \tag{1.11}$$

and for all $s \in (0, 1/2)$ and $x \in \overline{B}_s$ the following representation formula holds

$$u(x) = \int_{B_s} g_s(x, y)h(y)dy + \int_{\partial B_s} u(Q)d\omega_s^x(Q), \tag{1.12}$$

where $g_s(\cdot,\cdot)$ and $d\omega_s^x$ are respectively the Green's function and harmonic measure for S on $\overline{B}_s$. Then, there exists a family of points $\{x_s \backslash s \in (0, 1/4)\}$ with $|x_s| = s/2$ so that

$$\lim_{s\to 0} \frac{u(x_s)}{s} = 0.$$

Proof. Observe that from (1.12) and the arguments in [3], it follows that for each fixed $s \in (0, 1/4)$ the function

$$w_s(x) = u(x) - \int_{B_s} g_s(x, y)h(y)dy$$

is a good solution to $Sw = 0$ on B_s, and from Harnack's Inequality for good solutions [3], we conclude that there is a constant C depending on λ and n so that

$$u(x) \le \int_{B_s} g_s(x, y)h(y)\, dy - C \int_{B_s} g_s(0, y)h(y)\, dy \text{ for all } x \in B_{s/2}. \tag{1.13}$$

We introduce the following functions on B_s:

$$v_s(x) = \int_{B_s} g_s(x, y)h^+(y)dy \quad \text{and} \quad v_s^m(x) = \int_{B_s} g_s(x, y)h_m^+(y)dy,$$

where $h^+ = \max\{h, 0\}$, $h_m^+ = h^+$ whenever $h^+ \le m$ and identically zero otherwise. Here m denotes a nonnegative integer.

We claim that the following properties hold for each fixed $s \in (0, 1/4)$:

(i) The sequence$\{v_s^m\}$ converges uniformly to v_s on $B_{s/2}$.

(ii) There exists a sequence of points $\{x_s^m\}$ with $|x_s^m| = s/2$ so that $v_s^m(x_s^m) \le v_s^m(0)$.

If we assume the claims we can find a subsequence of the above sequence of points which we will denote without loss of generality as $\{x_s^m\}$ converging to some point x_s with $|x_s| = s/2$, and from the uniform convergence of the sequence $\{v_s^m\}$ we conclude that $v_s(x_s) \le v_s(0)$. This together with (1.13) imply that for some constant C depending on λ and n

$$u(x_s) \le C \int_{B_s} g_s(0, y)|h(y)|\, dy.$$

From the above inequality, the fact that $g_s(0.y) \le g_1(0.y)$, and the estimate (see [6], [1] and [2])

$$\left\{\int_{B_s} g_s(x, y)^{\frac{n}{n-1}}\, dy\right\}^{\frac{n-1}{n}} \le Cs \text{ for all } x \in B_s, \tag{1.14}$$

where C depends only on λ and n, we get

$$\frac{u(x_s)}{s} \leq C\left[\int_{B_s} g_1(0.y)|h(y)|^2\,dy\right]^{1/2}$$

and from (1.11) the right-hand side of the above inequality goes to zero as s tends to zero.

Therefore, to finish the proof of the lemma we need to prove the last two claims. In order to do so, we introduce a sequence $\{S^k\}$ of elliptic operators with smooth coefficients on $\mathbf{R}^n$ whose coefficients are regularizations of the coefficients of the operator S and observe that [3] v_s^m is the good solution on B_s to the following problem

$$\begin{cases} Sw = -h_m^+ & \text{on } B_s \\ w = 0 & \text{on } \partial B_s \end{cases}$$

If $v_s^{m,k}$ denotes the solution to the above problem with S replaced by S^k, we know [3] that for each fixed $x \in (0,1/2)$ and $m > 0$ the sequence $\{v_s^{m,k}\}$ converges uniformly to v_s^m on B_s.

We will start with the first claim. Observe that from (1.11), (1.14) and the fact that $h_m^+ \leq |h|$ we conclude that there is a constant C depending on λ and n so that

$$\|v_s^m\|_{L^\infty(B_s)} \leq s\,C\,T \quad \text{for all } m \geq 0. \tag{1.15}$$

We will show that there exists $\alpha \in (0,1)$ and C depending on λ and n so that

$$\sup_{x,y\in B_{s/2}} \frac{|v_s^m(x) - v_s^m(y)|}{|x-y|^\alpha} \leq C(s^{1-\alpha} + s^{-\alpha})T \text{ for all } m \geq 0. \tag{1.16}$$

But (1.15) and (1.16) imply that any subsequence of $\{v_s^m\}$ has a uniformly convergent subsequence $\{\overline{B}_{s/2}\}$. From the dominated convergence theorem and (1.11) is easy to see that $\{v_s^m\}$ converges pointwise to v_s. These imply that $\{v_s^m\}$ converges uniformly to v_s on $\overline{B}_{s/2}$.

Let $B_{2r}(z)$ be a ball contained in B_s, $M_t^{m,s} = \sup_{B_t(z)} v_s^m$, $m_t^{m,s} = \inf_{B_t(z)} v_s^m$ and $g_{B_t(z)}(\cdot,\cdot)$ denote the Green's function for S on $B_t(z)$ Observe that the function

$$v_s^m(x) - m_{2r}^{m,s} - \int_{B_{2r}(z)} g_{B_{2r}(z)}(x,y)h_m^+(y)\,dy$$

is a good solution to $Su = 0$ on $B_{2r}(z)$ which is nonnegative on $\partial B_{2r}(z)$, and from the Strong Maximum principle for good solutions [3] it is nonnegative on $B_{2r}(z)$. Hence, from the Harnack's Inequality for good solutions, (1.14) and the fact that $g_{B_{2r}(z)}(\cdot,\cdot) \leq g_s(\cdot,\cdot)$ we conclude

$$M_r^{m,s} - m_{2r}^{m,s} \leq C\{m_r^{m,s} - m_{2r}^{m,s} + rT\}.$$

Applying the same argument to $M_{2r}^{m,s} - v_s^m(\cdot)$ we obtain

$$M_{2r}^{m,s} - m_r^{m,s} \leq C\{M_{2r}^{m,s} - M_r^{m,s} + rT\}$$

and adding up the last two inequalities we get

$$\operatorname{Osc}_{B_r(z)} v_s^m \le \frac{C-1}{C+1} \operatorname{Osc}_{B_{2r}(z)} v_s^m + CrT,$$

and it is very well known (see [5], chapter 8) that the above inequality implies (1.6).

We will now prove the second claim by contradiction. If we assume that it does not hold, we would have that $v_s^m(x) - v_s^m(0) > 0$ for all $x \in \partial B_{s/2}$. Then, for each $\varepsilon > 0$ we can find $k_0 = k_0(\varepsilon, s, m)$ so that $v_s^{m,k}(x) - v_s^{m,s}(0) + \varepsilon > 0$ on $\partial B_{s/2}$ for all $k \ge k_0$. Since $S^k(v_s^{m,k}) \le 0$ on $B_{s/2}$, we would conclude from the Strong Maximum principle for strong solutions that the last inequality holds over all of $B_{s/2}$. Applying lemma 1.8 to $v_s^{m,k}(\cdot) - v_s^{m,k}(0) + \varepsilon$ on $B_{s/2}$ we would obtain that for all $\sigma \in (0,1)$

$$\left[r^{-n} \int_{B_{\sigma s/2}} [v_s^{m,k}(y) - v_s^{m,k}(0) + \varepsilon]^p \, dy\right]^{1/p} \le C\varepsilon.$$

Letting first k tend to ∞ and then ε tend to zero in the above inequality we would get that v_s^m is identically constant on $B_{s/2}$ which is a contradiction. This finishes the proof of the lemma.

2. Proof of the Theorem. We will first prove theorem 0.1 when the set E reduces to one point x_0.

THEOREM 2.1. *Let L be a time independent elliptic operator as in (0.1) satisfying (0.2) and assume that the matrix $A(x)$ is continuous on $\mathbf{R}^n$ except possibly at x_0. Then the Dirichlet problem (0.5) is well posed for L on $R_r(x_0, a)$ for all $a \in \mathbf{R}$ and $r > 0$.*

Proof. From the ideas in [3] it suffices to show uniqueness when f is supported away from the line $\mathcal{L} = \{(x,t) \in \mathbf{R}^{n+1} : x = x_0\}$ and $\varphi \in C(\partial R_r(x_0, a))$. We may also assume without loss of generality that $x_0 = 0$ and that f has support contained in $\mathbf{R}^{n+1} \setminus (B_{\sigma r} \times \mathbf{R})$ for some $\sigma \in (0,1)$.

Let $w, v \in C(\overline{R}_r(0,a))$ be two possibly different good solutions to (0.5) with f and φ as before. Setting $u = w - v$ we have that u is a strong solution to $Lu = 0$ on $R_r(0,a) \setminus \mathcal{L}$ with $u = 0$ on $\partial R_r(0,a)$. If u is not identically zero it must attain its nonzero minimum at some point along the segment of line $R_r(0,a) \cap \mathcal{L}$ and we may assume without loss of generality that this minimum is attained at the point $(0,a)$ with $a = 0$. Since we are assuming that f is supported away from the set $R_r(0.0) \cap \mathcal{L}$, it follows from Lemma 1.6 and [3] that there exists a measurable function h on $B_{\sigma r/2}$ so that

$$u(x,0) = \int_{B_s} g_s(x,y)h(y)\,dy + \int_{\partial B_s} u(Q,0)\,d\omega_s^x(Q) \text{ for all } x \in B_s \text{ and } s \in (0, \sigma r/2),$$

where $g_s(x,y)$ and $d\omega_s^x$ are respectively the Green's function and harmonic measure for the operator S evaluated at x on B_s and the function h satisfies

$$\sup_{x \in B_{\sigma r/2}} \int_{B_{\sigma r/2}} g_{\sigma r}(x,y)h^2(y)\,dy < \infty.$$

From these and by dilation it follows that the function $u(\sigma r, 0) - u(0., 0)$ satisfies the conditions in Lemma 1.9. Thus, we conclude that there exists a family of points $\{x_s : s \in (0, \sigma r/4)\}$ with $|x_s| = s/2$ so that

$$\lim_{s \to 0} \frac{u(x_s, 0) - u(0,0)}{s} = 0. \tag{2.1}$$

On the other hand, for each unit vector ϑ in $\mathbf{R}^n$ the function $u(\cdot,\cdot) - u(0,0)$ satisfies the conditions of Lemma 1.7 with $B_r(x_0)$, n and y are replaced respectively by $B_{r/2}((r/2)\vartheta, 0))$, $n+1$ and $(0,0)$. Hence, we get

$$u(x,0) - u(0,0) \geq C\frac{|x|}{r} \inf_{R_{3r/4} \setminus R_{r/8}} [u(\cdot,\cdot) - u(0,0)] \text{ for } |x| \leq r/8. \tag{2.2}$$

From (2.1) and (2.2) it follows that there exists a point $(z,t) \in R_{3r/4} \setminus R_{r/8}$ with $u(z,t) = u(0,0)$. But this is a contradiction because u is a strong solution to $Lu = 0$ on $R_r \setminus \mathcal{L}$ and cannot attain its minimum at an interior point of $R_r \setminus \mathcal{L}$ unless it is a constant. Since $u = 0$ on $\partial R_r(0,a)$, it implies that u must be identically zero on R_r and proves the lemma.

THEOREM 2.2. *Let L be a time independent elliptic operator as in (0.1) satisfying (0.2) and assume that the matrix $A(x)$ is continuous on $\mathbf{R}^n$ except possibly for a countable set of points E with at most one cluster point $x_0 \in \mathbf{R}^n$. Then, the Dirichlet problem (0.5) is well posed for L on $R_r(x_0, a)$ for all $a \in \mathbf{R}$ and $r > 0$.*

Proof. Without loss of generality we may assume that $x_0 = 0$ and that $\varphi \in C(\partial R_r(0,a))$ and f is supported in $\mathbf{R}^{n+1} \setminus (B_{\sigma r} \times \mathbf{R})$ for some $\sigma \in (0,1)$ away from the line $\mathcal{L} = \{(x,t) \in \mathbf{R}^{n+1} : x = 0\}$. Let $w, v \in C(\overline{R}_r(0,a))$ be two possibly different good solutions to (0.5) with f and φ as before. Setting $u = w - v$ we have from Theorem 2.1 and Theorem 6 in [3] that w is a good solution to $Lu = 0$ on any open set Ω contained in $R_r(0,a) \setminus \mathcal{L}$ with $u = 0$ on $\partial R_r(0,a)$. If u is not identically zero it must attain its nonzero minimum at some point along the segment of line $R_r(0,a) \cap \mathcal{L}$ and we may assume without loss of generality that this minimum is attained at the point $(0,a)$ with $a = 0$. From Lemmas 1.6 and 1.9 we still can find a family of points $\{x_s : x_s \in (0, \sigma r/4)\}$ with $|x_s| = s/2$ so that (2.1) holds. To obtain (2.2) in the proof of Theorem 2.1 we used the fact that u was a strong solution to $Lu = 0$ on $B_{r/2}(((r/2)\vartheta, 0))$. In this case u is only a good solution to $Lu = 0$ on $B_{r/2}(((r/2)\vartheta, 0))$; but we showed that under these conditions Lemma 1.7 still holds. Thus, we also get (2.2) in this case and arrive at the same contradiction.

The final conclusion in Theorem 0.1 follows from the localization argument of Theorem 6 in [3].

REFERENCES

[1] ALEKSANDROV, A. D., Majorization of solutions of second order linear equations, Vestnick Leningrad Univ., 21, 5–25 (1966). English translation in AMS Trans. (2) (1968), 120–143.

[2] BAKELMAN, I. YA, Theory of quasilinear elliptic equations, Siberian Math. J., 2(1961), 179–186.

[3] Cerutti, M. C., Escauriaza, L., Fabes, E. B., Uniqueness in the Dirichlet problem for some elliptic operators with discontinuous coefficients. To appear, Annali di Matematica Pura ed Applicata.

[4] Fabes, E. B., Stroock, D., The L^p-integrability of Green's function and fundamental solutions for elliptic and parabolic equations, Duke Math. J. 51(1984), no. 4, 977–1016.

[5] Gibarg, D. and Trudinger, N. S., Elliptic Partial Differential Equations of Second Order, Springer-Verlag-New York-Berlin, 1983.

[6] Pucci, C., Limitazioni per soluzioni di equaziones ellittiche, Annali di Mat. Pura ed Appl. 161 (1966), 15–30.

[7] Safanov, MN. V., Harnack's inequality for elliptic equations and the Hölder property of their solutions, Zap. Nauchn. Sem. Leningrad. Odtel. Mat. Inst. Sreldov (Lomi), 96 (1980), 272–287. English translation in J. Soviet May. 21 (1983), no. 5.

THE SPECTRAL RADIUS OF THE CLASSICAL LAYER POTENTIALS ON CONVEX DOMAINS

EUGENE FABES,* † MARK SAND,* † AND JIN KEUN SEO* †

Introduction. Let D denote a bounded Lipschitz domain in R^n. For almost every (with respect to surface measure $d\sigma$) $Q \in \partial D$ the exterior normal N_Q at Q exists. The solution u to the Dirichlet problem,

$$\Delta u = 0 \text{ in } D, \quad u|_{\partial D} = g,$$

with $g \in L^2(\partial D, d\sigma)$ can be represented in the form of the classical double layer potential

$$u(X) = \frac{1}{\omega_n} \int_{\partial D} \frac{N_Q \circ (Q - X)}{|X - Q|^n} [((1/2)I + K)^{-1} g](Q) d\sigma(Q).$$

Here $X \in D$, ω_n is the area of the unit sphere $\{X \in R^n : |X| = 1\}$ in R^n, and

$$Kf(P) = \lim_{\epsilon \to 0} \frac{1}{\omega_n} \int_{|P-Q|>\epsilon} \frac{N_Q \circ (Q - P)}{|P - Q|^n} f(Q) d\sigma(Q),$$

the limit understood in the sense of $L^2(\partial D, d\sigma)$. The invertibility of $\frac{1}{2}I + K$ on $L^2(\partial D, d\sigma)$ was shown by G. Verchota in [6].

Similarly the solution u, unique up to constants, of the Neumann problem

$$\Delta u = 0 \text{ in } D, \quad \frac{\partial u}{\partial N}|_{\partial D} = g$$

with $g \in L_0^2(\partial D, d\sigma) \equiv \{g \in L^2(\partial D, d\sigma) : \int_{\partial D} g d\sigma = 0\}$, can be represented in the form of a single layer potential

$$u(X) = \begin{cases} \dfrac{1}{2\pi} \displaystyle\int_{\partial D} \log \frac{1}{|X - Q|} \quad ((-1/2)I + K^*)^{-1} g(Q) d\sigma(Q), & n = 2 \\ \dfrac{1}{(n-2)\omega_n} \displaystyle\int_{\partial D} \frac{1}{|X - Q|^{n-2}} ((-1/2)I + K^*)^{-1} g(Q) d\sigma(Q), & n \geq 3 \end{cases}$$

where K^* is the adjoint of the operator K defined above. The invertibility of $-\frac{1}{2}I + K^*$ on $L_0^2(\partial D, d\sigma)$ was also proved in [6].

As it stands the only seemingly noncomputable or nonapproximable parts of the above representations are those involving $(\frac{1}{2}I + K)^{-1}$ and $(-\frac{1}{2}I + K^*)^{-1}$. For nonsmooth domains the structure of these inverses are not well understood. It is the purpose of this paper to shed light on this matter at least in the case of convex domains in R^n. In fact our main result is

*Partially supported by NSF Grant Number DMS 90-01411.
†School of Mathematics, University of Minnesota, Minneapolis, MN 55455.

THEOREM (2.6). *If D is a bounded Lipschitz and convex domain in R^n, $n \geq 2$, then the spectral radius of K^* on $L_0^2(\partial D, d\sigma) \equiv \{f \in L^2(\partial D, d\sigma) : \int_{\partial D} f d\sigma = 0\}$ is $< \frac{1}{2}$. In particular*

$$(-1/2I + K^*)^{-1} = -2\sum_{j=0}^{\infty}(2)^j K^{*j}$$

where the series converges absolutely in the operator norm on $L_0^2(\partial D, d\sigma)$

1. On the resolvent set of K^* for Lipschitz domains. Recall that

$$K^* f(P) = \lim_{\epsilon \to 0} \frac{1}{\omega_n} \int_{\partial D} \frac{N_P \circ (P - Q)}{|P - Q|^n} f(Q) d\sigma(Q).$$

The following results on the eigenvalues and resolvent set of K^* are known.

THEOREM (1.1). *Suppose D is a bounded Lipschitz domain in R^n, $n \geq 2$. The eigenvalues of K^*, as an operator on $L^2(\partial D, d\sigma)$, are real and lie in the interval $(-\frac{1}{2}, \frac{1}{2}]$.*

A proof of the above theorem can be found in O.D. Kellogg's classical book ([3]) for the case of smooth domains. The proof, however, remains valid for Lipschitz domains once we know K^* is bounded on $L^2(\partial D, d\sigma)$. For a detailed discussion in the Lipschitz case see [1]. For dimensions $n \geq 3$ Kellogg's proof as described in [1] extends to the case of special unbounded Lipschitz domains, namely domains above the graph of a globally defined and Lipschitz continuous function.

DEFINITION 1. A domain $D \subset R^n$ is called a special Lipschitz domain if there is a Lipschitz continuous function $\varphi : R^{n-1} \to R$ such that

$$D = \{(x, y) : x \in R^{n-1}, y > \varphi(x)\}.$$

THEOREM (1.2). *Suppose D is a special Lipschitz domain in R^n, $n \geq 3$, then the eigenvalues of K^*, as an operator on $L^2(\partial D, d\sigma)$, are real and lie in the interval $(-\frac{1}{2}, \frac{1}{2}]$.*

We will see shortly that for special Lipschitz domains in R^n, $n \geq 3$ the eigenvalues of K^*, as an operator on $L^2(\partial D, d\sigma)$, lie in the interval $(-\frac{1}{2}, \frac{1}{2})$. The fact that $\frac{1}{2}$ is not an eigenvalue is actually due to the ideas of Verchota as described in [6].

THEOREM (1.3). *Assume D is a bounded Lipschitz domain in R^n, $n \geq 2$. If λ is real and $|\lambda| > \frac{1}{2}$ then $\lambda I - K^*$ is invertible on $L^2(\partial D, d\sigma)$.*

A proof of Theorem 1.3 can be found in [1]. By an easy extension of the ideas in [1] we will extend Theorem 1.3 to the case of special Lipschitz domains in R^n, $n \geq 3$, and with these same ideas, all due to Verchota, we will be able to include the end points in the unbounded case.

THEOREM (1.4). *Suppose D is a special Lipschitz domain in R^n, $n \geq 3$. If λ is real and $|\lambda| \geq \frac{1}{2}$, then $\lambda I - K^*$ is invertible on $L^2(\partial D, d\sigma)$.*

Proof. We begin with cases $\lambda = \frac{1}{2}$ and $\lambda = -\frac{1}{2}$. Set $u(X) = \mathcal{S}(f)(X)$, the single layer potential of f, with $f \in L^2(\partial D, d\sigma)$ and having compact support. Let e_i denote the ith standard basis vector of R^n. Applying the divergence theorem we have

$$\int_{\partial D} (e_n \circ N)|\nabla u|^2 = \int_D D_n(|\nabla u|^2)dX$$
$$= 2\int_D \sum_{i=1}^n (D_i D_n u) D_n u dX = 2\int_{\partial D} \frac{\partial u}{\partial N} D_n u d\sigma$$

Writing $e_n = (e_n \circ N)N + \tau$ we see that

$$\int_{\partial D} (e_n \circ N)[(\frac{\partial u}{\partial N})^2 + (\frac{\partial u}{\partial \tau})^2]d\sigma = 2\int_{\partial D} (\frac{\partial u}{\partial N})^2 e_n \circ N d\sigma + 2\int_{\partial D} \frac{\partial u}{\partial N}\frac{\partial u}{\partial \tau} d\sigma.$$

Hence

$$\int_{\partial D} (e_n \circ N)(\frac{\partial u}{\partial N})^2 d\sigma = \int_{\partial D} (e_n \circ N)(\frac{\partial u}{\partial \tau})^2 d\sigma - 2\int_{\partial D} \frac{\partial u}{\partial N}\frac{\partial u}{\partial \tau} d\sigma$$

The last equality implies there exists a constant $C_m > 0$ depending only on $m = \sup_{R^{n-1}} |\nabla\varphi|$ $(D = \{(x,y) : y > \varphi(x)\})$ such that

$$\frac{1}{C_m}\int_{\partial D} (\frac{\partial u}{\partial \tau})^2 d\sigma \leq \int_{\partial D} (\frac{\partial u}{\partial N})^2 d\sigma \leq C_m \int_{\partial D} (\frac{\partial u}{\partial \tau})^2 d\sigma.$$

As a consequence,

$$\frac{1}{C_m}\int_{\partial D} (\frac{\partial u}{\partial \tau})^2 d\sigma \leq \int_{\partial D} [(-1/2I + K^*)f]^2 d\sigma \leq C_m \int_{\partial D} (\frac{\partial u}{\partial \tau})^2 d\sigma.$$

If we apply the same method to the domain $R^n \backslash \bar{D}$ and to $u = \mathcal{S}(f)$ we obtain

$$\frac{1}{C_m}\int_{\partial D} (\frac{\partial u}{\partial \tau})^2 d\sigma \leq \int_{\partial D} [((1/2)I + K^*)f]^2 d\sigma \leq C_m \int_{\partial D} (\frac{\partial u}{\partial \tau})^2 d\sigma.$$

Let's remember that since $u = \mathcal{S}f$ is continuous across ∂D tangential derivatives from either side of D are the same. Hence for some other positive constant $\tilde{C}_m$ depending only on m we have

$$\frac{1}{\tilde{C}_m}\int_{\partial D} [((-1/2)I + K^*)f]^2 d\sigma \leq \int_{\partial D} [((1/2)I + K^*)f]^2 d\sigma$$
$$\leq \tilde{C}_m \int_{\partial D} [((-1/2) + I + K^*)f]^2 d\sigma. \tag{1.4.1}$$

Inequality (1.4.1) implies $(-\frac{1}{2}I+K^*)$ and $(\frac{1}{2}I+K^*)$ have the same kernel on $L^2(\partial D, d\sigma)$, and so this kernel must only include $f \equiv 0$. The argument given in [6] for the closed range of $(-\frac{1}{2}I+K^*)$ and of $(\frac{1}{2}I+K^*)$ in the case of a bounded Lipschitz domain can be used to conclude the same for these operators in the case of a special Lipschitz domain $D \subset R^n$, $n \geq 3$.

To show the surjectivity on $L^2(\partial D, d\sigma)$ of $-\frac{1}{2}I+K^*$ and $\frac{1}{2}I+K^*$ we use the technique found in [4]. In the case ∂D is the graph of a Lipschitz function φ the surjectivity on $L^2(\partial D, d\sigma)$ is the same as the surjectivity on $L^2(R^{n-1}, dx)$ of $-\frac{1}{2}I+K^*$ and $\frac{1}{2}I+K^*$ where

$$K^*f(x) = \frac{1}{\omega_n} \int_{R^{n-1}} \frac{\varphi(x)-\varphi(z)-\nabla\varphi(x)\circ(x-z)}{[|x-z|^2+(\varphi(x)-\varphi(z))^2]^{n/2}} f(z)dz.$$

For $t \geq 0$ set

$$K_t^*f(x) = \frac{1}{\omega_n} \int_{R^{n-1}} \frac{t(\varphi(x)-\varphi(z)-\nabla\varphi(x)\circ(x-z))}{[|x-z|^2+t^2(\varphi(x)-\varphi(z))^2]^{n/2}} f(z)dz.$$

We have already shown that $-\frac{1}{2}I+K_t^*$ and $\frac{1}{2}I+K_t^*$ are Fredholm operators for each $t \geq 0$. The mapping $t \to K_t^*$ is continuous from $[0,\infty)$ into the space of bounded linear operators on $L^2(R^{n-1}, dx)$. Since $(-\frac{1}{2}I+K_t^*)|_{t=0} = -\frac{1}{2}I$ we conclude that the index of $-\frac{1}{2}I+K_t^*$ is zero for each $t \geq 0$. Since $-\frac{1}{2}I+K^* \equiv -\frac{1}{2}I+K_1^*$ is injective it is also surjective. The same result holds for $\frac{1}{2}I+K^*$.

Now we will prove the invertibility of $\lambda I - K^*$ for λ real and $|\lambda| > 1/2$. Again set $u = \mathcal{S}f$ with $f \in L^2(\partial D, d\sigma)$ and having compact support. Once again the divergence theorem gives

$$\begin{aligned}\int_{\partial D} (e_n \circ N)|\nabla u|^2 d\sigma &= \int_D div(e_n|\nabla u|^2)dX \\ &= 2\int_{\partial D} \frac{\partial u}{\partial N} D_n u d\sigma\end{aligned}$$

For $u = \mathcal{S}f(x)$, $X \in D$,

$$D_n u|_{\partial D} = (\frac{-e_n \circ N}{2})f + K_n f$$

where

$$K_n f(p) = \lim_{\epsilon \to 0} \frac{1}{\omega_n} \int_{|P-Q|>\epsilon} \frac{e_n \circ (P-Q)}{|P-Q|^n} f(Q)d\sigma(Q),$$

the limit taken in $L^2(\partial D, d\sigma)$. Hence

$$\int_{\partial D} |e_n \circ N|[((1/2)I - K^*)f]^2 d\sigma \leq 2\int_{\partial D} ((1/2)I - K^*)f(-\frac{e_n \circ N}{2}I + K_n)f d\sigma.$$

We rewrite $\frac{1}{2}I - K^*$ as $(\frac{1}{2} - \lambda)I + (\lambda I - K^*)$ and obtain

$$(\lambda - 1/2)^2 \int_{\partial D} |e_n \circ N| f^2 d\sigma - 2(\lambda - 1/2) \int_{\partial D} |e_n \circ N| f(\lambda I - K^*) f d\sigma$$

$$+ \int_{\partial D} |e_n \circ N| [(\lambda I - K^*) f]^2 d\sigma \leq 2(1/2 - \lambda) \int_{\partial D} \frac{|e_n \circ N|}{2} f^2 d\sigma +$$

$$+2(-1/2 + \lambda) \int_{\partial D} f K_n f d\sigma + 2 \int_{\partial D} (\lambda I - K^*) f (\frac{|e_n \circ N|}{2} I + K_n) f d\sigma.$$

Since $K_n^* = -K_n$ on $L^2(\partial D, d\sigma)$ we have $\int_{\partial D} f K_n f d\sigma = 0$. From the last inequality it is easy to there exists $C_m > 0$ depending only on $m \geq sup|\nabla\varphi|$ and dimension such that

$$[(\lambda - 1/2)^2 + (\lambda - 1/2)] \int_{\partial D} |e_n \circ N| f^2 d\sigma \leq C_m \|(\lambda I - K^*) f\|_{L^2(\partial D, d\sigma)} \|f\|_{L^2(\partial D, d\sigma)}.$$

Since $(\lambda - \frac{1}{2})^2 + \lambda + \frac{1}{2} = \lambda^2 - \frac{1}{4}$ and $|\lambda| > \frac{1}{2}$ we conclude

$$\int_{\partial D} f^2 d\sigma \leq C_{m,n,\lambda} \int_{\partial D} [(\lambda I - K^*) f]^2 d\sigma$$

with $C_{m,n,\lambda}$ depending only m, n, and λ.

So $(\lambda I - K^*)$ has closed range on $L^2(\partial D, d\sigma)$ and the argument used to establish the surjectivity of $-\frac{1}{2}I + K^*$ can be applied to show the same for $\lambda I - K^*$.

This concludes the proof of Theorem (1.4). □

2. The spectral radius of K^* on convex domains. Theorems (1.1)-(1.4) give us considerable information on the eigenvalues and resolvent of K^* as an operator on $L^2(\partial D, d\sigma)$. When the kernel of K^* is nonnegative, as in the case of convex domains, a classical result in functional analysis combines with the above theorems to complete the picture of the spectral radius of K^* on $L^2(\partial D, d\sigma)$.

LEMMA (2.1). *Assume T is a bounded linear operator on L^2 of a measure space (Ω, μ) such that $Tf \geq 0$ (a.e.μ) whenever $f \geq 0$ (a.e. μ). Then the spectral radius of T belongs to the spectrum of T .*

For a proof of Lemma (2.1) see [2] or [5]. When we combine Lemma (2.1) with Theorems (1.3) and (1.4) we obtain

THEOREM (2.2). i) *Assume D is a bounded Lipschitz and convex domain in R^n, $n \geq 2$. Then the spectral radius of K^* on $L^2(\partial D, d\sigma)$ equals $\frac{1}{2}$.*

ii) If D is a special Lipschitz domain of R^n, $n \geq 3$, which is convex, then the spectral radius of K^* on $L^2(\partial D, d\sigma)$ is $< \frac{1}{2}$.

Proof. i) Lemma (2.1) and Theorem (1.3) imply that the spectral radius of K^* as an operator on $L^2(\partial D, d\sigma)$ is $\leq \frac{1}{2}$. Since $\frac{1}{2}I - K^*$ maps $L^2(\partial D, d\sigma)$ into $L^2_o(\partial D, d\sigma)$ we conclude that the spectral radius is $\frac{1}{2}$.
ii) In this case Lemma (2.1) combines with Theorem (1.4) to give that the spectral radius of K^* on $L^2(\partial D, d\sigma)$ is strictly less than $\frac{1}{2}$.

We will now begin the somewhat more difficult task of showing that for bounded, convex, and Lipschitz domains of R^n, $n \geq 2$, the spectral radius of K^* as an operator on $L^2_o(\partial D, d\sigma)$ is strictly less than $\frac{1}{2}$.

LEMMA (2.3). **(Localization Lemma)** *Let D be a bounded Lipschitz domain in R^n, $n \geq 2$. Fix a complex number λ and assume for each point $P \in \partial D$ there are positive numbers $r_{p,\lambda}$ and $C_{p,\lambda}$, depending on P and λ, such that for each boundary ball $\Delta_r \equiv B_r(P) \cap \partial D$, with $\frac{1}{3}r_{p,\lambda} \leq r \leq r_{p,\lambda}$, we have*

$$\|f\|_{L^2(\Delta_r,d\sigma)} \leq C_{P,\lambda}\|(\lambda I - K^*)(fX_{\Delta_r}\|_{L^2(\Delta_r,d\sigma)}$$

for all $f \in L^2(\Delta_r, d\sigma)$. ($X_E$ denotes the characteristic function of the set E.) If λ is not an eigenvalue of K^* on $L^2(\partial D, d\sigma)$ then $\lambda I - K^*$ has closed range on $L^2(\partial D, d\sigma)$.

Proof. Assume $(\lambda I - K^*)(f_j) \to g$ in $L^2(\partial D, d\sigma)$. If a subsequence of the f_j's converges weakly in L^2 to some $f \in L^2(\partial D, d\sigma)$, it is easy to conclude $(\lambda I - K^*)f = g$. In fact for any $h \in L^2(\partial D, d\sigma)$,

$$\int_{\partial D} (\lambda I - K^*)f\bar{h}d\sigma = \lim_j \int_{\partial D} f_j\overline{(\bar{\lambda} I - K)h}d\sigma$$
$$= \lim_j \int_{\partial D} (\lambda I - K^*)f_j\bar{h}d\sigma = \int_{\partial D} g\bar{h}d\sigma.$$

Hence $(\lambda I - K^*)f = g$.

If then $\lim_j \|f_j\|_{L^2(\partial D,d\sigma)} = \infty$ we divide each f_j by its L^2-norm and obtain a new sequence which we again denote by $\{f_j\}$ such that $\|f_j\|_{L^2(\partial D,d\sigma)} = 1$ and $(\lambda I - K^*)f_j \to 0$ in $L^2(\partial D, d\sigma)$. We will arrive to a contradiction of this situation.

Since $\|f_j\|_{L^2(\partial D,d\sigma)} = 1$ we may assume the sequence converges weakly in L^2. Also since $\lambda I - K^*$ is injective, the argument in the first paragraph of this proof allows us to conclude that the sequence converges weakly to zero.

Fix $P \in \partial D$ and set $\Delta_s \equiv \Delta_s(P) \equiv B_s(P) \cap \partial D$ and take $\frac{1}{3}r_{p,\lambda} \leq r < s \leq r_{p,\lambda}$. From the hypotheses

$$\int_{\Delta_r} |f_j|^2 d\sigma \leq C \int_{\Delta_r} |(\lambda I - K^*)(f_jX_{\Delta_r})|^2 d\sigma$$
$$\leq C \int_{\Delta_r} |(\lambda I - K^*)(f_j)|^2 d\sigma +$$
$$+ C \int_{\Delta_r} |K^*(f_jX_{\Delta_s \setminus \Delta_r})|^2 d\sigma + C \int_{\Delta_r} |K^*(f_jX_{\partial D \setminus \Delta_s})|^2 d\sigma.$$

Hence

$$\int_{\Delta_r} |f_j|^2 d\sigma \leq \epsilon_j(r,s) + C \int_{\Delta_s \setminus \Delta_r} |f_j|^2 d\sigma \tag{2.3.1}$$

where C depends on P and λ and for fixed r, s, and P, $\lim_j \epsilon_j = 0$. Now choose a large integer N and partition $[\frac{1}{2} r_{p,\lambda}, r_{p,\lambda}]$ by the numbers $\frac{1}{2} r_{p,\lambda} = r_0 < r_1 < \cdots < r_N = r_{p,\lambda}$. From (2.3.1) for $i = 0, \ldots, N-1$,

$$\int_{\Delta_{r_i}} |f_j|^2 d\sigma \leq \epsilon(j, r_i, r_{i+1}) + C \int_{\Delta_{r_{i+1}} \setminus \Delta_{r_i}} |f_j|^2 d\sigma.$$

Adding in i from $i = 0, \ldots, N-1$,

$$N \int_{\Delta_{\frac{1}{2} r_{p,\lambda}}} |f_j|^2 d\sigma \leq \sum_{i=0}^{N-1} \epsilon(j, r_i, r_{i+1}) + C \int_{\Delta_{r_{p,\lambda}} \setminus \Delta_{\frac{1}{2} r_{p,\lambda}}} |f|^2 d\sigma.$$

Hence $\overline{\lim}_{j\to\infty} \int_{\Delta_{\frac{1}{2} r_{p,\lambda}}} |f_j|^2 d\sigma = 0$ and so $\lim_j \int_{\partial D} |f_j|^2 d\sigma = 0$. This is a contradiction since $\int_{\partial D} |f_j|^2 d\sigma = 1$ for each j .

THEOREM (2.4). *If D is a bounded, convex, Lipschitz domain in R^n, $n \geq 2$, and $\lambda \in \mathbf{C} \setminus R$ satisfies $|\lambda| = \frac{1}{2}$, then $\lambda I - K^*$ has closed range on $L^2(\partial D, d\sigma)$.*

Proof. We will show that when D is convex the hypotheses of the Localization Lemma (2.3) are valid.

We begin with the case of the dimension $n \geq 3$. Fix $P \in \partial D$. There exists $r_p > 0$ and a convex and Lipschitz continuous function $\varphi : R^{n-1} \to R$ such that for $0 < r \leq 2r_p$,

$$D \cap B_r(P) = \{(x,y) : x \in R^{n-1}, y > \varphi(x)\} \cap B_r(P).$$

Let K^*_φ denote the operator corresponding to K^* but for the special Lipschitz domain

$$D_\varphi = \{(x,y) : x \in R^{n-1}, y > \varphi(x)\}.$$

The convexity of φ implies $K^*_\varphi f \geq 0$ when $f \geq 0$. From Theorem (2.2) part (ii) the spectral radius of K^*_φ is $< \frac{1}{2}$ on $L^2(\partial D_\varphi, d\sigma)$. In particular there exists an integer $M > 0$ such that K^{*M}_φ has operator norm strictly less than $\frac{1}{2}$ on $L^2(\partial D, d\sigma)$. The nonnegativity of the kernels immediately imply that the operator norm of $(X_{\Delta_r(P)} K^* X_{\Delta_r(P)})^M$ on $L^2(\Delta_r, d\sigma)$ is $\leq$ the operator norm of K^*_φ on $L^2(\partial D_\varphi, d\sigma)$ for each $r \leq 2r_p$. Hence there exists $C_p > 0$ such that for $|\lambda| = \frac{1}{2}$

$$\|f\|_{L^2(\Delta_r, d\sigma)} \leq C_p \|(\lambda I - K^*)(f X_{\Delta_r})\|_{L^2(\Delta_r, d\sigma)}$$

for all $f \in L^2(\Delta_r, d\sigma)$. If now $\lambda \in \mathbf{C}\backslash R, \lambda$ is not an eigenvalue of K^* and the hypotheses of the Localization Lemma are all valid. So $\lambda I - K^*$ has closed range on $L^2(\partial D, d\sigma)$.

Next we turn to the remaining case when the dimension $n = 2$. In this case we can prove that for each $P \in \partial D$ there exists $r_p > 0$ and $0 \le \eta_p < \frac{1}{2}$ such that for all $f \in L^2(\Delta_{r_p}(P), d\sigma)$,

$$\|K^*(fX_{\Delta_{r_p}})\|_{L^2(\Delta_{r_p}(P),d\sigma)} \le \eta_p \|f\|_{L^2(\Delta_{r_p}(P),d\sigma)}.$$

We will show how to obtain the above estimate by studying the kernel of K^* in local coordinates. For fixed P there exists $r_p > 0$ such that

$$\Delta_{r_p}(P) = \{(x, \varphi(x)) : x \in R^{n-1}\} \cap B_{r_p}(P)$$

with $\varphi : R \to R$ Lipschitz continuous and convex. We may also assume P has coordinates $(0, \varphi(0)) = (0, 0)$.

In local coordinates the kernel of K^* has the form

$$\frac{1}{2\pi\sqrt{1+\varphi'(x)^2}} \quad \frac{\varphi(x) - \varphi(z) - \varphi'(x)(x-z)}{(x-z)^2 + (\varphi(x) - \varphi(z))^2}.$$

Set $\varphi'(0+) = \lim_{x\to 0+} \frac{\varphi(x)}{x}$ and $\varphi'(0-) = \lim_{x \to 0-} \frac{\varphi(x)}{x}$. If x and z have the same sign, say $x > 0$ and $z > 0$, then the above kernel is pointwise bounded by the nonnegative kernel

$$\frac{\varphi'(x) - \varphi'(z)}{x - z}.$$

Since both $\varphi'(x)$ and $\varphi'(z)$ are near to $\varphi'(0+)$ in the L^∞ sense when x and z are nonnegative and near 0, we conclude

$$\left\| \int_0^s \frac{\varphi'(x) - \varphi'(z)}{x - z} f(z)dz \right\|_{L^2((0,s),dx)} \le \epsilon(s) \|f\|_{L^2((0,s),dx)}$$

where $\epsilon(s) \to 0$ as $s \to 0+$. A similar result holds for $x < 0$ and $s < 0$.

In the case $x > 0$ and $z < 0$ then the kernel of K^* in local coordinates can be written as

$$\frac{1}{2\pi\sqrt{1+\varphi'(0+)^2}} \quad \frac{\varphi'(0+) - \varphi'(0-))z}{(x-z)^2 + (\varphi'(0+)x - \varphi'(0-)z)^2} + E(x,z)$$

where

$$\left\| \int_{-s}^0 |E(x,z)| f(z) dz \right\|_{L^2((0,s)dx)} \le \epsilon(s) \|f\|_{L^2((-s,0),dx)}$$

where $\epsilon(s) \to 0$ as $s \to 0+$. The operator norm of the operator

$$\int_{-\infty}^0 \frac{z}{(x-z)^2 + (\varphi'(0+)x - \varphi'(0-)z)^2} f(z)dz$$

as an operator from $L^2((-\infty, 0), dx)$ into $L^2((0,\infty), dx)$ is computable and the result is

$$\|K^* f X_{\Delta_s}\|_{L^2(\Delta_s, d\sigma)} \le [\eta + \epsilon(s)] \|f\|_{L^2(\Delta_s, d\sigma)}$$

where $\eta < \frac{1}{2}$ and $\epsilon(s) \to 0$ as $s \to 0+$.

THEOREM (2.5). *Suppose D is a bounded convex and Lipschitz domain of R^n, $n \geq 2$. If $\lambda \in \mathbf{C}\backslash R$ and $|\lambda| = \frac{1}{2}$ then $\lambda I - K^*$ is invertible on $L^2(\partial D, d\sigma)$.*

Proof. From Theorem (2.4) $\lambda I - K^*$ has closed range. Also from Theorem (1.1) it is injective. Hence there is a constant $C > 0$ such that for all $f \in L^2(\partial D, d\sigma)$,

$$\|f\|_{L^2(\partial D, d\sigma)} \leq C\|(\lambda I - K^*)f\|_{L^2(\partial D, d\sigma)}. \tag{2.5.1}$$

To show $\lambda I - K^*$ is surjective pick $g \in L^2(\partial D, d\sigma)$ and take $\lambda_j \in \mathbf{C}$, $|\lambda_j| > \frac{1}{2}$, with λ_j converging to λ. From Theorem (2.2) part (i) there exists $f_j \in L^2(\partial D, d\sigma)$ such that $(\lambda_j I - K^*)f_j = g$. If a subsequence of the f_j's converges weakly in $L^2(\partial D, d\sigma)$ to f then, as in the first part of the proof of Lemma (2.3) we easily conclude that

$$(\lambda I - K^*)f = g.$$

If, however, $\|f_j\|_{L^2(\partial D, d\sigma)} \to \infty$ we divide each f_j by its L^2-norm and obtain a new sequence, $\{\tilde{f}_j\}$, such that

$$\|\tilde{f}_j\|_{L^2(\partial D, d\sigma)} = 1 \quad \text{and} \quad \|(\lambda_j I - K^*)\tilde{f}_j\|_{L^2(\partial D, d\sigma)} \to 0.$$

However from 2.5.1

$$\begin{aligned} 1 = \|\tilde{f}_j\|_{L^2(\partial D, d\sigma)} &\leq C\|(\lambda I - K^*)\tilde{f}_j\|_{L^2(\partial D, d\sigma)} \\ &\leq C|\lambda - \lambda_j| + C\|(\lambda_j I - K^*)\tilde{f}_j\|_{L^2(\partial D, d\sigma)} \end{aligned}$$

and this produces clearly a contradiction.

THEOREM (2.6). *Suppose D is a bounded, convex and Lipschitz domain in R^n, $n \geq 2$. Then the spectral radius of K^* on $L_0^2(\partial D, d\sigma)$ is $< \frac{1}{2}$.*

Proof. From Theorem (2.2) part (i) if $|\lambda| > \frac{1}{2}$ then $\lambda I - K^*$ is invertible on $L^2(\partial D, d\sigma)$. Since $\int_{\partial D} (\lambda I - K^*)f d\sigma = (\lambda - \frac{1}{2}) \int_{\partial D} f$ we see that $\lambda I - K^*$ is invertible on $L_0^2(\partial D, d\sigma)$ for $|\lambda| > \frac{1}{2}$. This same argument, using now Theorem (2.5), implies $\lambda I - K^*$ is invertible on $L_0^2(\partial D, d\sigma)$ for $|\lambda| = \frac{1}{2}$, $\lambda \in \mathbf{C}\backslash R$. The invertibility of $\frac{1}{2}I - K^*$ and $\frac{1}{2}I + K^*$ on $L_0^2(\partial D, d\sigma)$ was already established in [6] for Lipschitz domains.

This completes the proof of Theorem (2.6). □

REFERENCES

[1] L. ESCAURIAZA, E. FABES, G. VERCHOTA, *On a regularity theorem for weak solutions to transmission problems with internal Lipschitz boundaries*, to appear in the Proceedings of the AMS.

[2] S. KARLIN, *Positive operators*, Journal of Mathematics and Mechanics 8, No. 6 (1959), pp. 907-937.

[3] O.D. KELLOGG, *Foundations of Potential Theory*, Dover, New York 1953.

[4] C.E. KENIG, *Recent progress on boundary-value problems on Lipschitz domains*, A.M.S. Proceedings of Symposia in Pure Mathematics, vol. 43 (1985), pp. 175-205.

[5] M.A. KRASNOSEL'SKII ET AL, *Approximate Solution of Operator Equations*, Walters - Noordhoff, Groningen, 1972.

[6] G. VERCHOTA, *Layer potentials and regularity for the Dirichlet problem for Laplace's equation in Lipschitz domains*, Journal of Functional Analysis 59 (1984), pp. 572-611.

UNIQUE CONTINUATION FOR DEGENERATE ELLIPTIC EQUATIONS

NICOLA GAROFALO* **

1. A famous result, first proved in $\mathbb{R}^2$ by Carleman [C] in 1939, states that if $V \in L^\infty_{\text{loc}}(\mathbb{R}^N)$ and u is a solution to $\Delta u = Vu$ in a connected open set $D \subset \mathbb{R}^N$, then u cannot vanish to infinite order at a point $x_0 \in D$ unless $u \equiv 0$ in D. We are interested in analogous results when the (elliptic) Laplacian in $\mathbb{R}^N$ is replaced by a subelliptic operator of the type

$$\mathcal{L} = \sum_{j=1}^{N-1} X_j^2, \tag{1.1}$$

where $X_1, \ldots, X_{N-1}$ are smooth vector fields satisfying Hörmander's condition for hypoellipticity [H]:

$$\operatorname{rank} \operatorname{Lie}[X_1, \ldots, X_{N-1}] = N \text{ at every point.} \tag{1.2}$$

Unfortunately, there is no analogue of Carleman's result for (1.1), even when $V \in C^\infty$. This is a consequence of the following:

THEOREM (Bahouri [B]). *Let $X_1, \ldots, X_{N-1}$ be C^∞ vector fields in an open set $D \subset \mathbb{R}^N$ satisfying* (1.2). *Suppose, moreover, that*

(i) The vector space generated by $X_1, \ldots, X_{N-1}$ has dimension $N-1$ at every point of D;

(ii) There exists a point $x_0 \in D$ such that in a neighborhood of it

$$\varepsilon_N \wedge d\varepsilon_N \neq 0, \quad \varepsilon_N \wedge (d\varepsilon_N)^2 = 0$$

where ε_N is the one–form that to every vector field X associates

$$\varepsilon_N(X) = X_1 \wedge X_1 \wedge \cdots \wedge X_{N-1} \wedge X.$$

Then there exists an open set Ω, with $x_0 \in \Omega \subset D$, and a function $V \in C^\infty(\Omega)$ such that the equation $-\mathcal{L}u + Vu = 0$ admits a nontrivial solution $u \in C^\infty(\Omega)$ vanishing in an open subset of Ω.

When $N = 3$ or 4 the conclusion of Theorem 1.1 holds without assumption (ii). Thus when $N = 3$ or 4 every operator (1.1) for which (1.2) and (i) hold fails to have

---

*Supported by the NSF, grant DMS–9096158
**Department of Mathematics, Purdue University, West Lafayette, IN 47907

the unique continuation property for some $V \in C^\infty$. In particular, the construction in [B] implies that for the operator in $\mathbf{R}^3$

$$\mathcal{L} = \left(\frac{\partial}{\partial x} + 2y\frac{\partial}{\partial t}\right)^2 + \left(\frac{\partial}{\partial y} - 2x\frac{\partial}{\partial t}\right)^2 \tag{1.3}$$

there exists a neighborhood of the origin Ω, and a $V \in C^\infty(\Omega)$ such that

$$\operatorname{supp} V \subset \Omega \cap \{x \geq 0\}, \quad V \text{ is flat at } \{x = 0\}, \tag{1.4}$$

for which the equation $-\mathcal{L}u + Vu = 0$ has a nontrivial solution $u \in C^\infty(\Omega)$ flat at $\{x = 0\}$ and supported in $\Omega \cap \{x \geq 0\}$. Now $\mathcal{L}$ in (1.3) is an important operator: It is the subelliptic Laplacian on the three dimensional Heisenberg group $\mathbf{H}^1$. More in general, for every $n \in \mathbf{N}$ let $\mathbf{H}^n$ denote the Lie group whose underlying manifold is $\mathbf{C}^n \times \mathbf{R}$ with group law

$$(z,t) \circ (z',t') = \left(z + z', t + t' + 2\operatorname{Im}\sum_{j=1}^{n} z_j\overline{z'_j}\right). \tag{1.5}$$

$(\mathbf{H}^n, \circ)$ is the $(2n+1)$–dimensional *Heisenberg group*. A basis for the Lie algebra of left–invariant vector fields with respect to (1.5) is provided by

$$X_j = \frac{\partial}{\partial x_j} + 2y_j\frac{\partial}{\partial t}, \quad Y_j = \frac{\partial}{\partial y_j} - 2x_j\frac{\partial}{\partial t}, \quad j = 1, \dots, n, \quad \frac{\partial}{\partial t}, \tag{1.6}$$

and the *subelliptic Laplacian* on $\mathbf{H}^n$ is defined by

$$\Delta_{\mathbf{H}^n} = \sum_{j=1}^{n} (X_j^2 + Y_j^2). \tag{1.7}$$

If $A = (a_{ij})$ is the $(2n+1) \times (2n+1)$ matrix defined by $a_{ij} = \delta_{ij}$ for $i, j = 1, \dots, 2n$, $a_{2n+1,j} = 2y_j$ for $j = 1, \dots, n$, $a_{2n+1,n+j} = -2x_j$ for $j = 1, \dots, n$, $a_{2n+1,2n+1} = 4|z|^2$, $a_{ji} = a_{ji}$ for all i, j, then

$$\Delta_{\mathbf{H}^n} = \operatorname{div}(A\nabla). \tag{1.8}$$

An easy computation shows that $\det A \equiv 0$, thus proving that $\Delta_{\mathbf{H}^n}$ is not elliptic at any point. However, as it is easily checked $[X_j, Y_k] = -4\delta_{jk}\frac{\partial}{\partial t}$ for $j, k = 1, \dots, n$, so that $\Delta_{\mathbf{H}^n}$ is hypoelliptic, see [H]. In fact, see (1.23) below, $\Delta_{\mathbf{H}^n}$ is real–analytic hypoelliptic like Laplace's operator Δ in $\mathbf{R}^n$. Therefore, if u solves $\Delta_{\mathbf{H}^n} u = 0$ in a connected open set $D \subset \mathbf{H}^n$, then u cannot vanish to infinite order at $(z_0, t_0) \in D$, unless $u \equiv 0$ in D. As a consequence of a theorem of Bony [Bo] if V is real analytic then $-\Delta_{\mathbf{H}^n} + V$ has the unique continuation property. Can one hope for any positive results for nonanalytic V's? Recently, E. Lanconelli and I have provided an affirmative answer to this question, see [GLan1] and [GLan2]. In order to state the results I need to introduce some notation.

There is a natural family of non–isotropic dilations associated with $\Delta_{\mathbf{H}^n}$, namely

$$\delta_\lambda(z,t) = (\lambda z, \lambda^2 t), \quad \lambda > 0, \quad (z,t) \in \mathbf{H}^n. \tag{1.9}$$

It is readily recognized that

$$\Delta_{\mathbf{H}^n} \circ \delta_\lambda = \lambda^2 \delta_\lambda \circ \Delta_{\mathbf{H}^n}. \tag{1.10}$$

A function u on $\mathbf{H}^n$ is said homogeneous of degree $k \in \mathbf{R}$ if

$$u \circ \delta_\lambda = \lambda^k u, \quad \lambda > 0. \tag{1.11}$$

Let X be the smooth vector field

$$X = \sum_{j=1}^{n} \left(x_j \frac{\partial}{\partial x_j} + y_j \frac{\partial}{\partial y_j} \right) + 2t \frac{\partial}{\partial t}. \tag{1.12}$$

Then Euler's formula holds, i.e., u satisfies (1.11) iff

$$Xu = ku, \tag{1.13}$$

which means that X is the generator of the group $\{\delta_\lambda\}_{\lambda>0}$. Let $dz\,dt$ denote Lebesgue measure on $\mathbf{H}^n$ ($dz\,dt$ is a left–and right–invariant Haar measure on $\mathbf{H}^n$). Under the rescaling (1.9) $dz\,dt$ changes by a factor λ^{2n+2} so that the number

$$Q = 2n + 2 \tag{1.14}$$

is the homogeneous dimension of $\mathbf{H}^n$. A remarkable function on $\mathbf{H}^n$ is given by

$$d(z,t) = \left(|z|^4 + t^2\right)^{\frac{1}{4}}. \tag{1.15}$$

Since d is homogeneous of degree one with respect to (1.9), (1.13) gives

$$Xd = d. \tag{1.16}$$

In the sequel, for a function u on $\mathbf{H}^n$ we let

$$|\nabla_{\mathbf{H}^n} u|^2 = \sum_{j=1}^{n} \left[(X_j u)^2 + (Y_j u)^2 \right]. \tag{1.17}$$

A function which has a special geometric significance is

$$|\nabla_{\mathbf{H}^n} d|^2 = \frac{|z|^2}{d(z,t)^2} \stackrel{\text{def}}{=} \psi(z,t). \tag{1.18}$$

ψ is homogeneous of degree zero, so

$$X\psi = 0, \tag{1.19}$$

and moreover $0 \leq \psi \leq 1$, $\psi(z,0) \equiv 1$, $\psi(0,t) \equiv 0$. With A as in (1.8) it is not difficult to recognize that

$$A\nabla d = \frac{1}{d}(\psi X + \varphi T), \tag{1.20}$$

where I have set

$$\varphi(z,t) = \frac{t}{d(z,t)^2} \tag{1.21}$$

and

$$T = \sum_{j=1}^{n} \left(y_j \frac{\partial}{\partial x_j} - x_j \frac{\partial}{\partial y_j} \right). \tag{1.22}$$

A remarkable fundamental solution for $-\Delta_{\mathbf{H}^n}$ with pole at the origin was discovered by Folland [F]

$$\Gamma(z,t) = \frac{C_Q}{d(z,t)^{Q-2}}, \tag{1.23}$$

where $C_Q > 0$ is a suitable number depending only on Q in (1.14). Let us introduce the balls

$$B_r = \{(z,t) \in \mathbf{H}^n | d(z,t) < r\}, \quad r > 0. \tag{1.24}$$

By means of (1.24) it is clear that

$$B_r = \left\{ (z,t) \in \mathbf{H}^n | \Gamma(z,t) > \frac{C_Q}{r^{Q-2}} \right\}. \tag{1.25}$$

Since $Td = 0$ the vector field T is tangential to the ball B_r. Also, using $Td = 0$ and (1.16) it is easy to deduce (1.18) from (1.20).

Finally, an easy computation shows that in the coordinates (z,t) (1.7) can be written

$$\Delta_{\mathbf{H}^n} = \Delta_z + 4|z|^2 \frac{\partial^2}{\partial t^2} + 4\frac{\partial}{\partial t}T, \tag{1.26}$$

with T given by (1.22).

One of the main results in [GLan2] is given by the following

THEOREM 1.1. *Let $R_0 > 0$ be fixed and let V be a measurable function in B_{R_0} such that there exist $C > 0$ and $f : (0, R_0) \to \mathbf{R}^+$ increasing for which*

$$\int_0^{R_0} \frac{f(r)}{r} dr < \infty \tag{1.27}$$

and

$$|V(z,t)| \le C \frac{f(d(z,t))}{d(z,t)^2} \psi(z,t), \quad \text{a.e. } (z,t) \in B_{R_0}, \tag{1.28}$$

where ψ is as in (1.18). Let u be a solution in B_{R_0} of

$$\Delta_{\mathbf{H}^n} u = Vu. \tag{1.29}$$

Suppose there exist $C_1 > 0$ and $g : (0, R_0) \to \mathbf{R}^+$ increasing and satisfying (1.27) such that

$$|tTu(z,t)| \le C_1 g\left(d(z,t)\right) |z|^2 |u(z,t)|, \text{ a.e. } (z,t) \in B_{R_0}. \tag{1.30}$$

Then there exist $r_0 = r_0(Q, C, C_1, f, g) > 0$ and $K = K(Q, C, C_1, f, g, u) > 0$ such that

$$\int_{B_{2r}} u^2 \psi dz\, dt \le K \int_{B_r} u^2 \psi dz\, dt \tag{1.31}$$

for every $r \in \left(0, \frac{r_0}{2}\right)$.

COROLLARY 1.1. *Under the assumptions of Theorem 1.1 if u vanishes to infinite order at the origin in the sense that for every $k \in \mathbf{N}$*

$$\int_{B_r} u^2 \psi dz\, dt = 0(r^k) \text{ as } r \to 0,$$

then $u \equiv 0$ in B_{r_0}, where r_0 is the number whose existence is claimed in Theorem 1.1.

The dependence of the constants r_0 and K in the statement of Theorem 1.1 on the various parameters involved is quite explicit. Specifically, r_0 can be chosen such that

$$f(r_0) < C^{-1} \left(\frac{Q-2}{2}\right)^2 \text{ and } g(r_0) < 2C_1^{-1},$$

while

$$K = 2^Q \exp\left\{2 \log 2 \max\left(1, N(r_0)\right) \left[1 + \exp\left(M \int_0^{R_0} \frac{f(t) + g(t)}{t} dt\right)\right]\right\},$$

where $M > 0$ is a constant depending only on Q, C, C_1, f, g, and

$$N(r_0) = r_0 \frac{\displaystyle\int_{B_{r_0}} \left[|\nabla_{\mathbf{H}^n} u|^2 + Vu^2\right] dz\, dt}{\displaystyle\int_{\partial B_{r_0}} u^2 \frac{\psi}{|\nabla d|} dH_{Q-2}}.$$

Another result in [GLan2] is concerned with weaker assumptions both on the potential V and on the differential inequality (1.30). In what follows let $V = V^+ - V^-$ where V^+, V^- are respectively the positive and the negative part of V.

THEOREM 1.2. *Suppose there exist $C > 0$ and a dimensional constant $\delta = \delta_Q > 0$ such that*

$$0 \leq V^+(z,t) \leq \frac{C}{d(z,t)^2}\psi(z,t), \quad 0 \leq V^-(z,t) \leq \frac{\delta}{d(z,t)^2}\psi(z,t), \text{ a.e. } (z,t) \in B_{R_0}.$$

Suppose moreover that there exists $C_1 > 0$ such that

$$|tTu(z,t)| \leq C_1 |z|^2 |u(z,t)|, \quad \text{a.e. } (z,t) \in B_{R_0}.$$

There exists $r_0 = r_0(Q, C, C_1, \delta) > 0$ such that if as $r \to 0^+$

$$\int_{B_r} u^2 \psi dz\, dt = 0\left(\exp\left[-Ar^{-\alpha}\right]\right)$$

for some $A, \alpha > 0$, then must be $u \equiv 0$ in B_{r_0}.

The proof of Theorems 1.1, 1.2 is modelled on the method developed in [GL1], [GL2] for elliptic equations. Yet, in the present subelliptic context several difficulties arise of a rather subtle geometric nature. It is hoped that, once the model operator $\Delta_{\mathbf{H}^n}$ will be fully understood, the approach in [GLan2] will carry to more general operators such as (1.1). Here is a list of the main ingredients involved in the proof of Theorems 1.1, 1.2.

1) Representation formulas for functions on $\mathbf{H}^n$ in terms of weighted averages on the level sets (1.25) of the fundamental solution (1.23).
2) Uncertainty inequality for $\mathbf{H}^n$.
3) First variation estimates for the energy integral associated to (1.29).
4) A frequency function on $\mathbf{H}^n$ and its growth properties on balls of infinitesimal radii.

We have some work in progress which aims to extend the results on [GLan2] to operators such as (1.1). Using the results in [RS], [S], [NSW] Lanconelli and I [GLan3] have been able to establish 1) and 2) above in the general context of (1.1) and to carry over part of 3) and 4).

2. In this section I want to discuss the issue of Carleman estimates for some degenerate elliptic operators related to (1.7). Consider the natural action of the torus $\mathbf{T}$ on $\mathbf{H}^n$ given by

$$\varphi_\theta(z,t) = \left(e^{i\theta}z, t\right), \quad \theta \in [0, 2\pi], \ (z,t) \in \mathbf{H}^n. \tag{2.1}$$

It is not difficult to recognize that a function u on $\mathbf{H}^n$ is invariant with respect to (2.1), i.e., $u \circ \varphi_\theta = u$ for every $\theta \in [0, 2\pi]$, iff $Tu \equiv 0$. In virtue of (1.26) this remark shows that if u is a solution to $\Delta_{\mathbf{H}^n} u = 0$ and u is invariant with respect to the action (2.1), then

$$\Delta_z u + 4|z|^2 \frac{\partial^2 u}{\partial t^2} = 0. \tag{2.2}$$

This one less degree of freedom expressed by invariance with respect to (2.1) in a way trivializes the complex geometry of $\Delta_{\mathbf{H}^n}$. Thus, for instance, left invariance with respect to (1.5) is lost when passing from (1.26) to (2.2). Nonetheless, from the viewpoint of pde's (2.2) is an interesting operator which has several features in common with $\Delta_{\mathbf{H}^n}$. (2.2) (and its variants) was first studied by Grushin [Gr1], [Gr2], who established its hypoellipticity.

I want to consider for $z \in \mathbf{R}^n$, $t \in \mathbf{R}^m$ the model operator in $\mathbf{R}^{n+m}$

$$P_\alpha = \Delta_z + |z|^{2\alpha}\Delta_t, \quad \alpha > 0, \tag{2.3}$$

which is elliptic for $z \neq 0$ and degenerates on the manifold $\{0\} \times \mathbf{R}^m$. As for (1.7) there is a natural family of non–isotropic dilations attached to P_α, namely

$$\delta_\lambda(z,t) = \left(\lambda z, \lambda^{\alpha+1} t\right), \quad \lambda > 0, \quad (z,t) \in \mathbf{R}^{n+m}. \tag{2.4}$$

As in (1.14) one can define a homogeneous dimension related to (2.4)

$$Q = n + m(\alpha + 1). \tag{2.5}$$

This number plays an important role in the local analysis of P_α at points of the manifold $\{0\} \times \mathbf{R}^m$. Off this set the operator is locally uniformly elliptic and the homogeneous dimension coincides with the topological dimension $N = m + n$. Henceforth, I will write $\mathbf{R}^N$ instead of $\mathbf{R}^{n+m}$.

In the light of Bahouri's cited result Carleman estimates fail for (1.7). In spite of the degeneracy of (2.3), which becomes increasingly stronger with $\alpha \to \infty$, Carleman estimates for P_α which allow to prove unique continuation from the manifold $\{0\} \times \mathbf{R}^m$ do exist, instead. Provided that the notion of vanishing be suitably calibrated on the natural geometry of P_α. This is part of the content of the paper [G]. To explain the results consider the function on $\mathbf{R}^N$

$$d(z,t) = d_\alpha(z,t) = \left(|z|^{2(\alpha+1)} + (\alpha+1)^2|t|^2\right)^{\frac{1}{2(\alpha+1)}}. \tag{2.6}$$

In [G] I prove that with Q as in (2.5) a fundamental solution $\Gamma(z,t)$ of $-P_\alpha$ in (2.3) with pole at $(0,0) \in \mathbf{R}^N$ is given by

$$\Gamma(z,t) = \frac{C_{\alpha,Q}}{d(z,t)^{Q-2}}, \tag{2.7}$$

where $C_{\alpha,Q} > 0$ is a suitable constant depending only on α, Q. When $n = 2k$, $k \in \mathbf{N}$, and $m = \alpha = 1$, this remarkable fundamental solution reduces (up to rescaling in t), to that found by Folland, see (1.23), (1.15). Consider then the singular weight function $\exp(\Gamma^\beta)$, $\beta > 0$, where Γ is as in (2.7), or what is the same $\exp(d^{-\beta})$, with d as in (2.6). By analogy with (1.24), (1.25) denote by B_r the level sets of the function d in (2.6), i.e.,

$$B_r = \left\{(z,t) \in \mathbf{R}^N | d(z,t) < r\right\}. \tag{2.8}$$

Consider the degenerate gradient of a function u defined by

$$|\nabla_\alpha u|^2 = |\nabla_z u|^2 + |z|^{2\alpha}|\nabla_t u|^2. \tag{2.9}$$

With d as in (2.6) a computation yields

$$|\nabla_\alpha u|^2 = \frac{|z|^{2\alpha}}{d(z,t)^{2\alpha}} \stackrel{\text{def}}{=} \psi_\alpha(z,t). \tag{2.10}$$

One should compare (2.10) with (1.18). In what follows we fix $R_0 < 1$ and consider functions $u \in C_0^\infty(B_{R_0} \backslash \{(0,0)\})$. For such u's we have for each $\beta > 0$

$$\exp(2d^{-\beta})u^2 = o(1) \text{ as } d \to 0.$$

We also require that for each $\beta > 0$

$$\psi_\alpha^{-\frac{1}{2}} \exp(d^{-\beta}) P_\alpha u \in L^2(B_{R_0}), \tag{2.11}$$

with ψ_α as in (2.10). Then in [G] the following Carleman estimate is established.

THEOREM 2.1. *Let $u \in C_0^\infty(B_{R_0} \backslash \{(0,0)\})$, $R_0 < 1$, and satisfying (2.11). Then there exist β_0, $C_0 > 0$ depending only on α, Q such that*

$$\int_{\mathbf{R}^N} \exp(2d^{-\beta}) \left[\psi_\alpha d^{-2\beta-2}u^2 + |\nabla_\alpha u|^2\right] dz\, dt$$
$$\leq C_0 \left(\beta^{-2} + R_0^\beta\right) \int_{\mathbf{R}^N} \psi_\alpha^{-1} d^{\beta+2} \exp(2d^{-\beta})(P_\alpha u)^2 dz\, dt$$

for every $\beta \geq \beta_0$.

As a consequence of Theorem 2.1 we obtain the following theorem of unique continuation. Let us fix $R > 0$ and consider in B_R solutions to the following differential inequality

$$|P_\alpha u(z,t)| \leq C_1 \frac{\psi_\alpha(z,t)}{d(z,t)^2}|u(z,t)| + C_2 \frac{\psi_\alpha(z,t)^{\frac{1}{2}}}{d(z,t)}|\nabla_\alpha u(z,t)|, \tag{2.12}$$

where $C_1, C_2 > 0$ are suitable constants. Let $M(R)$ denote the class of those measurable functions in B_R such that for every $\beta > 0$

$$\|\exp(d^{-\beta})u\|_{L^\infty(B_r)} = o(1) \text{ as } r \to 0, \text{ and } \exp(d^{-\beta})|\nabla_\alpha u| \in L^2(B_R).$$

THEOREM 2.2. *Suppose that u satisfies (2.12) in B_R for some constants $C_1, C_2 > 0$, and that $u \in M(R)$. Then there exists $r_0 = r_0(\alpha, Q, C_1, C_2) > 0$ such that $u \equiv 0$ in B_{r_0}.*

In [P] Protter used the singular weight $\exp(|x|^{-\beta})$, $\beta > 0$, to prove a Carleman estimate for Laplace's equation. Our approach to Theorem 2.1 is modelled on that in [P], although we must take in due account the degenerate geometry of P_α. For instance, a crucial ingredient in the proof of Theorem 2.1 is the following remarkable integral identity. Let X_α be the smooth vector field in $\mathbf{R}^N$ which generates the group of dilations (2.4)

$$X_\alpha = \sum_{i=1}^n z_i \frac{\partial}{\partial z_i} + (\alpha+1) \sum_{j=1}^m t_j \frac{\partial}{\partial t_j}.$$

Then in [G] we prove

THEOREM 2.3. *Let $D \subset \mathbb{R}^N$ be a bounded, piecewise C^1 domain and let $u \in C^2(D) \cap C^1(\overline{D})$. Then*

$$(2.13)\qquad 2\int_{\partial D}(A_\alpha \nabla u \cdot \nu)X_\alpha u\, dH_{N-1} - \int_{\partial D}|\nabla_\alpha u|^2 X_\alpha \cdot \nu dH_{N-1} = (2-Q)\int_D |\nabla_\alpha u|^2 dz\, dt + 2\int_D X_\alpha u P_\alpha u\, dz\, dt.$$

In (2.13) we have let $A_\alpha = (a_{ij})$ with $a_{ij} = \delta_{ij}$, $i,j = 1,\dots,n$, $a_{ij} = \delta_{ij}|z|^{2\alpha}$ for $i,j = n+1,\dots,n+m$, $a_{ij} = a_{ji} = 0$ if $i = n+1,\dots,n+m$, $j = 1,\dots,n$. Also, ν denotes the outer unit normal to ∂D and Q is as in (2.5). As a consequence of Theorem 2.3 we obtain the following positivity result.

COROLLARY 2.1. *Let u be such that $X_\alpha u$, $P_\alpha u$, $|\nabla_\alpha u| \in L^2(\mathbb{R}^N)$, and suppose that* supp u *is compact. Then*

$$(2.14)\qquad \int_{\mathbb{R}^N} X_\alpha u\, P_\alpha u\, dz\, dt = \frac{Q-2}{2}\int_{\mathbb{R}^N} |\nabla_\alpha u|^2 dz\, dt.$$

One should observe that in virtue of (2.5) the right hand side of (2.14) is always larger than or equal to zero. For a nontrivial u the right hand side of (2.14) equals zero only when $n = m = 1$ and $\alpha = 0$, in which case P_α becomes the Euclidean Laplacian in $\mathbb{R}^2$.

I close this note with a conjecture. I believe that the fundamental solution (2.7) might prove useful to establish an $L^p - L^q$ Carleman estimate which yield strong unique continuation from a point of $\{0\} \times \mathbb{R}^m$ for $-P_\alpha + V$ with $V \in L^r_{\text{loc}}$. With Q as in (2.5) I conjecture that if $V \in L^{Q/2}(B_{R_0})$ for some $R_0 > 0$, and if u solves $|P_\alpha u| \le |V|\,|u|$ in B_{R_0}, there exists $r_0 = r_0(\alpha, Q, \|V\|_{L^{Q/2}}) > 0$ such that if $\int_{B_r} u^2 \psi dz\, dt = 0(r^k)$ for every $k \in \mathbb{N}$ as $r \to 0$, then $u \equiv 0$ in B_{r_0}. This conjecture constitutes the appropriate analogue of Jerison and Kenig's celebrated $L^{\frac{n}{2}}$-result for the Schrödinger operator $H = -\Delta + V$, see [JK].

Bibliography

[B] H. BAHOURI, *Non-prolongement unique des solutions d'opérateurs, "Somme de carrés"*, Ann. Inst. Fourier, Grenoble, 36(4) (1986), pp. 137–155.

[Bo] J. M. BONY, *Principe du maximum, inégalité de Harnack et unicité du problème de Cauchy pour les opérateurs elliptiques dégénérés*, Ann. Inst. Fourier, Grenoble, 19, 1 (1969), pp. 277–304.

[C] T. CARLEMAN, *Sur un problème d'unicité pour les systéms d'equations aux derivées partielles a deux variables independentes*, Ark. Mat., 26B (1939), pp. 1–9.

[F] G. B. FOLLAND, *A fundamental solution for a subelliptic operator*, Bull. Amer. Math. Soc., 79(2) (1973), pp. 373–376.

[G] N. GAROFALO, *Unique continuation for a class of elliptic operators which degenerate on a manifold of arbitrary codimension*, preprint.

[GL1] N. GAROFALO AND F. H. LIN, *Monotonicity properties of variational integrals, A_p weights and unique continuation*, Indiana Univ. Math. J., 35(2) (1986), pp. 245–268.

[GL2] ——————, *Unique continuation for elliptic operators: A geometric–variational approach*, Comm. Pure Appl. Math, 40 (1987), pp. 347–366.

[GLan1] N. Garofalo and E. Lanconelli, *Zero–order perturbations of the subelliptic Laplacian on the Heisenberg group and their uniqueness properties*, Bull. Amer. Math. Soc., 23(2) (1990), pp. 501–511.

[GLan2] ——————, *Frequency functions on the Heisenberg group, the uncertainty principle and unique continuation*, Ann. de l'Inst. Fourier, 40(2) (1990), pp. 313–356.

[GLan3] ——————, *Work in progress.*

[GR1] V. V. Grushin, *On a class of hypoelliptic operators*, Math. USSR Sbornik, 12(3) (1970), pp. 458–476.

[GR2] ——————, *On a class of hypoelliptic pseudodifferential operators degenerate on a submanifold*, Math. USSR Sbornik, 13(2) (1971), pp. 155–186.

[H] L. Hörmander, *Hypoelliptic second order differential equations*, Acta Math., 119 (1967), pp. 147–171.

[NSW] A. Nagel, E. M. Stein, and S. Wainger, Acta Math., 155 (1985), pp. 103–148 paper Balls and metrics defined by vector fields, I. Basic properties.

[P] M. H. Protter, *Unique continuation for elliptic equations*, Trans. Amer. Math. Soc., 95(1) (1960), pp. 81–91.

[RS] L. P. Rotschild and E. M. Stein, *Hypoelliptic differential operators and nilpotent groups*, Acta Math., 137 (1976), pp. 247–320.

[S] A. Sanchez–Calle, *Fundamental solutions and geometry of the sum of squares of vector fields*, Invent. Math., 78 (1984), pp. 143–160.

SHARP ESTIMATES FOR HARMONIC MEASURE IN CONVEX DOMAINS

DAVID JERISON*

Key words. harmonic measure, convex, Monge-Ampère

INTRODUCTION

In this note we will prove estimates for harmonic measure in convex and convex C^1 domains. It is not hard to show that in a convex domain, surface measure belongs to the Muckenhoupt class A_1 with respect to harmonic measure (Lemma 3). If the boundary of the domain is also of class C^1, then it follows from [JK1] that the constant in the A_1 condition tends to 1 as the radius of the ball tends to 0 (Lemma 7′). Our main estimates (Theorems A and B) are of the same type. The novelty is that they are not calculated with respect to balls, but rather with respect to "slices" formed by the intersection of the boundary with an arbitrary half-space. In addition to proving Theorems A and B we will also explain how these estimates are related to an approach to regularity for the Monge-Ampère equation due to L. Caffarelli and to a problem of prescribing harmonic measure as a function of the unit normal.

These results were announced in [J1]. Details of the original proof of *a priori* inequalities and regularity for the prescribed harmonic measure problem from [J1] will be presented here. The proof requires Caffarelli's rather difficult $W^{2,p}$ estimates for the Monge-Ampère equation [C3]. However, motivated by [J1] (see also Remark 2), Caffarelli has strengthened his estimate [C2] so that Theorem B and the $W^{2,p}$ estimates are no longer needed for our application to prescribing harmonic measure. This simplification will be explained in the final section of the paper. The full simplified proof of regularity appears in [J2], along with existence and uniqueness results.

1. Estimates for harmonic measure. Let B_r be the ball of radius r about 0 in $\mathbf{R}^N$. Let Ω be a convex, open subset of $\mathbf{R}^N$. We will assume that Ω is normalized so that $B_1 \subset \Omega \subset B_T$. Denote by $d\omega$ harmonic measure for Ω at 0, in other words,

$$u(0) = \int_{\partial\Omega} f \, d\omega$$

where u is the harmonic function with boundary values f, and f is any continuous function on the boundary. Let $d\sigma$ denote surface measure on $\partial\Omega$. By Dahlberg's theorem $d\omega$ is mutually absolutely continuous with $d\sigma$, and so we can define the Radon-Nikodym derivative $h = d\omega/d\sigma$. Let $H = \{x \in \mathbf{R}^N : (x - x_0) \cdot \theta \geq 0\}$ be a half-space with boundary $P = \{x \in \mathbf{R}^N : (x - x_0) \cdot \theta = 0\}$. We will suppose that the normal θ is chosen so that $0 \notin H$. (We make this choice for notational convenience only: we want the definition of $\frac{1}{2}F$ that follows to be the right one. On

*Massachusetts Institute of Technology, Cambridge, MA 02139

the other hand, the case $0 \in H$ is a trivial one. It follows that the slice F contains more than "half" of $\partial\Omega$ so that the conclusions of Theorem 1 are easy to obtain.) Denote by Π the orthogonal projection onto P. Let $E = P \cap \Omega$, let z be the center of mass of E, and denote $\frac{1}{2}E = \{\frac{1}{2}(x - z) + z : x \in E\}$. Let $F = H \cap \partial\Omega$ and $\frac{1}{2}F = \{x \in \partial\Omega : \Pi x \in \frac{1}{2}E\}$.

THEOREM A. *Let Ω be a convex, open subset of $\mathbf{R}^N$. Suppose that $B_1 \subset \Omega \subset B_T$. There is a constant C depending only on N and T such that for every slice $F = H \cap \partial\Omega$,*

$$\omega(F) \leq C\omega\left(\frac{1}{2}F\right) \tag{1}$$

$$\max_F h \leq C\frac{1}{\sigma(F)}\int_F h\, d\sigma. \tag{2}$$

(Here and elsewhere we abbreviate by $\max h$ the essential supremum of h.)

THEOREM B. *Suppose, in addition to the assumptions of Theorem A, that $\partial\Omega$ is of class C^1. Then for any $\epsilon > 0$ there exists $\rho > 0$ depending only on ϵ, T, N, and the C^1 modulus of continuity of the boundary such that with the notations above,* diameter$(E) \leq \rho$ *implies*

$$\max_F h \leq (1+\epsilon)\frac{1}{\sigma(F)}\int_F h\, d\sigma.$$

Part (1) of Theorem A is proved in [J2]. We will have to repeat the proof here in order to obtain the stronger statement (2). The proof is not entirely self-contained because we assume the key lemma (Lemma 4) proved in [J2]. However, the reader can also consult the proof of Lemma 8 which recapitulates the proof of Lemma 4 in the small constant case. Let G denote Green's function for Ω with pole at 0. Since T controls the Lipschitz constant of the region, various estimates for Lipschitz domains are valid with constants depending only on N and T.

For $x \in \partial\Omega$ denote a surface ball of radius r by $\Lambda(x,r) = B(x,r) \cap \partial\Omega$ where $B(x,r) = \{y \in \mathbf{R}^N : |y - x| < r\}$.

LEMMA 1 (DAHLBERG'S COMPARISON THEOREM; SEE [D],[JK]). *For every $r < 1/2$ every $x \in \partial\Omega$ and every point $y \in \Omega$ such that $|x-y| = r$ and* distance$(y, \partial\Omega) >$ *$r/4T$,*

$$c^{-1}|G(y)|r^{N-2} \leq \omega(\Lambda(x,r)) \leq c|G(y)|r^{N-2}.$$

where c is a constant depending only on N and T.

LEMMA 2 (CARLESON LEMMA; SEE [HW],[JK]). *Every positive harmonic function u defined in $B(x,2r) \cap \Omega$ and vanishing on $B(x,2r) \cap \partial\Omega$ satisfies*

$$\max_{\bar{B}(x,r)\cap\bar{\Omega}} u \leq Cu(y),$$

where y is defined in Lemma 1 and C depends only on N and T.

Next, we have a lemma that does depend on convexity, not just the Lipschitz constant.

LEMMA 3. *The density $h = d\omega/d\sigma$ satisfies*

$$h(z) \leq C\omega(\Lambda(x,r))/\sigma(\Lambda(x,r))$$

for almost every $z \in \Lambda(x,r)$. The constant C depends only on N and T.

Recall that $d\sigma \in A_1(d\omega)$ means that $f = d\sigma/d\omega$ satisfies

$$\frac{1}{\omega(\Lambda(x,r))}\int_{\Lambda(x,r)} f\,d\omega \leq C \operatorname*{essinf}_{z\in\Lambda(x,r)} f(z).$$

This is exactly the condition in Lemma 3 since $f = 1/h$. It should be compared to Dahlberg's estimate (valid for all Lipschitz domains):

$$\frac{1}{\sigma(\Lambda(x,r))}\int_{\Lambda(x,r)} h^2\,d\sigma \leq C\frac{\omega(\Lambda(x,r))^2}{\sigma(\Lambda(x,r))^2}. \tag{3}$$

This "reverse Schwarz inequality" is equivalent to $d\sigma \in A_2(d\omega)$.

To prove Lemma 3 it suffices to consider $r < 1/4$. Let $\Omega(z,s) = B(z,s)\cap\Omega$, and let $H(z,s)$ be the hemisphere containing $\Omega(z,s)$ formed by slicing the ball $B(z,s)$ with a tangent plane to $\partial\Omega$ at z. Let y be the point of $\partial\Omega(z,r)$ furthest from $\partial\Omega$. Then, by Lemma 1 and Harnack's inequality,

$$|G(y)| \leq Cr\omega(\Lambda(x,r))/\sigma(\Lambda(x,r)).$$

By Lemma 2,

$$\max_{\Omega(z,r)} |G| \leq Cr\omega(\Lambda(x,r))/\sigma(\Lambda(x,r)).$$

(Here, as elsewhere, the constant C may change from line to line, but it depends at most on N and T.) Using the maximum principle, one can compare $|G|$ to a harmonic function on $H(z,r)$ with constant value on the spherical part of the boundary and zero on the flat part. It follows that

$$|G(y')| \leq C|y'-z|\omega(\Lambda(x,r))/\sigma(\Lambda(x,r)),$$

for all $y' \in \Omega(z,r/2)$. It then follows from Lemma 1 (with y' in place of y) that

$$\omega(\Lambda(z,s))/\sigma(\Lambda(z,s)) \leq C\omega(\Lambda(x,r))/\sigma(\Lambda(x,r)),$$

for all $0 < s < r$. Thus, by the Lebesgue differentiation theorem,

$$h(z) \leq C\omega(\Lambda(x,r))/\sigma(\Lambda(x,r))$$

almost everywhere, as desired.

LEMMA 4 (SEE [J1] AND [J2] LEMMA D). *Let x and y belong to Ω, $x \notin B_{1/2}$. There is a constant C depending only on N, $c > 0$, and T such that for any t, $c \le t \le 1$,*

$$|G(x)| \le C|G(tx + (1-t)y|.$$

REMARK 1. *Let $M > 1$. Let 0 be the center of mass of a convex set E in $\mathbf{R}^n$. If $x \in E$, $y \in \partial E$, and $p = tx + (1-t)y \in (1 - \frac{1}{M})E$, then $t \ge c > 0$ for a constant c depending only on n and M.*

The remark is proved by observing that the value of t is unchanged by the transformation $(x_1, x_2, \dots, x_n) \mapsto (a_1x_1, a_2x_2, \dots, a_nx_n)$. The lemma of F. John [dG] implies that after such a transformation we may assume that $B_1 \subset \Omega \subset B_n$, in which case the assertion is easy to check.

We can now complete the proof of (1) and (2). Define

$$\delta = \max_{x \in F} \text{distance}(x, E).$$

The John lemma implies that there are similar ellipsoids E_1 and E_2 of comparable size such that $E_1 \subset E \subset E_2$. There is a constant C depending on T such that the axes of E_1 all exceed δ/C. Choose points $p_j \in \frac{1}{2}E_1$, $j = 1, \dots, N_1$ such that $B(p_j, \delta/C_1^2)$ are mutually disjoint for different values of j and $\bigcup_k B(p_j, \delta/C_1) \supset \frac{1}{4}E_1$.

Consider $x \in E$ and choose $y \in \partial E$ so that y is on the ray from x to p_j. Using the Remark 1 with $M = 2$ we find that $p_j = tx + (1-t)y$ with $1 \le t > c$ for some dimensional constant $c > 0$. It follows from Lemma 4 that $|G((1-\delta)x)| \le C|G((1-\delta)p_j)|$. In particular, the values of $|G((1-\delta)p_j)|$ are comparable for all j to the value, say, $v = |G((1-\delta)p_1)|$. Then for any $z \in F$ we find $|G((1-\delta)\Pi z)| \le Cv$. Therefore, by Lemmas 1 and 3, $\max_F h \le Cv/\delta$. On the other hand,

$$\begin{aligned}\int_{\frac{1}{2}F} h\, d\sigma &\ge \sum_{j=1}^{N} \frac{\omega(\Lambda(\Pi^{-1}p_j, \delta/C_1^2)}{\sigma(\Lambda(\Pi^{-1}p_j, \delta/C_1^2)}\delta^{N-1} \\ &\ge C^{-1}(v/\delta)\delta^{N-1}N_1 \ge C^{-1}(v/\delta)\sigma(F).\end{aligned}$$

In all we have

$$\max_F h \le C\frac{1}{\sigma(F)}\int_{\frac{1}{2}F} h\, d\sigma$$

from which both (1) and (2) follow.

2. The C^1 case. In order to prove Theorem B, we need to develop a more precise analogue of each of the lemmas used to prove Theorem A. We will change notations slightly since we will be dealing with regions above a graph. Let r satisfy $0 < r < 1/4$. Suppose that $0 \in \partial\Omega$ and that $\Omega \cap B_{2r} = \{(x_0, x) \in \mathbf{R} \times \mathbf{R}^n : x_0 > \phi(x) \text{ and } |(x_0, x)| < 2r\}$. Here, $n = N - 1$ and ϕ is a convex function. The former origin will be denoted X^0. For $x \in \mathbf{R}^n$, denote $\Lambda(x, s) = \{(\phi(y), y) : |y - x| < s\}$. In place of Lemma 1 we have

LEMMA 5. *For any $\lambda > 0$ and any $\eta > 0$ there exist positive numbers α and δ such that if $|\nabla\phi(x)| \leq \delta$ for all $|x| < 2r$, then*

$$|G(X) - x_0 \frac{\omega(\Lambda(x,s))}{\sigma(\Lambda(x,s))}| \leq \eta x_0 \frac{\omega(\Lambda(x,s))}{\sigma(\Lambda(x,s))},$$

whenever $X = (x_0, x)$, $|x| < \alpha^2 r$, $\lambda\alpha^2 r \leq s \leq \alpha^2 r$, and $\lambda\alpha^2 r \leq x_0 \leq \alpha^2 r$.

Proof. We will prove Lemma 5 using estimates in [JK1]. Denote $A_0 = \Lambda(0, \alpha r)$. Denote $a = \omega(A_0)/\sigma(A_0)$. [JK1, Lemma 2.6] implies that for any $\epsilon > 0$, one can choose α and δ sufficiently small depending on λ, so that for all s and x, $|a - \omega(\Lambda(x,s))/\sigma(\Lambda(x,s))| < \epsilon a$. Define the kernel function by $k(X,Q) = (d\omega^X/d\omega)(Q)$, the Radon-Nikodym derivative of harmonic measure at X with respect to harmonic measure at X^0. Recall that G is Green's function with pole at X^0. Denote by G^X Green's function with pole at X. With the coordinates $X = (x_0, x_1, \ldots, x_n)$ denote by ∂_0 the partial derivative in the x_0 direction. The main step in the proof is to show that $\partial_0 G(X)$ is very close to $-a$. By Green's formula

$$\begin{aligned}\partial_0 G(X) + \partial_0 G^X(X^0) & \int_\Omega (\partial_0 G)\delta_X - (\partial_0 \delta_{X^0})G^X \\ &= \int_\Omega (\partial_0 G)\Delta G^X - (\Delta\partial_0 G)G^X \\ &= \int_{\partial\Omega} (\partial_0 G)(Q)\frac{\partial G^X}{\partial\nu}(Q)\, d\sigma(Q) \\ &= \int_{\partial\Omega} (\partial_0 G)(Q)k(X,Q)h(Q)\, d\sigma(Q).\end{aligned}$$

(The use of Green's formula in Lipshitz domains is justified, for example, in [JK2].) For $j \geq 1$, denote $A_j = \partial\Omega \cap (B(0, 2^j\alpha r)\backslash B(0, 2^{j-1}\alpha r))$. For $Q \in A_0$, $Q = (\phi(y), y)$, one easily computes that $\partial_0 G(Q) d\sigma(Q) = -h(\phi(y), y)dy$, and therefore

$$\int_{A_0} k(X,Q)\partial_0 G(Q)h(Q)\, d\sigma(Q) = -\int_{|x|<\alpha r} k(X,Q)h(Q)^2 dy \equiv -I_0.$$

I_0 will be the main term. For $j \geq 1$, define

$$I_j = \int_{A_j} k(X,Q)h(Q)^2\, d\sigma(Q).$$

Then $|\partial_0 G(X) + I_0| \leq |\partial_0 G^X(X^0)| + \sum_{j=1}^\infty I_j$. We begin by showing that the right hand side is negligible compared to a.

For $Z \in B(X^0, 1/4)$ and $Y \in \partial B(X^0, 1/2)$, $C^{-1} \leq |G^Y(Z)| \leq C$. Therefore, $|G^Y(Z)| \leq |G^Y(X^0)|$, and hence, by the maximum principle, $|G^X(Z)| \leq C|G^Y(X^0)|$. But the symmetry of Green's function implies $G^X(X^0) = G(X)$, so we see that

$$\max_{Z\in B(X^0,1/4)} |G^X(Z)| \leq C|G(X)|.$$

It follows from a direct calculation of the derivative of the Poisson kernel for the ball that $|\partial_0 G^X(X^0)| \leq C|G(X)|$. But by Lemma 1, $|G(X)| \leq Cx_0\omega(A_0)/\sigma(A_0) = Cx_0 a$.

Recall Dahlberg's boundary Harnack principle:

LEMMA 6 ([D],[JK]). *Let $Z \in \partial\Omega$, and $0 < t < 1/2$. Let u and v be positive harmonic functions in $B(Z, 2t) \cap \Omega$ which vanish continuously on $B(Z, 2t) \cap \partial\Omega$. Choose $Y \in B(Z, 2t) \cap \Omega$ so that* dist $(Y, \partial\Omega) > r/T$. *Then*

$$u(X) \leq C\frac{u(Y)}{v(Y)}v(X) \quad \text{for all } X \in B(Z,t) \cap \Omega.$$

For $j \geq 1$, define $\Lambda_j = B(0, 2^j\alpha r) \cap \partial\Omega$, and choose X_j so that $2^{j-1}\alpha r < |X_j| < 2^j\alpha r$ and dist $(X_j, \partial\Omega) > 2^j\alpha r/8T$. (If $2^j\alpha r > 4T$, so that X_j does not exist, then A_j is empty and $I_j = 0$.) We will apply Lemma 6 to $u(X) = k(X, Q)$ and $v(X) = -G(X)$. Note that for $Q \in A_{j+2}$, $u(X)$ vanishes as X tends to Λ_{j+1}. It follows from [JK] or [JK1, Lemma 1.10] that $C^{-1} \leq k(X_j, Q)/k(X_j, Q') \leq C$ for all Q and Q' in Λ_{j+2}. Also,

$$\int_{\Lambda_{j+2}} k(X_j, Q)h(Q)\, d\sigma(Q) \leq \int_{\partial\Omega} k(X_j, Q)h(Q)\, d\sigma(Q) = 1.$$

Hence, $u(X_j) = k(X_j, Q) \leq C/\omega(\Lambda_{j+2})$ for all $Q \in \Lambda_{j+2}$. Lemma 1 implies $v(X_j) = |G(X_j)| \geq C^{-1}(2^j\alpha r)\omega(\Lambda_{j+2})/\sigma(\Lambda_{j+2})$. Therefore, by Lemma 6 for $\Omega \cap B(0, 2^{j+1}\alpha r)$,

$$k(X, Q) \leq C\frac{\sigma(\Lambda_{j+2})}{\omega(\Lambda_{j+2})^2}(2^j\alpha r)^{-1}|G(X)|.$$

for all $Q \in A_{j+2}$. Next, recall again that $|G(X)| \leq Cx_0a$. Therefore, using (3) for Λ_{j+2} and $x_0 \leq \alpha^2 r$,

$$\begin{aligned} I_{j+2} &= \int_{A_{j+2}} k(X, Q)h(Q)^2\, d\sigma(Q) \\ &\leq C\frac{\sigma(\Lambda_{j+2})}{\omega(\Lambda_{j+2})^2}\int_{A_{j+2}} h(Q)^2\, d\sigma(Q)(2^j\alpha r)^{-1}x_0a \\ &\leq C(2^j\alpha r)^{-1}x_0a \leq C2^{-j}\alpha a. \end{aligned}$$

Summing the geometric series, we obtain $||\partial_0G(X) + I_0| \leq C\alpha a$, for a constant C depending only on T and N.

Next we wish to show that I_0 is close to a. Let

$$I_0' = \int_{|y|<\alpha r} P_{x_0}(x - y)h(\phi(y), y)^2\, dy,$$

where $P_t(z) = c_Nt(|z|^2 + t^2)^{-N/2}$ is the Poisson kernel for the upper half-space. By [JK1, Lemma 2.10], for any $\epsilon > 0$ $|I_0 - I_0'/a| \leq \epsilon I_0'/a$, for sufficiently small α and sufficiently small δ depending on both ϵ and α. Moreover, by [JK1, Lemmas 2.11 and 2.12] one can choose α and then δ so that $|I_0' - a^2| \leq \epsilon a^2$. Thus for any $\epsilon > 0$ and any $\lambda > 0$ we can choose $\alpha > 0$ and then $\delta > 0$ so that $|\partial_0G(X) + a| \leq \epsilon a$, provided X satisfies the limitations of the lemma. On the other hand, even if the limitation on x_0 is omitted we still have $|G(X)| \leq Cx_0\omega(\Lambda(x, x_0))/\sigma(\Lambda(x, x_0))$. In particular,

$0 \le |G(\lambda\alpha^2 r, x)| \le C\lambda\alpha^2 ra$. Therefore, integrating in the vertical direction, we obtain

$$|G(t,x)| \le C\lambda\alpha^2 ra + (1+\epsilon)(t-\lambda\alpha^2 r)a \le (1+\epsilon+C\sqrt{\lambda})ta,$$

for $\alpha^2 r > t > \sqrt{\lambda}\alpha^2 r$. Similarly,

$$|G(t,x)| \ge (1-\epsilon)(t-\lambda\alpha^2 r)a \ge (1-\epsilon-\sqrt{\lambda})ta,$$

for $\alpha^2 r > t > \sqrt{\lambda}\alpha^2 r$. This implies Lemma 5 with $\eta = \epsilon + C\sqrt{\lambda}$ and $\sqrt{\lambda}$ replacing λ.

LEMMA 7. *For any $\eta > 0$ and any $\lambda > 0$ there exist $\alpha > 0$ and $\delta > 0$ such that with the notations of Lemma 5,*

$$h(\phi(z), z) \le (1+\eta)\frac{\omega(\Lambda(x,s))}{\sigma(\Lambda(x,s))}$$

for any $z \in \Lambda(x,s)$.

This lemma is an easy consequence of Lemma 5. One makes a comparison, using the maximum principle, with the linear (harmonic) function which vanishes on the flat part of the boundary of a hemisphere (modified on a very small set near the bottom of the curved part of the boundary). The bounds on size near the bottom are taken care of by the Carleson lemma. The details are left to the reader.

Lemma 7 can be restated in a somewhat less cumbersome form as

LEMMA 7$'$. *Let Ω be a C^1 convex domain. For any $\epsilon > 0$ there exists $\rho > 0$ such that for $0 < r < \rho$,*

$$\max_{\Lambda} h \le (1+\epsilon)\omega(\Lambda)/\sigma(\Lambda),$$

for any surface ball Λ of radius r.

LEMMA 8. *Let ϕ be a convex function on $\mathbf{R}^n$ satisfying $\phi(0) = 0$. Let $\Omega = \{(x_0, x) : x_0 > \phi(x)\}$. For any $\alpha > 0$, there exists $C > 0$ such that if $0 < \eta < C^{-1}$ and u is a positive harmonic function in $B_2 \cap \Omega$ that vanishes on $B_2 \cap \partial\Omega$ satisfying*

$$X \in \Omega \quad \text{and} \quad |u(X) - x_0| \le \eta x_0$$

for all $X = (x_0, x) \in \bar{B}_2$ such that $\eta \le x_0 \le 2$, then

$$tu(X) \le u(tX)/(1 - C\eta), \text{ for } \alpha \le t \le 1 \text{ and } X \in \bar{B}_1 \cap \bar{\Omega}.$$

Proof. Convexity of Ω implies that the function $w(X) = u(tX)$ is defined in $B_2 \cap \Omega$. The estimate for u implies $w(X) \ge tu(X) - 2\eta t x_0 \ge t(u(X) - 4\eta)$ for all $X \in \bar{B}_2 \cap \bar{\Omega}$ such that $tx_0 \ge \eta$. By Carleson's lemma and the maximum principle applied as in Lemma 3, $u(X) \le C_1 x_0$ for all $X \in \bar{B}_{3/2} \cap \bar{\Omega}$. In particular, if

$X \in \bar{B}_{3/2} \cap \bar{\Omega}$ and $tx_0 \leq \eta$, $t(u(X) - C_1\alpha^{-1}\eta) \leq 0 \leq w(X)$. Hence, $w(X) \geq t(u(X) - C_1\alpha^{-1}\eta)$ for all $X \in \bar{B}_{3/2} \cap \bar{\Omega}$. Let v be the harmonic function in $B_{3/2} \cap \Omega$ satisfying $v = C_1\alpha^{-1}\eta$ on $\Omega \cap \partial B_{3/2}$ and $v = 0$ on $B_{3/2} \cap \partial\Omega$. By the maximum principle, $w \geq t(u - v)$ on $\bar{B}_{3/2} \cap \bar{\Omega}$. By the boundary Harnack principle, there is a constant C depending on $C_1\alpha^{-1}$ such that $v(X) \leq C\eta u(X)$ for all $X \in \bar{B}_1 \cap \bar{\Omega}$. Therefore, $w(X) \geq t(1 - C\eta)u(X)$.

Let us now conclude the proof of Theorem B. Let γ be the inradius of E and let s be the diameter of E. Fix a large constant M. Denote

$$E^* = \{X : \operatorname{dist}(X, E) \leq M\gamma\}, \quad E' = (1 - 1/M)E,$$

where the dilation of E is taken with respect to the center of mass. We will show that there exists $\rho > 0$ depending on ϵ and M such that for $s < \rho$,

$$\max_{F^*} h \leq (1 + \epsilon)\frac{\sigma(\Lambda)}{\omega(\Lambda)},$$

where $F^* = \Pi^{-1}E^*$ and $\Lambda = B(\Pi^{-1}X, r)$ for any $X \in E'$ and any r such that $\gamma/M < r < M\gamma$. Indeed, let $X_0 \in E'$ and $Z \in E$. Choose $Y \in \partial E$ so that the ray from Z to X_0 leaves E at Y. We will suppose that the origin is at Y. Lemma 5 implies that for sufficiently small ρ, the hypothesis of Lemma 8 is valid for the function

$$u(X) = -\frac{1}{s}G(Y + sX)\frac{\sigma(\Lambda(Y, s))}{\omega(\Lambda(Y, s))}$$

in the region $Y + sX \in B(Y, 2s) \cap \Omega$. Define t by $X_0 = tZ + (1 - t)Y$. Then since $X_0 \in E'$, Remark 1 implies $t \geq \alpha$ for some $\alpha > 0$ depending on M. Define $\mathbf{e}$ as the unit vector perpendicular to the plane of E. The conclusion of Lemma 8 for the point $X = s^{-1}(Z - Y + M\gamma\mathbf{e})$ is that

$$|tG(Z + M\gamma\mathbf{e})| \leq |G(X_0 + tM\gamma\mathbf{e})|/(1 - C\eta).$$

By Lemma 5, we can choose $\rho > 0$ sufficiently small depending on η, M, α and the C^1 modulus of continuity of $\partial\Omega$ such that

$$|G(X_0 + tM\gamma\mathbf{e})| \leq (1 + \eta)tM\gamma\omega(\Lambda)\sigma(\Lambda) \quad \text{and}$$

$$M\gamma\omega(\Lambda')/\sigma(\Lambda') \leq |G(Z + M\gamma\mathbf{e})|(1 + \eta),$$

where $\Lambda = \Lambda(\Pi^{-1}X_0, r)$, $M^{-1}\gamma < r < M\gamma$ and $\Lambda' = \Lambda(\Pi^{-1}Z, M\gamma)$. Moreover, Lemma 7 shows that

$$\max_{\Lambda'} h \leq (1 + \eta)\omega(\Lambda')/\sigma(\Lambda').$$

Combining all these inequalities, we find

$$\max_{F^*} h \leq \frac{(1 + \eta)^2}{(1 - C\eta)}\frac{\omega(\Lambda)}{\sigma(\Lambda)} = (1 + \epsilon)\frac{\omega(\Lambda)}{\sigma(\Lambda)}.$$

The remainder of the proof is a routine covering argument. Denote $F' = \Pi^{-1}E'$ and $H = \max_{F^*} h$. Let $\Lambda_J = \{X \in \Lambda : h(X) \leq H(1 - J\epsilon)\}$. Then

$$1 \leq \frac{(1+\epsilon)}{H}\int_\Lambda h\,d\sigma/\sigma(\Lambda) \leq (1+\epsilon)\left\{\frac{\sigma(\Lambda\backslash\Lambda_J)}{\sigma(\Lambda)} + (1-J\epsilon)\frac{\sigma(\Lambda_J)}{\sigma(\Lambda)}\right\}$$
$$\leq 1 + \epsilon - J\epsilon\frac{\sigma(\Lambda_J)}{\sigma(\Lambda)}.$$

Thus $\sigma(\Lambda_J) \leq \sigma(\Lambda)/J$. Choose a covering for F' by surface balls Λ^i satisfying

$$\bigcup_i \Lambda^i \supset F' \quad \text{and} \quad \sum_i \sigma(\Lambda^i) \leq C\sigma(F').$$

Then

$$\int_F h \geq \int_{F'} h\,d\sigma - \int_{\cup_i\Lambda^i} h\,d\sigma$$
$$\geq (1-J\epsilon)H(\sigma(F') - \sum_i \sigma(\Lambda^i_J))$$
$$\geq (1-J\epsilon)H(\sigma(F') - C\sigma(F')/J)$$
$$\geq (1-J\epsilon)H(1-C/J)(1-1/M)^n\sigma(F).$$

We can choose M, J, large and then $\epsilon > 0$ so that the final factor on $\sigma(F)$ is arbitrarily close to 1. This ends the proof of Theorem B.

3. Monge-Ampère equations. We will now show how the estimates of Theorems A and B lead to *a priori* inequalities in a problem of prescribing harmonic measure. We take a convex region Ω with $0 \in \Omega$. Consider the Gauss map g defined almost everywhere $d\sigma$ taking a point $X \in \partial\Omega$ to the unit normal θ at X. Since harmonic measure and surface measure are mutually absolutely continuous, the Gauss map defines a measure $g_*(d\omega)$ on the unit sphere S^n in $\mathbf{R}^{n+1}$. The prescribed harmonic measure problem is: given a measure $d\mu$ of total mass 1 on the unit sphere, find a convex region containing the origin such that $g_*(d\omega) = d\mu$. In $\mathbf{R}^2$, this problem is solved by a continuous version of the Schwarz-Christoffel formula for conformal mapping of polygons ([J1],[J2]). Suppose that $\partial\Omega$ is C^∞ and strongly convex (i.e., the Hessian of the defining function can be chosen to be positive definite at all boundary points.) Then $g_*(d\omega) = Rd\theta$ for a positive density $R \in C^\infty(S^n)$. The solution in higher dimensions depends on the *a priori* inequality

$$\|\partial\Omega\|_{C^{k+2,\alpha}} \leq C(\|R\|_{C^{k,\alpha}(S^n)}, \max 1/R), \tag{4}$$

for some function $C(.,.)$.

Let us compare this problem with the Minkowski problem: given a measure $d\nu$ on the unit sphere, find a convex domain Ω so that the push-forward of surface measure, $g_*(d\sigma)$ equals $d\nu$. In the smooth case, $g_*(d\sigma) = (1/K)d\theta$, where K is the Gauss curvature. *A priori* inequalities of the form

$$\|\partial\Omega\|_{C^{k+2,\alpha}} \leq C(\|K\|_{C^{k,\alpha}(S^n)}, \max 1/K)$$

were proved by L. Caffarelli [C1], using earlier estimates of Alexandrov, Nirenberg, Calabi, Pogorelov, Cheng and Yau [CY]. The point here is that Caffarelli obtained the best possible gain of two derivatives.

Caffarelli's estimate is a regularity estimate for the Monge-Ampère equation. To see this, let us write down the equation for Gauss curvature in terms of a function expressing $\partial\Omega$ locally as a graph. Let ϕ be a convex function on $\mathbf{R}^n$. Suppose that $\Omega \cap B = \{(x_0, x) : x_0 > \phi(x)\} \cap B$ for some ball B. Denote the Hessian matrix of ϕ by

$$[\phi_{ij}(x)] = \left[\frac{\partial^2 \phi(x)}{\partial x_i \partial x_j}\right]$$

Then

$$\det\, [\phi_{ij}(x)] = (1 + |\nabla\phi(x)|^2)^{(n+2)/2} K(\theta)$$

where $\theta = (-1, \nabla\phi(x))/(\sqrt{1 + |\nabla\phi(x)|^2}$, the outer unit normal. Thus regularity estimates for the Monge-Ampère equation translate into estimates for the boundary in terms of Gauss curvature (as a function of the unit normal).

Caffarelli also addressed the question of regularity in the borderline case $k = 0$ and $\alpha = 0$. He proved

THEOREM 1 [C2]. *Suppose that Ω is a convex domain and $B_1 \subset \Omega \subset B_T$ and the Gauss curvature satisfies*

$$C^{-1} \leq K \leq C. \tag{5}$$

Then $\partial\Omega$ is C^1 and strictly convex and the modulus of continuity of the first derivative and the modulus of strict convexity (defined below) depend only on C, T, and the dimension.

A domain Ω is strictly convex if there are no line segments in the boundary. This is not as stringent as the notion of strong convexity, which says that the function ϕ has positive definite Hessian. To define the modulus of strict convexity, consider a "slice" F. Denote by δ the largest distance from a point of F to the plane P defining the slice. As above, denote by s the diameter of the convex set $E = P \cap \Omega$. The modulus of strict convexity is a function $f(\delta)$ which tends to zero as δ tends to zero and for which $s \leq f(\delta)$ for every F.

The equation associated to prescribing the density R of harmonic measure with respect to the unit normal is very similar to the Minkowski equation. If we identify the boundary of Ω with the unit sphere and use the informal notations $d\sigma$, $d\omega$, and $d\theta$ without bothering with the g_* we have $h = d\omega/d\sigma$, $K = d\theta/d\sigma$, and $R = d\omega/d\theta$. Thus $K = (d\omega/d\sigma)(d\theta/d\omega) = h/R$, and the equation for the harmonic measure problem is

$$\det\, [\phi_{ij}(x)] = (1 + |\nabla\phi(x)|^2)^{(n+2)/2} h(\phi(x), x)/R(\theta). \tag{6}$$

In other words, the new feature of the problem is the factor h. The scheme for proving (4) is to prove estimates for ϕ and deduce estimates for h. Then one returns to (6) to prove further estimates for ϕ, and so on. In contrast to the Minkowski problem, our problem is dilation invariant. So we normalize Ω so that the unit ball is that largest ball centered at 0 contained in Ω. Next, it is rather easy to deduce that $\Omega \subset B_T$ for some T depending only on dimension and $\max 1/R$ [J2]. This is in contrast to the analogous estimate in the Minkowski problem which is quite tricky [CY]. However, bounds on T are not sufficient to give a positive lower bound on h. Instead we must make do with the bound of Theorem A.

REMARK 2. *Theorem 1 is valid with the weaker hypothesis in place of (5) that*

$$\max_F K \leq \frac{C}{\sigma(F)} \int_{\frac{1}{2}F} K \, d\sigma.$$

Now we can take our first step toward proving (4). Since $K = h/R$, and since we are assuming that R is bounded above and below by positive constants, Theorem A implies that the hypothesis in Remark 2 is valid. Therefore, we can conclude that the modulus of strict convexity and C^1 modulus of continuity are controlled by the right hand side of (4). The proof of the remark is essentially the same as the proof of [C2]. The only difference is to replace Lemma 2 in [C2] by the following lemma.

LEMMA 9. *Let ϕ be a convex function on B_1, the unit ball in $\mathbf{R}^n$. Suppose that $\phi \leq 0$ on ∂B_1, $\det(\phi_{ij}) \geq h \geq 0$ on B_1, and*

$$\int_{B_{1/2}} h(x)dx \geq 1.$$

Then there is a dimensional constant c such that

$$\min_{B_1} \phi \leq -c < 0.$$

Proof. In place of the Gauss map we consider the closely related mapping $x \mapsto \nabla\phi(x)$.

$$\operatorname{vol} \nabla\phi(B_{1/2}) \geq \int_{B_{1/2}} h(x)dx \geq 1.$$

Therefore, the image $\nabla\phi(B_{1/2})$ cannot be contained in a ball of small radius: there exists a dimensional constant c_n and a point $y \in B_{1/2}$ such that $|\nabla\phi(y)| \geq c_n > 0$. Moreover, by convexity, the tangent plane to the graph of ϕ at y stays below zero on ∂B_1. Therefore, $\phi(y) \leq -c_n/2$.

The remaining details are exactly as in [C2]. The argument there proceeds by contradiction, but it is an easy matter to change it to a direct quantitative argument giving *a priori* estimates. While we are not going to repeat [C2] here, it is worthwhile to explain that the slices F are the same as level sets of ϕ minus a linear function:

$$F = \{(\phi(x), x) : \phi(x) - (ax + b) < 0\}.$$

On the other hand, $\det \phi_{ij}$ is unchanged if we subtract a linear function from ϕ. Moreover, this determinant is simply rescaled by a constant when x undergoes a transformation $(x_1, \ldots, x_n) \mapsto (a_1 x_1, \ldots, a_n x_n)$. Thus any estimate can be rescaled to one at unit scale. This is why one need only consider the unit ball in Lemma 9.

Next, we have another regularity theorem due to Caffarelli.

THEOREM 2 [C3]. *Suppose that $\partial\Omega$ is strictly convex and C^1 that $C^{-1} < K < C$ and that K is continuous. Then $\partial\Omega$ is of class $W^{2,p}$ for any $p < \infty$.*

(In brief, a strictly convex function ϕ for which $\det \phi_{ij}$ is positive and continuous is necessarily of class $W^{2,p}$.)

Once again, we need to extend this result to allow for a right hand side in the Monge-Ampère equation which vanishes.

THEOREM 2$'$. *Suppose that $\partial\Omega$ is strictly convex and for any $\epsilon > 0$ there exists $\rho > 0$ such that if F is a slice of $\partial\Omega$ and* diameter$(F) < \rho$, *then*

$$\max_F K \leq \frac{(1+\epsilon)}{\sigma(F)} \int_F K \, d\sigma.$$

Then $\partial\Omega$ is of class $W^{2,p}$ for any $p < \infty$.

Let us now show how to complete the proof of (4). We have proven so far that $\partial\Omega$ is strictly convex and C^1 with control on the corresponding moduli in terms of positive upper and lower bounds for R. It follows that Theorem B applies, and h satisfies the conclusion there. Furthermore, $R \circ g$ is a continuous on $\partial\Omega$. Therefore, the conclusion of Theorem B is also valid with h replaced by $K = h/R$. In other words, the hypothesis of Theorem 2$'$ is satisfied. Thus $\partial\Omega$ is of class $W^{2,p}$, for all $p < \infty$. In particular, it is of class $C^{1,\alpha}$ for some $\alpha > 0$. Once we have $C^{1,\alpha}$ regularity for the boundary, standard estimates tell us that we have a positive lower bound for h. Furthermore, the $C^{k+1,\alpha}$ norm of the boundary controls the $C^{k,\alpha}$ norm of h. The sharp regularity estimates now follow from Caffarelli's Monge-Ampère estimates for Hölder classes in [C1].

The proof of Theorem 2$'$ is the same as in [C3] except that Lemma 4 of [C3] is replaced by

LEMMA 10. *Let E be a convex subset of $\mathbf{R}^n$ such that $B_1 \subset E \subset B_n$. Suppose that ϕ is a convex function in E such that $\det \phi_{ij} = h \leq 1$ in E and $\phi = 0$ on ∂E. Suppose further that*

$$\frac{1}{|E|} \int_E h \, dx \geq 1 - \epsilon,$$

where $|E|$ denotes the volume of E. Define w as the (convex) solution to $\det w_{ij} = 1$ in E and $w = 0$ on ∂E. Then

$$\max_E |\phi - w| \leq \epsilon |E|^{1/n}.$$

Once again a linear change of variables in x–space changes the Hessian determinant in a very simple way, so that estimates with the normalization $B_1 \subset E \subset B_n$ imply a uniform estimate for all sets E corresponding to slices of $\partial\Omega$. Moreover, the inequality on h in the hypothesis to Lemma 10 corresponds exactly to the hypothesis for K in Theorem 2$'$.

Proof. $\det \phi_{ij} \leq \det w_{ij}$ implies $\phi \geq w$, by the maximum principle [GT]. Denote by c_{ij} the cofactor matrix of w_{ij}, i.e., $c_{ij} w_{jk} = \delta_{ik} \det w_{pq} = \delta_{ik}$, with the convention that repeated indices are summed. Note that c_{ij} is positive definite and has determinant 1. Define a linear operator L by $Lv = c_{ij}v_{ij}$. Then $Lw = c_{ij}w_{ij} = \text{trace}(\delta_{ij}) = n$.

We claim that if $Lv = g$ and v_{ij} is positive definite, then

$$\det v_{ij} \le \left(\frac{g}{n}\right)^n.$$

Indeed, without loss of generality one can assume that c_{ij} is a diagonal matrix with diagonal entries $\lambda_1, \dots, \lambda_n$. Then $g = \lambda_1 v_{11} + \cdots + \lambda_n v_{nn}$. If we denote $\alpha_{ij} = \lambda_i^{1/2} v_{ij} \lambda_j^{1/2}$, then α_{ij} is positive definite, $\det \alpha_{ij} = \det v_{ij}$, and $\text{trace}\, \alpha_{ij} = g$. Our inequality now follows from the arithmetic-geometric mean inequality:

$$\det \alpha_{ij} \le \left(\frac{\text{trace}\, \alpha_{ij}}{n}\right)^n.$$

Let v be the solution to the problem $Lv = nh^{1/n}$ in E with $v = 0$ on ∂E. Let $\hat{v}$ be the convex envelope of v. Then $\hat{v} \le v$ and $\det \hat{v}_{ij} = (\det v_{ij})\chi$, where χ is the characteristic function of the set where $\hat{v} = v$. (This equation has to be taken in the viscosity sense; see [C3]). But the claim above implies $\det v_{ij} \le h = \det \phi_{ij}$. By the maximum principle [GT], $\phi \le \hat{v} \le v$. On the other hand, we can use the linear estimate Alexandrov-Pucci estimate [A,P]: since $L(w - v) = n(1 - h^{1/n})$,

$$\max_E |w - v| \le \left(\int_E (1 - h^{1/n})^n\right)^{1/n}.$$

Finally,

$$\int_E (1 - h^{1/n})^n \le \int_E 1 - h^{1/n} \le \int_E 1 - h \le \epsilon |E|.$$

Therefore, $v \le w + \epsilon|E|$, which ends the proof of Lemma 10 and hence of Theorem $2'$ and (4).

We conclude this note with a description of Caffarelli's simplification of this argument. He shows in [C4] that the doubling condition

$$\int_F K\, d\sigma \le C \int_{\frac{1}{2}F} K\, d\sigma,$$

alone suffices to obtain not only C^1 regularity and strict convexity, but also $C^{1,\alpha}$ regularity for some $\alpha > 0$. Thus instead of going through $W^{2,p}$ estimates to obtain $C^{1,\alpha}$ estimates, this can be done directly. Furthermore, these estimates are obtained without Theorem B. Indeed they only require the doubling condition (1) of Theorem A.

REFERENCES

[A] A. D. ALEXANDROV, *Uniqueness conditions and estimates of the solution of Dirichlet's problem*, Vestnik Leningr. Un.-ta., 13 (1963), pp. 5–29.

[C1] L. A. CAFFARELLI, *interior a priori estimates for solutions of fully non-linear equations*, Annals of Math., 130 (1989), pp. 189–213.

[C2] ———————, *A localization property of viscosity solutions to Monge-Ampère equations and their strict convexity*, Annals of Math., 131 (1990), pp. 129–134.

[C3] ———————, *Interior $W^{2,p}$ estimates for solutions of Monge-Ampère equations*, Annals of Math., 131 (1990), pp. 135–150.

[C4] ———————, *Some regularity properties of solutions to the Monge-Ampère equation*, Comm. on P. A. M. (to appear).

[CY] S.-Y. CHENG AND S.-T. YAU, *On the regularity of the solution of the n-dimensional Minkowski problem*, Comm. on P. A. M., 29 (1976), pp. 495–516.

[D] B. E. J. DAHLBERG, *Estimates for harmonic measure*, Arch. Rational Mech. Anal., 65 (1977), pp. 275–283.

[dG] M. DEGUZMAN, *Differentiation of Integrals in* $\mathbf{R}^n$, Lecture Notes 481, Springer-Verlag, New York.

[GT] D. GILBARG AND N. S. TRUDINGER, *Elliptic Partial Differential Equations of Second Order*, Second Edition, Springer-Verlag, New York, 1983.

[HW] R. HUNT AND R. WHEEDEN, *On boundary values of harmonic functions*, Transactions A. M. S., 132 (1968), pp. 307–322.

[J1] D. JERISON, *Harmonic measure in convex domains*, Bull. A. M. S., 21 (1989), pp. 255–260.

[J2] —————, *Prescribing harmonic measure on convex domains*, Inventiones Math. (to appear).

[JK] D. JERISON AND C. E. KENIG, *Boundary behavior of harmonic functions in nontangentially accessible domains*, Advances in Math., 46 (1982), pp. 80–147.

[JK1] ———————, *The logarithm of the Poisson kernel of a* C^1 *domain has vanishing mean oscillation*, Transactions A. M. S., 273 (1982), pp. 781–794.

[JK2] ———————, *Boundary value problems on Lipschitz domains*, in *Studies in Partial Differential Equations*, ed. W. Littman, M. A. A. Studies 23, 1982, pp. 1–68.

[P] C. PUCCI, *Limitazione per soluzioni de equazione ellitichi*, Annali Math. Pura Appl., 74 (1966), pp. 15–30.

ON THE POSITIVE SOLUTIONS OF THE FREE-BOUNDARY PROBLEM FOR EMDEN-FOWLER TYPE EQUATIONS

HANS G. KAPER*, MAN KAM KWONG*AND YI LI†,

1. Introduction. Let Ω be a smooth, bounded and connected domain in $\Re^n$. In this paper, we consider the following boundary value problem:

$$\begin{cases} \Delta u + f(u) = 0 & \text{in } \Omega, \\ u > 0 & \text{in } \Omega, \\ u = \frac{\partial u}{\partial n} = 0 & \text{on } \partial\Omega. \end{cases} \tag{1.1}$$

Here, n denotes the unit outer normal to $\partial\Omega$. (See [KK] for existence and uniqueness results for (1.1).) We prove the following two theorems.

THEOREM 1. *Let f be such that*

$$f(s) = f_1(s) + f_2(s), \tag{1.2}$$

where f_1 is nondecreasing and f_2 Lipschitz continuous. If $u \in C^2(\overline{\Omega})$ be a classical solution of (1.1), *then Ω is an open ball, $\Omega = B_R(x_0)$ say, in $\Re^n$ and u is radially symmetric about the center x_0. Furthermore,*

$$\frac{\partial u}{\partial r} < 0 \quad \text{for} \quad 0 < r \equiv |x - x_0| < R.$$

THEOREM 2. *Let $B_R(0)$ be a ball of radius $R > 0$. Let u be a classical solution of the boundary value problem,*

$$\begin{cases} \Delta u + f(u) = 0 & \text{in } B_R(0), \\ u > 0 & \text{in } B_R(0), \\ u = 0 & \text{on } \partial B_R(0). \end{cases} \tag{1.3}$$

If $f \in C^{0,1}_{loc}((0,\infty))$ and there exists an $s_0 > 0$ such that $f(s)$ is strictly decreasing in $[0, s_0]$, then u is radially symmetric about 0. *Furthermore,*

$$\frac{\partial u}{\partial r} < 0 \quad \text{for} \quad 0 < r < R.$$

We use the well-known moving-plane method, which was first proposed by Alexandrov. In 1971, Serrin used this method to prove the symmetry result for (1.1) in the case where $f(s)$ is real and constant. Since we are dealing with non-smooth functions like f_1, some stronger version of the Hopf near-boundary theorem has to be used. In fact, we use the moving-plane method, in combination with a result (Lemma 4) of Gidas, Ni and Nirenberg [GNN], to prove Theorem 1. To prove Theorem 2, we need to analyze the locations of possible minima of the difference $u - u^\lambda$ in order to continue the moving-plane process.

---

*Mathematics and Computer Science Division, Argonne National Laboratory, Argonne, IL 60439-4801.
This work was Supported by Applied Mathematical Sciences subprogram of the Office of Energy Research, U.S. Department of Energy, under Contract W-31-109-Eng-38
†University of Rochester, Rochester, N.Y. 14620.
Research supported in part by the National Science Foundation

2. Preliminaries. To prove Theorems 1 and 2, we need a series of technical lemmas, whose proofs can be found in [GNN], [H], [PW] and [S].

Let u be a nonnegative classical solution in Ω of the following differential inequality:

$$Lu \equiv a^{ij}(x)D_{ij}u + b^i(x)D_iu + c(x)u \leq 0 \quad \text{in} \quad \Omega, \tag{2.1}$$

where $(a^{ij}(x)) \geq \lambda I$ in Ω for some fixed $\lambda > 0$ and $a^{ij}, b^i, c \in L^\infty(\Omega)$.

LEMMA 1 (STRONG MAXIMUM PRINCIPLE). *If $u \geq 0$ in Ω and u vanishes at some point inside Ω, then $u \equiv 0$ in Ω.*

LEMMA 2 (HOPF BOUNDARY LEMMA). *If $x_0 \in \partial\Omega$, $u > 0$ in Ω, and $u(x_0) = 0$, then*

$$\frac{\partial u}{\partial n}(x_0) < 0$$

and

$$\frac{\partial u}{\partial n}(x_0) < 0$$

$$\lim_{x \to x_0, x \in \Omega} \frac{u(x) - u(x_0)}{|x - x_0|} > 0 \tag{2.2}$$

for any non-tangential limit.

LEMMA 3. ([GNN]) *Let Ω be a domain in $\Re^n$. Let $y_0 \in \partial\Omega$ and assume that, near y_0, $\partial\Omega$ consists of two transversally intersecting C^2-hypersurfaces $\varphi = 0$ and $\psi = 0$. Suppose that $\varphi, \psi < 0$ in Ω. Let u satisfy (2.1), $u > 0$ in Ω, and $u(y_0) = 0$. Assume that*

$$\ell(y_0) \equiv a^{ij}(y_0)D_i\varphi(y_0)D_j\psi(y_0) \geq 0$$

and, if $\ell(y_0) = 0$, assume furthermore that $a^{ij} \in C^2$ in some $\Omega_\varepsilon = \Omega \cap B_\varepsilon(y_0)$ $(\varepsilon > 0)$, and that $\nabla_t(\ell(y)) = 0$ at y_0 for any tangential derivatives ∇_t along $\{\varphi = 0\} \cap \{\psi = 0\}$. Then

$$\frac{\partial u}{\partial s} > 0 \quad \text{at} \quad y_0 \quad \text{if} \quad \ell(y_0) > 0,$$

$$\frac{\partial u}{\partial s} > 0 \quad \text{or} \quad \frac{\partial^2 u}{\partial s^2} > 0 \quad \text{at} \quad y_0 \quad \text{if} \quad \ell(y_0) = 0,$$

for any direction s entering Ω at y_0 transversally to the hypersurfaces $\varphi = 0$ and $\psi = 0$.

LEMMA 4. ([GNN]) *Let $x_0 \in \partial\Omega$ with $n_1(x_0) > 0$, where n_1 is the first component of n. Assume that $u > 0$ in Ω_ε, $u \equiv 0$ on $\partial\Omega \cap B_\varepsilon(x_0)$, and*

$$\Delta u + f(u) = 0 \quad \text{in } \Omega,$$

where f satisfies (1.2). Then there exists a $\delta > 0$ such that $D_1 u < 0$ in Ω_δ.

3. Proof of Theorem 1. Before we can use the moving-plane method, we introduce a few definitions. Let $e_1 = (1, 0, \ldots, 0)$ be the unit vector along the

x_1-axis, and let T_λ be the hyperplane $\{x_1 = \lambda\}$. Since Ω is bounded and smooth, $T_\lambda \cap \overline{\Omega} = \emptyset$ for large λ. Now, let λ decrease until T_λ touches $\overline{\Omega}$ at λ_0, say.

For $\lambda < \lambda_0$, let $\Sigma_\lambda^+ = \Omega \cap \{x_1 > \lambda\}$, and let Σ_λ^- be the reflection of Σ_λ^+ about the plane T_λ. Let x^λ be the reflection point of x about T_λ, i.e.,

$$x^\lambda = (2\lambda - x_1, x_2, \dots, x_n).$$

If $\lambda_0 - \lambda$ is small, Σ_λ^- will be inside Ω. But as λ decreases continuously, Σ_λ^- will be in Ω until one of the following occurs:

1. $\overline{\Sigma_\lambda^-}$ becomes internally tangent to $\partial\Omega$ at some point x_0 not on T_λ;
2. T_λ becomes orthogonal to $\partial\Omega$ at some point $y_0 \in \partial\Omega \cap T_\lambda$.

We let T_{λ_1} denote the plane that first reaches one of these two possibilities and call $\Sigma_{\lambda_1}^+$ the maximal cap.

LEMMA 5. *If*

$$u(x) \le u(x^\lambda) \text{ in } \Sigma_\lambda^+$$

for some $\lambda \in (\lambda_1, \lambda_0)$, *then*

$$\begin{cases} D_{e_1} u(x) < 0 & \text{on } \Omega \cap T_\lambda, \\ u(x) < u(x^\lambda) & \text{in } \Sigma_\lambda^+. \end{cases} \tag{3.1}$$

Proof. Let $v(x) = u(x^\lambda)$, $x \in \Sigma_\lambda^+$. Then

$$\Delta(v(x) - u(x)) + f_1(v(x)) + f_2(v(x)) - f_1(u(x)) - f_2(u(x)) = 0$$

and

$$v(x) - u(x) \ge 0 \quad \text{in } \Sigma_\lambda^+,$$

with

$$v(x) - u(x) \not\equiv 0 \text{ in } \Sigma_\lambda^+.$$

Now, $f_1(v(x)) \ge f_1(u(x))$ and $f_2(v(x)) - f_2(u(x)) = c_\lambda(x)(v(x) - u(x))$, where $c_\lambda(x)$ is bounded, because f_1 is nondecreasing and f_2 is Lipschitz. Hence,

$$\Delta(v(x) - u(x)) + c_\lambda(x)(v(x) - u(x)) \le 0, \tag{3.2}$$

$$v(x) - u(x) \ge 0 \quad \text{in} \quad \Sigma_\lambda^+, \tag{3.3}$$

and

$$v(x) - u(x) \not\equiv 0 \quad \text{in} \quad \Sigma_\lambda^+. \tag{3.4}$$

Then Lemma 1 implies that $v(x) - u(x) > 0$ in Σ_λ^+, while Lemma 2 gives us that

$$\left.\frac{\partial}{\partial x_1}(v(x) - u(x))\right|_{x_1=\lambda} > 0, \tag{3.5}$$

because $v(x) - u(x)|_{x_1=\lambda} \equiv 0$, which is the minimum. From (3.5), we obtain the inequality

$$\left. -\frac{\partial u}{\partial x_1} - \frac{\partial u}{\partial x_1}\right|_{x_1=\lambda} > 0,$$

which completes the proof of Lemma 5.

□

LEMMA 6. *Let u be a classical solution of* (1.1). *Then* (3.1) *holds for all* $\lambda \in (\lambda_1, \lambda_0)$.

Proof. From Lemma 4, we know that (3.2) and (3.3) hold for all $\lambda \in (\lambda_1, \lambda_0)$ with $\lambda_0 - \lambda$ sufficiently small.

Suppose the lemma is false. That is, suppose that there exists a $\lambda_2 \in (\lambda_1, \lambda_0)$ such that (3.1) holds for $\lambda \in (\lambda_2, \lambda_0)$ but not for $\lambda < \lambda_2$.

On the other hand, the continuity of u implies that

$$u(x) \leq u(x^{\lambda_2}), \quad \lambda \in \Sigma^+_{\lambda_2},$$

and since $\lambda_2 \in (\lambda_1, \lambda_0)$, Lemma 5 implies that (3.2) and (3.3) also hold for $\lambda = \lambda_2$.

Since $\lambda_2 > \lambda_1$, $n_1(x_0) > 0$ for each point $x_0 \in \partial\Sigma^+_{\lambda_2} \backslash (T_{\lambda_2} \cap \Omega)$. And hence Lemma 4 concludes that there exists $\varepsilon_{x_0} > 0$ such that

$$D_{e_1} u(x) < 0 \quad \text{in} \quad \Omega \cap B_{\varepsilon_{x_0}}(x_0).$$

Since $D_{e_1} u(x) < 0$ on $T_{\lambda_2} \cap \Omega$, we have that there exists some $\varepsilon > 0$, such that

$$D_{e_1} u(x) < 0 \quad \text{in } \Omega \cap \{x_1 > \lambda_2 - \varepsilon\}, \tag{3.6}$$

because $T_{\lambda_2} \cap \overline{\Omega}$ is compact.

Therefore, if (3.2) and (3.3) fail to hold in (λ_1, λ_2), we must have a sequence $\{\lambda^i\}$ such that

$$\lambda^i > \lambda_1 \quad \text{and} \quad \lambda^i \nearrow \lambda_2 \text{ as } i \to \infty,$$

with

$$u(x^i) \geq u(x^{i\lambda^i}) \quad \text{for some } x^i \in \Sigma^+_{\lambda_i}.$$

But Ω is bounded, so we can find a subsequence of $\{x^i\}$, say $\{x^i\}$ itself, which converges to some point $x_0 \in \overline{\Sigma^+_{\lambda_2}}$ as $i \to \infty$ with $u(x_0) \geq u(x_0^{\lambda_2})$. Therefore, $x_0 \in \partial\Sigma^+_{\lambda_2}$, because (3.1) holds for λ_2. Thus, we have either of two possibilities:

1. $x_0 \in \partial\Sigma^+_{\lambda_2} \backslash (T_\lambda \cap \overline{\Omega})$. But then $x_0 \in \partial\Omega$ with $x_0^{\lambda_2} \in \Omega$, since $\lambda_2 > \lambda_1$, which implies that $0 \geq u(x_0^{\lambda_2}) > 0$. This is impossible.
2. $x_0 \in T_{\lambda_2} \cap \overline{\Omega}$. Therefore, $x_0^{\lambda_2} = x_0$.

Since $\lambda^i > \lambda_1$, we have that the line segment P_i joining x^i and $x^{i\lambda^i}$ lies in Ω. Therefore, $u(x^i) \geq u(x^{i\lambda^i})$ implies that

$$D_{e_1} u(y^i) \geq 0 \quad \text{for some } y^i \in P_i.$$

But $x^i \to x_0$ and $x^{i\lambda^i} \to x_0^{\lambda_2} = x_0$, so P_i shrinks into the single point x_0. This gives us a contradiction with (3.6), because $y^i \in \Omega \cap \{x_1 > \lambda_2 - \varepsilon\}$ for i large enough. This completes the proof of Lemma 6.

□

LEMMA 6. *Let u be a classical solution of* (1.1). *Then* (3.1) *holds for all* $\lambda \in (\lambda_1, \lambda_0)$.

Proof. From Lemma 4, we know that (3.2) and (3.3) hold for all $\lambda \in (\lambda_1, \lambda_0)$ with $\lambda_0 - \lambda$ sufficiently small.

Suppose the lemma is false. That is, suppose that there exists a $\lambda_2 \in (\lambda_1, \lambda_0)$ such that (3.1) holds for $\lambda \in (\lambda_2, \lambda_0)$ but not for $\lambda < \lambda_2$.

On the other hand, the continuity of u implies that

$$u(x) \leq u(x^{\lambda_2}), \quad \lambda \in \Sigma^+_{\lambda_2},$$

and since $\lambda_2 \in (\lambda_1, \lambda_0)$, Lemma 5 implies that (3.2) and (3.3) also hold for $\lambda = \lambda_2$.

Since $\lambda_2 > \lambda_1$, $n_1(x_0) > 0$ for each point $x_0 \in \partial\Sigma^+_{\lambda_2} \backslash (T_{\lambda_2} \cap \Omega)$. And hence Lemma 4 concludes that there exists $\varepsilon_{x_0} > 0$ such that

$$D_{e_1} u(x) < 0 \quad \text{in} \quad \Omega \cap B_{\varepsilon_{x_0}}(x_0).$$

Since $D_{e_1} u(x) < 0$ on $T_{\lambda_2} \cap \Omega$, we have that there exists some $\varepsilon > 0$, such that

$$D_{e_1} u(x) < 0 \quad \text{in } \Omega \cap \{x_1 > \lambda_2 - \varepsilon\}, \tag{3.6}$$

because $T_{\lambda_2} \cap \overline{\Omega}$ is compact.

Therefore, if (3.2) and (3.3) fail to hold in (λ_1, λ_2), we must have a sequence $\{\lambda^i\}$ such that

$$\lambda^i > \lambda_1 \quad \text{and} \quad \lambda^i \nearrow \lambda_2 \text{ as } i \to \infty,$$

with

$$u(x^i) \geq u(x^{i\lambda^i}) \quad \text{for some } x^i \in \Sigma^+_{\lambda_i}.$$

But Ω is bounded, so we can find a subsequence of $\{x^i\}$, say $\{x^i\}$ itself, which converges to some point $x_0 \in \overline{\Sigma^+_{\lambda_2}}$ as $i \to \infty$ with $u(x_0) \geq u(x_0^{\lambda_2})$. Therefore, $x_0 \in \partial\Sigma^+_{\lambda_2}$, because (3.1) holds for λ_2. Thus, we have either of two possibilities:

1. $x_0 \in \partial\Sigma^+_{\lambda_2} \backslash (T_\lambda \cap \overline{\Omega})$. But then $x_0 \in \partial\Omega$ with $x_0^{\lambda_2} \in \Omega$, since $\lambda_2 > \lambda_1$, which implies that $0 \geq u(x_0^{\lambda_2}) > 0$. This is impossible.
2. $x_0 \in T_{\lambda_2} \cap \overline{\Omega}$. Therefore, $x_0^{\lambda_2} = x_0$.

Since $\lambda^i > \lambda_1$, we have that the line segment P_i joining x^i and $x^{i\lambda^i}$ lies in Ω. Therefore, $u(x^i) \geq u(x^{i\lambda^i})$ implies that

$$D_{e_1} u(y^i) \geq 0 \quad \text{for some } y^i \in P_i.$$

But $x^i \to x_0$ and $x^{i\lambda^i} \to x_0^{\lambda_2} = x_0$, so P_i shrinks into the single point x_0. This gives us a contradiction with (3.6), because $y^i \in \Omega \cap \{x_1 > \lambda_2 - \varepsilon\}$ for i large enough. This completes the proof of Lemma 6.

□

Proof of Theorem 1. By Lemma 6, (3.2) and (3.3) hold for all $\lambda \in (\lambda_1, \lambda_0)$. Let us discuss the following two possible cases.

Case 1. $\Sigma_{\lambda_1}^- \cup \Sigma_{\lambda_1}^+ \cup (T_{\lambda_1} \cap \Omega) = \Omega$.

Then Ω is symmetric about T_{λ_1}, in which case we have shown that

$$\text{(3.7)} \qquad \begin{cases} u(x) \le u(x^{\lambda_1}), & x \in \Sigma_{\lambda_1}^+, \\ & \qquad \forall\, \lambda \in (\lambda_1, \lambda_0), \\ D_{e_1} u < 0, & x \in \Sigma_{\lambda}^+, \end{cases}$$

or

$$\text{(3.8)} \qquad \begin{cases} u(x) = u(x^{\lambda_1}), & x \in \Omega \\ D_{e_1} u > 0, & \text{if } x \in \Omega \cap \Sigma_{\lambda_1}^-, \\ D_{e_1} u < 0 & \text{if } x \in \Omega \cap \Sigma_{\lambda_1}^+. \end{cases}$$

Case 2. $\Sigma_{\lambda_1}^- \cup \Sigma_{\lambda_1}^+ \cup (T_{\lambda_1} \cap \Omega) \subsetneq \Omega$.

Then $u(x^{\lambda_1}) - u(x) \ge 0$ in $\Sigma_{\lambda_1}^+$ and not identically zero. Therefore, the same argument as in Lemma 5 implies that

$$u(x^{\lambda_1}) > u(x) \quad \text{in } \Sigma_{\lambda_1}^+.$$

at some point $x_0 \notin T_{\lambda_1}$. As in the proof of Lemma 5, we find by letting $v(x) = u(x^{\lambda_1})$ that

$$\text{(3.9)} \qquad \begin{cases} \Delta(v(x) - u(x)) + c_{\lambda_1}(x)(v(x) - u(x)) \le 0 & \text{in } \Sigma_{\lambda_1}^+, \\ v(x) - u(x) > 0 & \text{in } \Sigma_{\lambda_1}^+, \\ v(x_0) - u(x_0) = 0. \end{cases}$$

Since $x_0 \notin T_{\lambda_1}$, $\Sigma_{\lambda_1}^+$ is smooth near x_0. Hence we may use Lemma 2 to conclude that

$$\frac{\partial}{\partial n}(v - u)(x_0) < 0,$$

which is in contradiction with the boundary condition $\frac{\partial v}{\partial n} = \frac{\partial u}{\partial n}(x_0) = 0$. Hence, T_{λ_1} must become orthogonal to $\partial\Omega$ at some point $y_0 \in \partial\Omega \cap T_{\lambda_1}$.

However, $u(x^{\lambda_1}) - u(x)$ satisfies (3.8) in $\Sigma_{\lambda_1}^+$, $y_0 \in \partial\Sigma_{\lambda}^+$, and, near y_0, $\partial\Sigma_{\lambda}^+$ consists of two transversally intersecting hypersurfaces $x_1 = \lambda_1$ and $\partial\Omega$, which become orthogonal at y_0. A simple computation shows that $\ell(y_0) = 0$ and, for any tangential direction t along $T_{\lambda_1} \cap \partial\Omega$ at y_0,

$$\nabla_t(\ell(y)) = 0 \quad \text{at } y_0,$$

which implies by Lemma 3 that for any s entering Ω at y_0 transversally to T_λ and $\partial\Omega$,

$$\frac{\partial(v-u)}{\partial s} > 0 \quad \text{or} \quad \frac{\partial^2(v-u)}{\partial s^2} > 0 \text{ at } y_0.$$

On the other hand, it follows from (1.1) that

$$(v-u)(y_0) = 0, \quad \nabla(v-u)(y_0) = 0, \quad \text{and } D^2(v-u)(y_0) = 0.$$

This again leads us to a contradiction, so it must be the case that

$$\Sigma_{\lambda_1}^- \cup \Sigma_{\lambda_1}^+ \cup (T_{\lambda_1} \cap \Omega) = \Omega.$$

On the other hand, since we can start moving the plane from the left to the right along the x_1-axis as well, we conclude that

$$\begin{cases} u(x) = u(x^{\lambda_1}), & x \in \Omega, \\ D_{e_1} u > 0 & \text{if} \quad x \in \Omega \cap \Sigma_{\lambda_1}^-, \\ D_{e_1} u < 0, & \text{if} \quad x \in \Omega \cap \Sigma_{\lambda_1}^+. \end{cases} \tag{3.10}$$

But equation (1.1) is rotationally invariant. Therefore Ω is symmetric in every direction. We thus find that Ω must be a ball, because it is connected. Then (3.10) gives the conclusions of Theorem 1.

4. Proof of Theorem 2. Gidas, Ni, and Nirenberg proved in [GNN] that the solutions of

$$\begin{cases} \Delta u + f(u) = 0 & \text{in } B_R(0), \\ u > 0 & \text{in } B_R(0), \\ u = 0 & \text{on } \partial B_R(0), \end{cases} \tag{4.1}$$

with

$$f(s) = f_1(s) + f_2(s), \tag{4.2}$$

where f_1 is Lipschitz continuous and f_2 nondecreasing, must be radially symmetric about 0 and, furthermore, $\frac{\partial u}{\partial r} < 0$ for $0 < r < R$. On the other hand, if a decomposition like (4.2) does not exist, in particular if f is not smooth, then it is an open problem whether positive solutions of (1.1) are radially symmetric. Actually, some examples given in [GNN, pp. 220] show that these cases could be very delicate.

In this part, we will try to treat a family of nonlinear terms f which are neither Lipschitz nor nondecreasing. Such situations arise, for example, in the study of free-boundary problems for Emden-Fowler type equations (see [KK]), where

$$f(u) = u^{1/p} - u^{1/q} \text{ with } 1 \leq p < q \leq \infty. \tag{4.3}$$

Remark 1. If $q = \infty$, then $f(s) = s^{1/p} - 1$ is an increasing function in s and therefore [GNN]'s result implies that u must be radially symmetric. Therefore, the difficult parts occur when $1 \leq p < q < \infty$. For such cases,

$$f(s) = \begin{cases} \text{strictly decreasing in } \left[0, (p/q)^{pq/q-p}\right], \\ \text{strictly increasing in } \left[(p/q)^{pq/q-p}, \infty\right). \end{cases}$$

Remark 2. Recently, new symmetry results have been obtained in [GL] and [LV] for equations on nonsmooth domains.

Proof of Theorem 2. First we define

$$\Lambda = \{\lambda \in (0,R) | u(x) < u(x^\lambda) \quad \text{if } x \in \Sigma_\lambda^+\}.$$

Because $u|_{\partial B_R} = 0$, there exists a $\lambda_0 \in (0,R)$ such that

$$u(B_R(0) \backslash B_{\lambda_0}(0)) \subset (0, s_0). \tag{4.4}$$

Step 1. $(\frac{1}{2}(\lambda_0 + R), R) \in \Lambda$.

For any $\lambda \in (\frac{1}{2}(\lambda_0 + R), R)$, f is a strictly decreasing function in the interval $[0, \max\{\sup_{\Sigma_\lambda^+} u, \sup_{S_\lambda^+} u^\lambda\}]$, where $u^\lambda = u(x^\lambda)$, because of (4.4), and

$$\begin{cases} \Delta(u^\lambda - u)(x) + f(u^\lambda(x)) - f(u(x)) = 0 & \text{in } \Sigma_\lambda^+, \\ u^\lambda - u = 0 & \text{on } \overline{T}_\lambda, \\ u^\lambda - u > 0 & \text{on } \partial\Sigma_\lambda^+ \setminus \overline{T}_\lambda. \end{cases}$$

Claim 1. If $u^\lambda - u \geq 0$ in Σ_λ^+, then $u^\lambda - u > 0$ in Σ_λ^+ and $\frac{\partial u}{\partial x_1} < 0$ on T_λ.

Suppose the claim is false, i.e. there exists a $y_0 \in \Sigma_\lambda^+$, such that $(u^\lambda - u)(y_0) = 0$. On the other hand, both u and u^λ are strictly positive in Σ_λ^+, so

$$\Delta(u^\lambda - u)(x) + \frac{f(u^\lambda(x)) - f(u(x))}{u^\lambda(x) - u(x)}(u^\lambda - u)(x) = 0,$$

where $\frac{f(u^\lambda(x)) - f(u(x))}{u^\lambda(x) - u(x)}$ is locally bounded, because $f \in C_{loc}^{0,1}((0,\infty))$.

Hence the strong maximum principle implies a contradiction. Therefore, if $u^\lambda - u \geq 0$ in Σ_λ^+, then $u^\lambda - u > 0$ there.

Claim 2. $u^\lambda - u \geq 0$ in Σ_λ^+.

Otherwise, because $u^\lambda - u \geq 0$ on $\partial\Sigma_\lambda^+$, $u^\lambda - u$ would have a strictly interior negative minimum, say at $y_0 \in \Sigma_\lambda^+$. But at y_0 we have $\Delta(u^\lambda - u)(y_0) \geq 0$ and, since $s_0 > u(y_0) > u^\lambda(y_0) > 0$ by (4.4), $f(u(y_0)) < f(u^\lambda(y_0))$. Therefore,

$$\Delta(u^\lambda - u)(y_0) + f(u^\lambda(y_0)) - f(u(y_0)) > 0,$$

a contradiction.

Thus Step 1 is proved.

Step 2. Λ is closed w.r.t. $(0,R)$.

If $\{\lambda^i\}$ is a sequence in Λ which converges to some λ in $(0,R)$, then, since

$$u(x) < u(x^{\lambda i}), \quad x \in \Sigma_{\lambda_i}^+,$$

letting $i \to \infty$, we find

$$u(x) \leq u(x^\lambda), \quad x \in \Sigma_\lambda^+.$$

But then 'Claim 1' in 'Step 1' shows that $u(x) < u(x^\lambda)$ in Σ_λ^+, i.e. $\lambda \in \Lambda$.

Step 3. Λ is open in $(0, R)$.

Suppose that Λ is not open. Then there exists a $\lambda \in \Lambda$ and a sequence $\{\lambda^i\} \in (0, R)$ s.t. $\lambda^i \to \lambda$ with $\lambda^i \notin \Lambda$. That is, for each i there exists $x^i \in \Sigma_{\lambda^i}^+$ with

$$0 > u(x^{i\lambda^i}) - u(x^i) = \min_{x \in \Sigma_{\lambda^i}^+} (u^{\lambda^i} - u)(x), \tag{4.5}$$

and the mean-value theorem implies

$$\frac{\partial u}{\partial x_1}(y^i) \geq 0 \text{ for some } y^i \in \overline{x^{i\lambda^i} x^i}. \tag{4.6}$$

Because $x^i \in B_R(0)$, there exists a subsequence, say $\{x^i\}$ itself, converging to a point $x_0 \in \overline{B}_R(0)$ and (4.5) implies that

$$u(x_0) \geq u(x_0^\lambda). \tag{4.7}$$

But $\lambda \in \Lambda$, therefore (4.7) could only occur on $\partial\Sigma_\lambda^+$. Hence $x_0 \in \partial\Sigma_\lambda^+$.

On the other hand,

$$u < u^\lambda \quad \text{on } \partial\Sigma_\lambda^+ \backslash \overline{T}_\lambda.$$

Therefore $x_0 \in \overline{T}_\lambda$.

In this case, since $x^i \to x_0 \in \overline{T}_\lambda$ and $\lambda^i \to \lambda$, we have $x^{i\lambda^i} \to x_0$. Therefore (4.6) implies that

$$\frac{\partial u}{\partial x_1}(x_0) \geq 0.$$

Hence, $x_0 \in \overline{T}_\lambda \cap \partial B_R(0)$, because

$$\frac{\partial u}{\partial x_1} < 0 \quad \text{in } \overline{T}_\lambda.$$

Now, since $x_0 \in \partial B_R(0)$ and $\lim_{i\to\infty} x^i = \lim_{i\to\infty} x^{i\lambda^i} = x_0$, we have $0 < u(x^{i\lambda^i}) < u(x^i) < s_0$ if i is large enough. Therefore, $f(u(x^i)) < f(u(x^{i\lambda^i}))$ for all i large enough.

But $\Delta(u^{\lambda^i} - u)(x^i) \geq 0$, since x^i is a minimum point of $u^{\lambda^i} - u$, so we reach a contradiction, because

$$\Delta(u^\lambda - u)(x^i) + f(u(x^{i\lambda^i})) - f(u(x^i)) = 0.$$

Therefore Λ is open.

Step 4. Since Λ is non-empty and both open and closed in $(0, R)$, it must be the case that $\Lambda = (0, R)$, so letting $\lambda \to 0$, we find that

$$u(x_1, \cdots, x_n) \leq u(-x_1, \cdots, x_n)$$

for $x \in \Sigma_0^+$. But both $B_R(0)$ and Δ are invariant under the symmetry group, so

$$u \quad \text{radially symmetric}$$

and

$$\frac{\partial u}{\partial r} < 0 \quad \text{for} \quad 0 < r < R,$$

by 'Claim 1'. Thus, the proof of Theorem 2 is complete.

REFERENCES

[GL] N. Garofalo and J. Lewis, *A Symmetry Result Related to Some Overdetermined Boundary Value Problems*, Amer. J. Math., 111 (1989), pp. 9–33.

[GNN] B. Gidas, W.-M. Ni and L. Nirenberg, *Symmetry and Related Properties via the Maximum Principle*, Comm. Math. Phys., 68 (1979), pp. 209–243.

[H] H. Hopf, *Lectures on Differential Geometry in the Large*, Stanford Univ., 1956.

[KK] H. G. Kaper and Man Kam Kwong, *Free Boundary Problems for Emden-Fowler Equations*, Differential and Integral Equations, 3 (1990), pp. 353–362.

[LV] J. Lewis and A. Vogel, *On Some Almost Everywhere Symmetry Theorems*, preprint.

[PW] M. Protter and H. Weinberger, *Maximum Principles in Differential Equations*, Prentice-Hall, 1967.

[S] J. Serrin, *A Symmetry Problem in Potential Theory*, Arch. Rational Mech. Anal., 43 (1971), pp. 304–318.

ABSOLUTE CONTINUITY OF PARABOLIC MEASURE

JOHN L. LEWIS*†‡ AND MARGARET MURRAY**†

1. Introduction. Let $\mathbf{R}$ be the real numbers and if $E \subseteq \mathbf{R}$, let $\overline{E}, \partial E, |E|$, denote the closure, boundary, and outer Lebesgue measure of E, respectively. Let

$$I = \{t : |t - t_0| \leq \lambda\}, \alpha I = \{t : |t - t_0| \leq \alpha\lambda\},$$

and if f is a function defined on I put

$$\|f\|_I = \sup\left\{\frac{|f(t) - f(s)|}{|t - s|^{1/2}} : s, t \in I\right\},$$

$$\|f\|_I^{\wedge} = \sup_{J \subseteq I}\left(|J|^{-1}\int_J\int_J \frac{(f(t) - f(s))^2}{(s - t)^2} ds dt\right)^{1/2},$$

where J is an interval. Extend f to $2I$ by defining, $f^*(t) = f(t_0 + \lambda)$, $t_0 + \lambda < t \leq t_0 + 2\lambda$, $f^*(t) = f(t)$, $t \in I$, and $f^*(t) = f(t_0 - \lambda)$, when $t_0 - 2\lambda \leq t < t_0 - \lambda$. Next if $0 < A < \infty$, put

$$D = D(I, A, f) = \left\{(x, t) : f^*(t) < x < \max_I f + 100A|I|^{1/2}, t \in 2I\right\}$$

$$Z = Z(I, A, f) = \left(f(t_0) + 10A|I|^{1/2}, t_0 + \frac{5}{8}|I|\right),$$

and if $\|f\|_I \leq 3A$, let $\tilde{\omega}[Z, F, D(I, A, f)] = \tilde{\omega}(Z, F)$, be the parabolic measure of the Borel set $F \subseteq \partial D$. That is, $\tilde{\omega}(Z, F)$, is the value at Z of the solution to the heat equation in D with boundary values 1 on F and 0 on $\partial D - F$, in the Brelot-Perron-Wiener sense. If $E \subseteq I$, put $\rho(E) = \{(f(t), t) : t \in E\}$, and let, $\omega(Z, E) = \tilde{\omega}(Z, \rho(E))$. We note that many authors have considered the Dirichlet problem for D and the mutual absolute continuity of ω and Lebesgue measure on I (see [1], [7], for references). In particular, Kaufman and Wu [6] were the first to show that $\|f\|_I < +\infty$, is not enough to insure the mutual absolute continuity of parabolic measure and Lebesgue measure on I. In fact they constructed a function f on $N = [-1, 1]$, with $\|f\|_N = 1$; and for which the corresponding D, ω, have the following property: There is a set $E \subset [-1, 1]$ of Hausdorff dimension $\alpha < 1$ such that $\omega([-1, 1] - E) = 0$. Lewis and Silver studied the mutual absolute continuity of these measures in [7]. They assumed f has compact support in $[-1, 1]$, and

$$|f(t) - f(s)| \leq \psi(|t - s|), s, t \in \mathbf{R}, \tag{1.1}$$

where ψ is an increasing function on $(0, \infty)$, with $\psi(0) = 0$, and

$$\int_0^\infty \tau^{-2}\psi(\tau)^2 d\tau = B < +\infty. \tag{1.2}$$

Under these assumptions they proved the following theorem.

*Mathematics Department, University of Kentucky, Lexington, Kentucky, 40506.
†Supported by NSF.
‡Supported by the Commonwealth of Kentucky through the Kentucky EPSCOR Program.
**Mathematics Department, VPI and SU, Blacksburg, Virginia, 24061.

THEOREM A. *Let $\Omega = \{(x,t) : x > f(t), t \in \mathbf{R}\}$ and define $\tilde{\omega}$ on $\partial\Omega$ relative to $(10,10)$. Then, $d\omega = h dt$ on $[-1,1]$, and for $1 < p < \infty$,*

$$(1.3) \qquad |J|^{-1} \int_J h^p dt \le \left[c(p,B)|J|^{-1} \int_J h dt \right]^p,$$

whenever J is an interval with $J \subseteq [-1,1]$.

In (1.3), as in the sequel, $c(p,B)$ denotes a positive constant depending only on p and B. Moreover, c denotes an absolute constant, not necessarily the same at each occurrence. From (1.3) and Hölder's inequality, we see that

$$(1.4) \qquad \omega(E)/\omega(J) \le c(p,B)(|E|/|J|)^{1-1/p}.$$

Also, we note that if $N = [-1,1]$, and $\|f\|_N \le A$, then ω is a doubling measure, i.e.,

$$(1.5) \qquad \omega(2I) \le c(A)\omega(I), I \subseteq [-1,1],$$

whenever I is an interval (see Lemma 1). From (1.3), (1.4), we see that ω satisfies a so called A_∞ condition with respect to Lebesgue measure on $[-1,1]$: Given H a Borel subset of the closed interval $J \subseteq [-1,1]$, there exits $\delta_1, \delta_2 > 0, 0 < \delta_1, \delta_2 < 1$ such that

$$(1.6) \qquad |H|/|J| \le \delta_1 \to \omega(H)/\omega(J) < \delta_2.$$

If a measure ω satisfies (1.5), (1.6), then it is said to be an A_∞ weight ($\omega \in A_\infty(|\cdot|)$) with respect to Lebesgue measure on $[-1,1]$. Coifman and Fefferman in [2] have shown that if $\omega \in A_\infty(|\cdot|)$, then there exists, $\alpha, c > 0$, such that

$$(1.7) \qquad |H|/|J| \le c[\omega(H)/\omega(J)]^\alpha.$$

Hence, $|\cdot| \in A_\infty(\omega)$. Also the above authors show that if $\omega \in A_\infty(|\cdot|)$, then (1.3), (1.4) are valid for some $p > 1$ with $c(p,B)$ replaced by $c(\delta_1, \delta_2)$.

We note that Theorem A is sharp in the following sense: Given ψ increasing on $(0,\infty)$ with $\psi(0) = 0$, $\psi(2r) \le 2\psi(r)$, $0 < r < \frac{1}{2}$, and $\int_0^1 \tau^{-2}\psi(\tau)^2 dr = +\infty$; there exists f with compact support in $[-1,1]$ for which the corresponding ω (defined as in Theorem A) and Lebesgue measure are mutually singular (see [7, Thm. 2]). Although Theorem A is essentially best possible in terms of a modulus of continuity condition, there is a weaker condition which insures that parabolic measure is an A_∞ weight with respect to Lebesgue measure. Before stating this condition, we note that the method of proof in Theorem A involves using classical double layer heat potentials. Thus, it has to be shown that a certain integral equation has a solution for all functions in a given Lebesgue class. To do this, certain singular integral estimates were made, using (1.1), (1.2). It then follows from a bootstrap type argument that the integral equation has a solution. Now, Russell Brown (oral communication) pointed out to the first author that the second author in [8] had considered singular integral operators with kernels involving functions which have fractional derivatives. Moreover, from a theorem of Strichartz [9], it was easily seen that the singular integral estimates in [8] were more general than in [7].

With these improved singular integral estimates, the technique in [7] can be used essentially unchanged to prove

THEOREM B. *Let f have compact support in $[-1,1]$ and define $\tilde{\omega}, \Omega$, relative to f as in Theorem A. There exists $\theta_0 > 0$ such that if*

$$\max\{\|f\|_N, \|f\|_N^{\wedge}\} \leq \theta_0, \tag{1.8}$$

then $dw = hdt$ on $[-1,1]$ and (1.3), (1.4) hold with $c(p,B)$ replaced by $c(p)$.

In this paper we use Theorem B to prove

THEOREM 1. *Let f be defined on the interval I and suppose that*

$$\max\{\|f\|_I, \|f\|_I^{\wedge}\} \leq A < +\infty. \tag{1.9}$$

Then $\omega(Z,\cdot,D(I,A,f))$ is an A_∞ weight with respect to Lebesgue measure on I. The constants in (1.6), (1.7), depend only on A.

From the A_∞ conclusion of Theorem 1 we see that ω and $|\cdot|$ are mutually absolutely continuous. Moreover, if $dw = hdt$, then from (1.3) we deduce that h is pth power integrable for some $p > 1$. If $q = p/(p-1)$, then from the argument in [7, Thm. 3], it now follows that an L_r Dirichlet problem for $D(I,A,f)$ has a unique solution, $r \geq q$. We note that Theorem 1 has a higher dimensional analogue, which we will obtain in a forthcoming paper.

We also prove

THEOREM 2. *Let f be defined on the closed interval I with $\|f\|_I < +\infty$, and*

$$\int_I \int_I \frac{(f(s)-f(t))^2}{(s-t)^2} ds dt < +\infty.$$

Then Lebesgue measure on I is absolutely continuous with respect to $\omega(Z,\cdot,D(I,A,f))$ on I.

Finally we prove

THEOREM 3. *Let f be defined on I with $\|f\|_I \leq A < +\infty$. If $\omega(Z,E,D(I,A,f)) > 0$, and $|E| = 0$, then E is contained ω almost everywhere in*

$$\left\{ t : \int_I \frac{(f(s)-f(t))^2}{(s-t)^2} ds = +\infty \right\}.$$

As motivation for the proof of Theorem 1, we note that Fefferman, Kenig, and Pipher considered a somewhat similar problem for elliptic operators in [4]. They use in an essential way, a comparison lemma for harmonic measures proved by Dahlberg, Kenig, and Jenison [3, Main lemma]. This comparison lemma has been generalized to the heat equation in Lip. $(\frac{1}{2}, 1)$ domains by Brown [1, Lemma 2.10]. Lemma 2 in section 2 is essentially a one space dimension statement of this lemma. In view of Lemma 2, our game plan is to construct h on I with (a) $\|h\|_I \le A$, (b) $\|h\|_I^\wedge \le (1-\epsilon)\|f\|_I^\wedge$ for some fixed $\epsilon > 0$, and (c) $f = h$ on a set of positive Lebesgue measure $\ge \epsilon|I|$. We then can apply the same argument to h, etc. Applying this argument repeatedly, we finally obtain a function with small $\| \ \|_I^\wedge$, to which Theorem B can be applied. Doing this and using Lemma 2, we get Theorem 1. We note that our proof is somewhat similar to the David buildup scheme (see [5, Ch. 8]). The proofs of Theorems 2 and 3 are similar.

2. Preliminary reductions. We shall need the following lemmas. In Lemmas 1-4, we write $\omega(\cdot,\cdot)$ for $\omega(\cdot,\cdot,D(I,A,f))$, when there is no chance of confusion.

LEMMA 1. *Let f be a function on I with $\|f\|_I \le 3A < +\infty$. Let $J = [t_1-\gamma, t_1+\gamma] \subseteq I, X = (x,t) \in D(I,A,f)$, with $t - t_1 > |x - f(t_1)|^2/100$, and $\frac{100\gamma}{99} \le t - t_1$. Then*

$$\omega(X,J) \le c(A)\omega(X, \frac{1}{2}J). \tag{2.1}$$

If E is a Borel set contained in J, and $Z(J) = (f(t_1) + 10A|J|^{1/2}, t_1 + \frac{5}{8}|J|)$, then

$$c(A)^{-1}\omega(Z(J),E) \le \frac{\omega(X,E)}{\omega(X,J)} \le c(A)\omega(Z(J),E). \tag{2.2}$$

For the proof of (2.1), see [10, Lemma 2.2]. (2.2) follows from (2.1) (see the proof of [1, Corollary 2.7]).

LEMMA 2. *Let f_1, f_2 be defined on $2I$ with $\|f_j\|_{2I} \le 3A < +\infty$, $j = 1,2$. Let $Z_j, j = 1,2$, be defined relative to f_j, I, as in section 1 and put*

$$D_j^*(I,A,f_j) = \{(x,t) : t \in 2I, \quad \text{and } f_j(t) < x < \min_I f_j + 100A|I|^{1/2}\},$$

$$\widetilde{D}_j = \widetilde{D}(I,A,f_j) = \{(x,t) : t \in 2I, \quad \text{and } f_j(t) < x < \max_I f_j + 100A|I|^{1/2}\}.$$

There exists $\alpha = \alpha(A)$, $c = c(A) > 0$, such that if $f_1 = f_2$ on a Borel set $E \subseteq I$, then

$$c(A)^{-1}\omega[Z_1,E,\widetilde{D}_1]^{1/\alpha} \le \omega[Z_2,E,D_2^*] \le c(A)\omega[Z_1,E,\widetilde{D}_1]^\alpha. \tag{2.3}$$

Lemma 2 can be proved by comparing both parabolic measures for E in (2.3) to the parabolic measures for E relative to a certain "parabolic sawtooth" contained in both domains. It then follows from Lemma 2.10 in [1], that Lemma 2 in true. A second way to prove Lemma 2 which avoids any regularity assumptions about the sawtooth, is to put $\hat{f} = \max(f_1,f_2)$, $f_- = \min(f_1,f_2)$. Repeating the argument in Lemma 2.10 of [1] with $D^*(I,A,\hat{f})$ playing the role of the sawtooth and $\widetilde{D}(I,A,f_-)$ the role of Ω, we get (2.3) with f_2 replaced by $\hat{f}$ and f_1 by f_-. We then deduce from the maximum principle for the heat equation that (2.3) is true for f_1, f_2. Next we prove

LEMMA 3. *Let f be defined on I. There exists $\theta_1 > 0$ such that if*

$$\max[\,\|f\|_I,\ \|f\|_I^{\wedge}\,] \le \theta_1,$$

then $\omega = \omega(Z,\cdot)$ satisfies (1.4) with $c(p,B)$ replaced by $c(p)$ whenever $1 < p < \infty$ and E is a Borel subset of the interval $J \subseteq I$.

Proof. Extend f to $4I$ by requiring that this extension be continuous on $4I$ and constant on each component of $4I - I$. We also denote this extension by f^*. Clearly, $\|f^*\|_I \le \theta_1$. Recall that $I = [t_0 - \lambda, t_0 + \lambda]$. If J is an interval, $J \subseteq 4I$, let

$$J = (4I - I) \cap J + I \cap J = J_1 + J_2.$$

If $L = \{(s,t) \in J_1 \times J_1 : |s-t| \ge |I|\}$, then

$$\begin{aligned}
\int_J\int_J \left(\frac{f^*(t)-f^*(s)}{s-t}\right)^2 dsdt &\le \int\int_L \frac{(f(t_0+\lambda)-f(t_0-\lambda))^2}{(s-t)^2} dsdt \\
&+ \int_{J_2}\int_{J_2} \left(\frac{f(s)-f(t)}{s-t}\right)^2 dsdt + 2\int_{J_1}\int_{J_2} \frac{(f^*(s)-f^*(t))^2}{(s-t)^2} dsdt \\
&\le c\theta_1^2|J| + 2\int_{J_1}\int_{J_2} \frac{(f^*(s)-f^*(t))^2}{(s-t)^2} dsdt.
\end{aligned} \tag{2.4}$$

If $t \le t_0 - \lambda$, $t \in J_1$, $s \in J_2$, then $|s-t| \ge |t - t_0 + \lambda|$, so

$$\begin{aligned}
\int_{J_2} \frac{(f^*(s)-f^*(t))^2}{(s-t)^2} ds &= \int_{J_2} \frac{(f^*(s)-f(t_0-\lambda))^2}{(s-t)^2} ds \\
&\le c\theta_1^2 \int_{J_2} |s-t|^{-1} ds \le c\theta_1^2 \log\left(\frac{8|J|}{|t-t_0+\lambda|}\right).
\end{aligned}$$

If $s \in J_2$, $t > t_0 + \lambda$, a similar inequality holds. Integrating these inequalities over J_1 we conclude first that

$$\int_{J_1}\int_{J_2} \frac{(f^*(s)-f^*(t))^2}{(s-t)^2} dsdt \le c\theta_1^2|J|,$$

and next from (2.4) that

$$\max\{\|f^*|_{4I}, \|f^*\|_{4I}^{\wedge}\} \le c\theta_1. \tag{2.5}$$

Let ϕ be an infinitely differentiable function on $\mathbf{R}$ with $\phi \equiv 1$ on $3I$, supp $\cdot\phi \subseteq 4I$, and $\|\phi'\|_\infty \le 1000|I|^{-1}$. Put

$$\bar{f}(t) = (4\lambda)^{-1/2}[(f^* - f(t_0))\phi](t_0 + 4\lambda t), \quad t \in \mathbf{R},$$

and note that $\text{supp}\cdot\bar{f} \subseteq [-1,1]$. An easy calculation using (2.5) shows that if $N = [-1,1]$, then

$$\max\{\ \|\bar{f}\|_N,\quad \|\bar{f}\|_N^{\wedge}\ \} \le c\theta_1.$$

Thus we can apply Theorem B to $\bar{f}$ provided $c\theta_1 \le \theta_0$. Fix θ_1 to be the largest number for which $c\theta_1 \le \theta_0$. From (1.4) of Theorem B, and (2.2) of Lemma 1 with $|I| \to \infty$, we deduce

$$\omega(Z^*, E^*, D) \le c(p)\left(\frac{|E|}{|J|}\right)^{1-1/p}.$$

Here, $D = \{(x,t) : x > \bar{f}(t)\}$, $E^* = \{t : t_0 + 4\lambda t \in E\}$, and $Z^* = \left(\bar{f}(t_2) + 10A\frac{\gamma^{1/2}}{(4\lambda)^{1/2}},\right.$ $\left. t_2 + \frac{5\gamma}{32\lambda}\right)$, where $t_1 = t_0 + 4\lambda t_2, \gamma, t_1$, as in Lemma 1. Since the heat equation is invariant under translations, and dilations of the form $(x,t) \to (dx, d^2t)$ when $d \in \mathbf{R}$, it follows from the above inequality and the maximum principle for the heat equation that

$$\omega[Z(J), E, D(I,A,f)] \le c(p)\left(\frac{|E|}{|I|}\right)^{1-1/p},$$

and there upon from (2.2) of Lemma 1 that Lemma 3 is valid. □

Finally we shall need

LEMMA 4. *Let f be defined on I with $\|f\|_I \le A$, $1 \le A < \infty$. There exists $\theta_j = \theta_j(A)$, $2 \le j \le 4$, $0 < \theta_j < 1$, such that if*

$$\int_I \int_I \left(\frac{f(s) - f(t)}{s - t}\right)^2 ds dt \le \theta_2 |I|, \tag{2.6}$$

then whenever F is a Borel subset of I with $|F|/|I| \ge 1 - \theta_3$, we have

$$\omega(Z, F) > \theta_4.$$

Proof. Observe from Lemma 3 that there exists $\theta_3 > 0$ with the property: If H is a Borel subset of I with $|H|/|I| \ge 1 - 2\theta_3$, $0 < \theta_3 < \frac{1}{4}$, then

$$\omega(Z, H, D(I,g)) \ge \frac{1}{2}\omega(Z, I, D(I,g)) \ge c$$

for g satisfying the hypotheses of Lemma 3 (apply (1.4) with $E = I - H, J = I$). With θ_3 now fixed, we shall prove Lemma 4 by constructing a function g, satisfying the hypotheses of Lemma 3, with $f \equiv g$ on a closed set $E \subseteq I$, and $|E|/|I| \ge 1 - \theta_3$. Using Lemma 3 and the fact that $|F \cap E| \ge (1 - 2\theta_3)|I|$, it then follows from Lemma 2 with $f = f_1$, $g = f_2$, and the above inequality with $H = F \cap E$, that Lemma 4 is valid. For the moment we allow θ_2 to vary. From (2.6) and the usual weak type estimates we find that

$$\alpha\left|\left\{t : \int_I \left(\frac{(f(s) - f(t)}{s - t}\right)^2 ds > \alpha\right\}\right| \le c\theta_2|I|.$$

Hence if α is large enough, there exists a closed set $E \subset I$, with $|E| \geq (1-\theta_3)|I|$, and

$$\int_I \left(\frac{f(s)-f(t)}{s-t}\right)^2 ds \leq c\theta_2, \quad t \in E. \tag{2.7}$$

If $0 < \rho < |I|$, then from (2.7)

$$\int_{\{s\in I:\rho/2<|s-t|<\rho\}} \left(\frac{f(s)-f(t)}{s-t}\right)^2 ds \leq c\theta_2, \quad t \in E.$$

Hence, from weak type estimates, for fixed $t \in E$

$$|f(s)-f(t)| \leq c(\theta_2 A^2 \rho^2)^{1/4},$$

when $\rho/2 < |s-t| < \rho$, except for s in a set of Lebesgue measure at most, $(\theta_2^{1/2}\rho)/A$. Since, $\|f\|_I \leq A$, it follows that

$$|f(s)-f(t)| \leq c(\theta_2 A^2 \rho^2)^{1/4}, \quad t \in E,$$

for all s, $\rho/2 < |s-t| < \rho$. We conclude from the arbitrariness of $\rho, 0 < \rho < |I|$, that

$$|f(s)-f(t)| \leq c(A^2\theta_2)^{1/4}|s-t|^{1/2}, \quad t \in E, s \in I. \tag{2.8}$$

Let $I - E = \cup I_k$, where $\{I_k\}$ are the components of $I - E$. Define g on I as follows: $g = f$ on E, and if $I_k = [a_k, b_k]$, then

$$g(t) = \frac{(f(b_k)-f(a_k))}{(b_k-a_k)}(t-a_k) + f(a_k), \quad a_k \leq t \leq b_k.$$

That is, g is linear on each I_k and equal to f at the endpoints of I_k. We claim that

$$\max\{\|g\|_I, \|g\|_I^{\wedge}\} \leq c(A^2\theta_2)^{1/4}. \tag{2.9}$$

(2.9) for $\|g\|_I$ follows easily from the definition of g and (2.8). To prove (2.9) for $\|g\|_I^{\wedge}$, suppose J is a subinterval of I. From (2.8) we deduce,

$$|g'| \leq c(A^2\theta_2)^{1/4}|I_k|^{-1/2}. \tag{2.10}$$

Thus,

$$\int_{I_k\cap J}\int_{I_k\cap J} \frac{(g(s)-g(t))^2}{(s-t)^2} dsdt \leq cA\theta_2^{1/2}|I_k \cap J|. \tag{2.11}$$

Moreover from (2.9) for $\|g\|_I$ we have with $d_k(t)$ the distance from $t \in I_k \cap J$ to $\mathbf{R} - (I_k \cap J)$,

$$\begin{aligned}
&\int\limits_{I_k\cap J}\int\limits_{J\cap(4I_k - I_k)} \frac{(g(s)-g(t))^2}{(s-t)^2} ds dt \\
&\le cA\theta_2^{1/2} \int\limits_{I_k\cap J} \left[\int\limits_{\{d_k(t)<|s-t|<4|J\cap 4I_k|\}} \frac{ds}{|s-t|} \right] dt \\
&\le cA\theta_2^{1/2} \int\limits_{|I_k\cap J|} \log\left(\frac{4|J\cap 4I_k|}{d_k(t)}\right) dt \\
&\le cA\theta_2^{1/2} |J \cap 4I_k|.
\end{aligned}$$

This inequality and (2.11) give

$$\int\limits_{I_k\cap J}\int\limits_{4I_k\cap J} \frac{(g(s)-g(t))^2}{(s-t)^2} ds dt \le cA\theta_2^{1/2} |4I_k \cap J|. \tag{2.12}$$

Similarly,

$$\int\limits_{I_k\cap J} \left(\int\limits_J \frac{(g(s)-g(t))^2}{(s-t)^2} ds \right) dt \le cA\theta_2^{1/2} |J|. \tag{2.13}$$

Now,

$$\begin{aligned}
\int\limits_J\int\limits_J \frac{(g(s)-g(t))^2}{(s-t)^2} ds dt &= \int\limits_{E\cap J}\int\limits_{E\cap J} \frac{(g(s)-g(t))^2}{(s-t)^2} ds dt \\
&+ 2\sum_k \int\limits_{I_k\cap J}\int\limits_{E\cap J} \frac{(g(s)-g(t))^2}{(s-t)^2} ds dt \\
&+ \sum_{k,l} \int\limits_{I_k\cap J}\int\limits_{I_l\cap J} \frac{(g(s)-g(t))^2}{(s-t)^2} ds dt \\
&= T_1 + T_2 + T_3.
\end{aligned} \tag{2.14}$$

Since $g = f$ on E we deduce from (2.7) that

$$T_1 \le c\theta_2 |J|. \tag{2.15}$$

To estimate T_2, let $G = \{I_k : I_k \subseteq J\}$. Since $\{I_k\}$ is made up of disjoint intervals we see that $\{I_k : I_k \cap J \neq \phi\} - G$ consists of at most two intervals. This fact and

(2.7)-(2.13) give

$$
\begin{aligned}
T_2 &\le cA\theta_2^{1/2}|J| + \sum_{I_k\in G}\int_{I_k}\left[\int_{(E-4I_k)\cap J}\frac{(g(s)-g(t))^2}{(s-t)^2}ds\right]dt\\
&\le cA\theta_2^{1/2}|J| + c\int_J\left[\int_{J\cap E}\left(\frac{f(s)-f(t)}{s-t}\right)^2 ds\right]dt\\
&\quad+\sum_{I_k\in G} c\int_{I_k}(f(t)-g(t))^2\left[\int_{\{|s-t|>2|I_k|\}}\frac{ds}{(s-t)^2}\right]dt\\
&\le cA\theta_2^{1/2}|J| + c\theta_2|J|\\
&\quad+cA\theta_2^{1/2}|J| = cA\theta_2^{1/2}|J|.
\end{aligned}\tag{2.16}
$$

Next let $G(k)=\{I_l : |I_l|\le|I_k|\}$. Then from (2.7)-(2.13), we deduce

$$
\begin{aligned}
T_3 &\le cA\theta_2^{1/2}|J| + \sum_{I_k\in G}\sum_{I_l\in G(k)}\int_{I_k}\int_{I_l}\frac{(g(s)-g(t))^2}{(s-t)^2}dsdt\\
&\le cA\theta_2^{1/2}|J| + 8\sum_{I_k\in G}\sum_{I_l\in G(k)}\left(\int_{I_k}\int_{I_l-4I_k}\frac{(f(s)-f(t))^2}{(s-t)^2}dsdt\right)\\
&\quad+8\sum_{I_k\in G}\sum_{I_l\in G(k)}\int_{I_k}\left(\int_{I_l-4I_k}\frac{(g(t)-f(t))^2}{(s-t)^2}ds\right)dt\\
&\quad+8\sum_{I_k\in G}\sum_{I_l\in G(k)}\int_{I_k}\left(\int_{I_l-4I_k}\frac{(g(s)-f(s))^2}{(s-t)^2}ds\right)dt\\
&= cA\theta_2^{1/2}|J| + P_1+P_2+P_3.
\end{aligned}\tag{2.17}
$$

To estimate P_1 choose $t_k\in\bar{I}_k\cap E$. Then from (2.8) we see that

(2.18)

$$
\begin{aligned}
&\int_{I_k}\left(\int_{I_l-4I_k}\frac{(f(s)-f(t)^2}{(s-t)^2}ds\right)dt\\
&\le 2\int_{I_k}\left(\int_{I_l-4I_k}\frac{(f(s)-f(t_k))^2}{(s-t)^2}ds\right)dt + 2\int_{I_k}\left(\int_{I_l-4I_k}\frac{(f(t)-f(t_k))^2}{(s-t)^2}ds\right)dt\\
&\le c\int_{I_k}\int_{I_l-4I_k}\frac{(f(s)-f(t_k))^2}{(s-t)^2}dsdt + cA\theta_2^{1/2}|I_k|^2\left(\int_{I_l-4I_k}\frac{ds}{(s-t_k)^2}\right).
\end{aligned}
$$

Summing first with respect to l in (2.18), using (2.7), integrating, and then summing with respect to k, we get

$$P_1\le cA\theta_2^{1/2}|J|.\tag{2.19}$$

Using (2.8) and (2.9) for $\|g\|_I$, we deduce

$$\begin{aligned}P_2 &\le cA\theta_2^{1/2} \sum_{I_k \in G} \sum_{I_l \in G(k)} |I_k| \int_{I_k} \left(\int_{I_1 - 4I_k} \frac{ds}{(s-t)^2} \right) dt \\ &\le cA\theta_2^{1/2} |J|.\end{aligned} \tag{2.20}$$

Finally to estimate P_3 note that the inside integral in P_3 is nonzero only if I_k lies at least $|I_l|$ distance from I_l, since $|I_l| \le |I_k|$. Using this fact and summing first with respect to k in P_3 it follows that

$$\begin{aligned}P_3 &\le \sum_{I_l \in G} \int_{I_l} \sum_{I_k \in G} \left(\int_{(I_k - 2I_l)} cA\theta_2^{1/2} \frac{|I_l|}{(s-t)^2} dt \right) ds \\ &\le \sum_{I_l \in G} cA\theta_2^{1/2} |I_l| \le cA\theta_2^{1/2} |J|.\end{aligned} \tag{2.21}$$

From (2.17)-(2.21), we find that

$$T_3 \le cA\theta_2^{1/2} |J|. \tag{2.22}$$

Combining (2.22), (2.16), and (2.15), we obtain from (2.14) that

$$\int_j \int_j \frac{(g(s) - g(t))^2}{(s-t)^2} dsdt \le cA\theta_2^{1/2} |J|. \tag{2.23}$$

From (2.23) we conclude that (2.9) is also valid for $\|g\|_{\hat{I}}$. If θ_2 is sufficiently small, we observe that Lemma 3 can be applied to g. With $\theta_2 > 0$ now fixed we see from our earlier remark that Lemma 4 is true.

3. Proof of Theorems 1–3. Theorem 1 is an easy consequence of the following lemma.

LEMMA 5. *Let f be defined on I with $\|f\|_I \le A < +\infty$, $A \ge 1$, and suppose that*

$$|I|^{-1} \int_I \int_I \left(\frac{(f(s) - f(t))}{s-t} \right)^2 dsdt \le a \le A^2. \tag{3.1}$$

Then there exists, $\delta = \delta(a, A)$, $\eta = \eta(a, A)$, $0 < \delta, \eta < 1$, such that whenever F is a Borel subset of I with $|F|/|I| \ge 1 - \delta$, then, $\omega(Z, F) \ge \eta$.

Proof. We prove Lemma 5 by an induction type argument on a. From Lemma 4 we see that Lemma 5 holds with $\delta = \theta_3(A)$, $\eta = \theta_4(A)$, provided $0 < a \le \theta_2(A)$. Suppose that we have proved Lemma 5 for $0 < a \le a_1 < A^2$, $a_1 \ge \theta_2(A)$. We shall need the following remark:

Remark. Let g be defined on I with $\|g\|_1 \leq A < +\infty$, and $\|g\|_I^\wedge \leq a_1^{1/2}$. Then we can apply the induction hypothesis to each closed subinterval $J \subseteq I$ and with f replaced by g. If H is a Borel subset of J with $|H|/|J| \leq \delta(a_1, A)$, then from the induction hypothesis with $F = J - H$, we have

$$\omega[Z_1(J), J - H, D(J, A, g)] \geq \eta(a_1, A),$$

where $Z_1 = Z(\cdot, A, g)$. Let $g^*, \tilde{g}$, denote the extension of g to $2I, 2J$, respectively. From the above inequality, and Lemma 2 with $g^* = f_2, \tilde{g} = f_1, J = I$, we see that

$$\omega[Z_1(J), J - H, D^*(J, A, g)] \geq c(A)\eta(a_1, A)^\beta,$$

for some $\beta = \beta(A) > 0$. Here, D^* is defined as in Lemma 2.

This inequality, Lemma 1, and the maximum principle for the heat equation yield

$$\omega[Z_1(I), J - H, D(I, A, g)] \geq c(a_1, A)\omega[Z_1(I), J, D(I, A, g)].$$

Hence

$$\omega[Z_1(I), H, D(I, A, g)] \leq [1 - c(a_1, A)]\omega[Z_1(I), J, D(I, A, g)]$$

and $\omega[Z_1(I), \cdot, D(I, A, g)]$ is an A_∞ weight with respect to Lebesgue measure on I.

To continue the proof of Lemma 5 we put, $\epsilon = \left(\frac{\theta_2}{c_1 A}\right)^{80}$, and shall show for c_1 large enough and $a_2 = \min[A, a_1(1 + \epsilon)]$, that Lemma 5 is valid for $a \leq a_2$, provided $\delta(a, A)$, $\eta(a, A)$ are defined suitably when $a_1 \leq a \leq a_2$. It then follows from induction that Lemma 5 is true. Let M be the family of all dyadic subintervals of I obtained by the usual bisection method. Let I', I'', be the intervals obtained from bisecting I once. If $J = I'$ or $J = I''$ and

$$\int_{(I-J)} \int_J \frac{(f(s) - f(t))^2}{(s - t)^2} ds dt \geq 20\epsilon a_1 |J|, \tag{3.2}$$

then from (3.1),

$$|I|(1 + \epsilon)a_1 \geq \int_I \int_I \left(\frac{f(s) - f(t)}{s - t}\right)^2 ds dt = \int_{(I-J)} \int_{(I-J)} \cdots + 2 \int_{(I-J)} \int_J \cdots + \int_J \int_J \cdot$$
$$\geq \int_{(I-J)} \int_{(I-J)} \cdots + 40\epsilon a_1 |J| + \int_J \int_J \cdots$$

Hence, either

$$\int_{I'} \int_{I'} \left(\frac{f(s) - f(t)}{s - t}\right)^2 ds dt \leq (1 - 18\epsilon)a_1 |I'|$$

or this inequality holds with I' replaced by I''. We can then take $\delta(a, A) = \frac{1}{2}\delta(a_1, A)$ for $a_1 \leq a \leq a_2$, and apply the induction assumption to conclude Lemma 5 is valid for $\omega(Z(J), \cdot, D(J, A, f))$ when either $J = I'$ or $J = I''$.

Using Lemmas 1 and 2, we then obtain Lemma 5 for $a_1 \le a \le a_2$. Thus, we assume (3.2) is false for both $J = I'$ and $J = I''$. Let $M' \subseteq M$, $M' = \{I_k\}$, be the collection of maximal closed dyadic subintervals of I such that (3.2) holds whenever $J \in M'$. We note that if $I_k \in M'$ is obtained from bisecting $\widetilde{I}_k$ (once), then

$$\int\limits_{I-\widetilde{I}_k}\int\limits_{\widetilde{I}_k} \frac{(f(s)-f(t))^2}{(s-t)^2}\,dtds \le 20\epsilon a_1|\widetilde{I}_k|, \tag{3.3}$$

since otherwise I_k would not be maximal. (3.3) implies as in the proof of (2.8) for $t \in I_k$, and $s \in I$ that

$$|f(s)-f(t)| \le c\epsilon^{1/10}\max[\,|I_k|^{1/2}, |s-t|^{1/2}\,]. \tag{3.4}$$

Indeed, from the usual weak type estimates, we deduce from (3.3) that

$$\int\limits_{I-\widetilde{I}_k} \frac{(f(s)-f(t))^2}{(s-t)^2}\,ds \le c(\epsilon a_1)^{1/2}A, \tag{3.5}$$

outside of a set $H \subseteq I_k$ with $|H| \le \frac{(\epsilon a_1)^{1/2}}{A}|I_k|$.

If $t_0 \in H$, then from (3.5), $\|f\|_I \le A$, and the definition of ϵ, we see for property chosen $t \in I_k - H$,

$$\begin{aligned}\int\limits_{I-\widetilde{I}_k} \frac{(f(s)-f(t_0))^2}{(s-t_0)^2}\,ds &\le 2\int\limits_{I-\widetilde{I}_k} \frac{(f(s)-f(t))^2}{(s-t)^2}\,ds \\ &\quad + cA(a_1\epsilon)^{1/2}|I_k|\int\limits_{I-\widetilde{I}_k}\frac{ds}{(s-t)^2} \le c(\epsilon a_1)^{1/2}A.\end{aligned} \tag{3.6}$$

So (3.6) holds whenever $t_0 \in I_k$. Using (3.6), the definition of ϵ, and essentially repeating the above argument, we obtain (3.4).

Put $E = I - \left(\bigcup_{I_k \in M'} I_k\right)$. Then (3.3) holds with $\widetilde{I}_k$ replaced by J whenever $J \cap E \ne \{\phi\}$ and $J \in M$. Now for almost every $t_0 \in I$, $0 = \lim_{\tau \to t_0} \frac{|f(\tau)-f(t_0)|}{|\tau - t_0|^{1/2}}$, as follows easily from (3.1) and $\|f\|_I \le A$. Using this fact, arguing as in (3.6), and taking a limit as $|J| \to 0$, we see for almost every $t_0 \in E$ that

$$\int\limits_I \frac{(f(s)-f(t_0))^2}{(s-t_0)^2}\,ds \le 20\epsilon a_1. \tag{3.7}$$

Since the integral in (3.7) is uniformly bounded on E, it follows from dominated convergence that (3.7) holds for every $t_0 \in E$. From (3.7) we find, as in the proof of (2.8) that

$$|f(s)-f(t)| \le c\epsilon^{1/10}|s-t|^{1/2}, \quad t \in E, \quad s \in I. \tag{3.8}$$

Next suppose that S is a finite subcollection of M' and L a closed subset of E with

$$\left(\sum_{I_k \in S} |I_k|\right) + |L| \geq \epsilon|I|. \tag{3.9}$$

Under this assumption we shall show that

$$\omega\left[Z, L \bigcup \left(\bigcup_{I_k \in S} I_k\right)\right] \geq \xi > 0 \tag{3.10}$$

for some $\xi = \xi(a_1, A)$. To do this define g on I by $g = f$ on E, and g is linear on each $I_k \in M'$ with $g = f$ at the endpoints of I_k. We claim that

$$\|g\|_I \leq c\epsilon^{1/10}, \quad \|g\|_I^\wedge \leq c\epsilon^{1/10}. \tag{3.11}$$

The proof of (3.11) for $\|g\|_I$ follows from (3.8), (3.4). The proof of (3.11) for $\|g\|_I^\wedge$ is similar to the proof of (2.9). Therefore, we shall only sketch the proof, indicating the necessary changes. The essential difference is that now E may be a long way from some I_k. This possibility, however, is compensated for by (3.3) and (3.4). From (3.4) and (3.8) we find that (2.11)-(2.14) hold with $A\theta_2^{1/2}$ replaced by, $\epsilon^{1/5}$. From (3.7), (3.3) we get (2.15), (2.16), with ϵa_1 replacing θ_2 in (2.15) and $\epsilon^{1/5}$ replacing $A\theta_2^{1/2}$ in (2.16). T_3 is estimated in the same way, as previously, except for the term involving P_1. In (2.17) we now estimate P_1 using (3.3) rather than $\{t_k\} \subseteq E$. Hence, (3.11) is true. From (3.11) we see that if c_1 is large enough, we can apply the induction hypothesis to g. Doing this, we obtain from our earlier remark that $\omega_1 = \omega(Z_1(I), \cdot, D(I, A, g))$ is an A_∞ weight with respect to Lebesgue measure on I. From (1.7), the fact that $f = g$ on L, and Lemmas 1-2 we conclude that if $|L| \geq \frac{1}{2}\epsilon$, then (3.10) holds. Otherwise, let $\{J_k\}$ be the components of $\bigcup_{I_k \in S} I_k$ and choose a subset T of $\{J_k\}$ such that

$$\bigcup_{I_k \in S} I_k \subseteq \bigcup_{J_k \in T} 16J_k,$$

and $J_k \cap 4J_l = \{\phi\}$, when $k \neq l$, $J_k, J_l \in T$. Let $\bar{g} = f$ on $\bigcup_{J \in T} J$ and $\bar{g} = g$ otherwise on I. Clearly, $\|\bar{g}\|_I \leq 3A$. Let

$$\bar{\omega} = \omega[Z(I, A, \bar{g}), \cdot, D(I, A, \bar{g})],$$
$$\omega = \omega[Z(I, A, f), \cdot, D(I, A, f)].$$

Then from Lemma 2 we deduce

$$c(A)\omega\left(\bigcup_{J \in T} J\right) \geq \bar{\omega}\left(\bigcup_{J \in T} J\right)^\beta \tag{3.12}$$

for some $\beta = \beta(A) \geq 1$. From the doubling property of $\bar{\omega}$ in (2.1) of Lemma 1, we find that if

$$V = \left[\bigcup_{J \in T} (2J - J) \right] \cap I$$

then

$$\bar{\omega}(V) \leq c(A)\bar{\omega}\left(\bigcup_{J \in T} J \right). \tag{3.13}$$

Since $\bar{g} = g$ on V, we can again use Lemma 2 to obtain

$$c(A)\bar{\omega}(V) \geq \omega_1(V)^{\beta} \tag{3.14}$$

Moreover, since $|V| \geq (\epsilon/64)|I|$, we can again use (1.7) to get, $\omega_1(V) \geq c(a_1, A)$. From (3.12)-(3.14) we now conclude that (3.10) is true.

Finally we use (3.10) to prove Lemma 5 for $a_1 \leq a \leq a_2$. Let $\delta(a, A) = \frac{\epsilon\delta(a_1, A)}{100}$, when $a_1 \leq a \leq a_2$. Next let S^* be the set of all $J \in M'$ such that

$$\int_J \int_J \frac{(f(s) - f(t))^2}{(s-t)^2} dsdt \leq a_1|J|.$$

Then from (3.1) and (3.2) we deduce,

$$\begin{aligned} 20\epsilon a_1 \sum_{J \in M'} |J| + \sum_{J \in M'} \int_J \int_J \frac{(f(s) - f(t))^2}{(s-t)^2} dsdt \\ \leq \int_I \int_I \frac{(f(s) - f(t))^2}{(s-t)^2} dsdt \leq (1+\epsilon) a_1 |I|. \end{aligned}$$

From this inequality we see for $0 < \epsilon < 1/100$, that either $|E| \geq 2\epsilon|I|$ or

$$\sum_{J \in M'} \int_J \int_J \frac{(f(s) - f(t))^2}{(s-t)^2} dsdt \leq (1 - 12\epsilon) a_1 |I|. \tag{3.15}$$

If (3.15) holds, then clearly

$$a_1 \left(\sum_{J \in M' - S^*} |J| \right) \leq (1 - 12\epsilon) a_1 |I|.$$

Hence if $|E| \leq 2\epsilon|I|$, then

$$\sum_{J \in S^*} |J| \geq 10\epsilon|I|. \tag{3.16}$$

Let F be as in Lemma 5 and observe from our choice of $\delta(a, A)$, $a_1 \leq a \leq a_2$, that if $|E| \geq 2\epsilon|I|$, then there exists a closed set $L \subseteq F \cap E$ with $|L| \geq \epsilon|I|$. Using (3.10), we see that Lemma 5 is valid in this case. If (3.16) holds, let S' be the subcollection of S^* with

$$(3.17) \qquad |F \cap J| \geq (1 - \delta(a_1, A))|J|, \quad J \in S'.$$

From our choice of $\delta(a, A)$, $a_1 \leq a \leq a_2$, and (3.16) we deduce for some finite subcollection S of S'

$$(3.18) \qquad \sum_{J \in S} |J| \geq \in |I|.$$

We now apply the induction hypothesis to f restricted to each $J \in S$. From the induction hypothesis, the definition of S^*, and (3.17), we deduce for $J \in S$,

$$\omega[Z(J), F \cap J, D(J, f)] \geq \eta(a_1, A).$$

From Lemma 1-2, and the maximum principle for the heat equation, it follows that

$$(3.19) \qquad \omega[Z(I), F \cap J, D(I, f)] \geq \eta_1(a, A)\omega[Z(I), J, D(I, f)],$$

where

$$c(A)\eta_1(a, A) = \eta(a_1, A)^{\beta}, \quad a_1 \leq a \leq a_2.$$

Summing (3.19), using (3.18) and (3.10), we get

$$\omega(F) \geq \eta_1(a, A)\omega\left(\bigcup_{J \in S} J\right) \geq \eta(a, A).$$

Hence, Lemma 5 is true when $|E| < 2\epsilon|I|$ and $a \leq a_2$. Since we have already considered the case when $|E| \geq 2\epsilon|I|$, we conclude by induction that Lemma 5 is true.

To prove Theorem 1 we use Lemma 5 on each subinterval $J \subseteq I$. Theorem 1 is then obtained, as in the remark for g in Lemma 5. We omit the details. To prove Theorem 2 suppose $|F| > 0$. Choose B large enough so that $\|f\|_I \leq B$ and there exists E closed, $E \subseteq F$, with

$$\int_I \frac{(f(s) - f(t))^2}{(s - t)^2} ds \leq B, \quad t \in E,$$

and $|E| \geq \frac{1}{2}|F|$. Put $g(t) = f(t)$, $t \in E$, and define g to be linear on each component of $I - E$ with $g = f$ at the endpoints of each component. Arguing as in the proof of Lemma 4, we find that

$$(3.20) \qquad \max\{\|g\|_I, \quad \|g\|_I^{\wedge}\} < +\infty.$$

From Theorem 1 and (1.7) with f replaced by g, we get

$$(3.21) \qquad \omega[Z(I, g), E, D(I, g)] \geq 0$$

Since $f = g$ on E we can use Lemmas 1-2, as previously, to conclude the above inequality also holds with g replaced by f. This completes the proof of Theorem 2. □

Finally to prove Theorem 3, suppose E is a closed subset of

$$\left\{t : \int_I \left(\frac{f(s)-f(t)}{s-t}\right)^2 ds \leq B < +\infty\right\}.$$

Define g as previously. Then (3.20) holds and we can again apply Theorem 1 to g. From (1.4) (for some $p > 1$) we see that if (3.21) is valid, then $|E| > 0$. From Lemmas 1-2, we now deduce that this conclusion is also valid with g in (3.21) replaced by f. Hence Theorem 3 is true.

REFERENCES

[1] R. Brown, *Area integral estimates for caloric functions*, Trans. Amer. Math. Soc. 315 (1989), pp. 565–589.

[2] R. Coifman and C. Fefferman, *Weighted norm inequalities for maximal functions and singular integrals*, Studia Math. 51 (1974), pp. 241–250.

[3] B. Dahlberg, C. Kenig, and D. Jenison, *Area integral estimates for elliptic differential equations with nonsmooth coefficients*, Ark Mat 22 (1984), pp. 97–108.

[4] R. Fefferman, C. Kenig, and J. Pipher, *The theory of weights and the Dirichlet problem for elliptic equations*, to appear.

[5] J. Journé, *Calderón-Zymund operators, pseudo-differential operators, and the Cauchy integral of Calderón*, Springer lecture notes in mathematics 994 (1983), Springer-Verlag.

[6] R. Kaufman and J.M. Wu, *Parabolic measure on domains of class* $Lip_{1/2}$, Compositio Mathematics 65 (1988), pp. 201–207.

[7] J. Lewis and J. Silver, *Parabolic measure and the Dirichlet problem for the heat equation in two dimensions*, Indiana Univ. Math. J. 37 (1988), pp. 801–839.

[8] M. Murray, *Commutators with fractional differentiation and BMO Sobolev spaces*, Indiana Univ. Math. J. 34 (1985), pp. 205–215.

[9] R. Strichartz, *Bounded mean oscillation and Sobolev spaces*, Indiana Univ. Math. J. 29 (1980), pp. 539–558.

[10] J.M. Wu, *On parabolic measures and subparabolic functions*, Trans. Amer. Math. Soc. 251 (1979), pp. 171–185.

SOME INEQUALITIES FOR THE DENSITY OF THE AREA INTEGRAL

CHARLES N. MOORE*

1. Introduction: results from probability. Let X_t be a continuous martingale starting at 0 and set $X^* = \sup\{|X_t| : t > 0\}$ and $S(X) = \sqrt{\langle X\rangle_\infty}$, where $\langle X\rangle_t$ is the quadratic variation process at time t. We then define a measure μ on $\mathbf{R}$ by $\mu(E) = d\langle X\rangle_t(\{t : X_t \in E\})$, where $d\langle X\rangle_t$ is the Riemann–Stieltjes measure on $[0,\infty)$ associated to the nondecreasing function $\langle X\rangle_t$. It is known that μ is absolutely continuous with respect to Lebesgue measure so that there exists a function $L(a)$, (called the local time), so that $\mu(E) = \int_E L(a)da$ for every Borel set E. More generally, for any Borel function f on $\mathbf{R}$:

$$\int_0^\infty f(X_t)\, d\langle X\rangle_t = \int_{\mathbf{R}} f(a)L(a)da \tag{1.1}$$

Take $f \equiv 1$ in (1.1) and we obtain:

$$S(X)^2 = \int_{\mathbf{R}} L(a)da = \int_{-X^*}^{X^*} L(a)da \tag{1.2}$$

where the last equality follows by noting that $L(a) = 0$ if $a \notin [-X^*, X^*]$. We now set $L^* = \sup\{L(a) : a \in \mathbf{R}\}$; L^* is called the maximal local time. Then (1.2) trivially yields $S(X)^2 \leq 2L^*X^*$, and this, the Cauchy–Schwarz inequality, and the Burkholder–Gundy inequality: $\|S(X)\|_p \approx \|X^*\|_p$ for $0 < p < \infty$, then gives $\|S(X)\|_p \leq C_p\|L^*\|_p$ for $0 < p < \infty$. The reverse inequality is more difficult and was shown by Barlow and Yor [4], [5]. In fact, even more is true. The following theorem is essentially to Bass [6] and independently, Davis [12], however, their statements of these results are different than appears here, but a careful analysis of their methods yields these results.

THEOREM A. *There exists constants c_1 and c_2 such that for all $\lambda > 0$ and all $0 < \varepsilon < 1$,*

(a) $P\{X^* > 2\lambda,\ L^* \leq \varepsilon\lambda\} \leq c_1 \exp\left(\frac{-c_2}{\varepsilon}\right) P\{X^* > \lambda\}$

(b) $P\{S(X) > 2\lambda,\ L^* \leq \varepsilon\lambda\} \leq c_1 \exp\left(\frac{-c_2}{\varepsilon^2}\right) P\{S(X) > \lambda\}$

(c) $P\{L^* > 2\lambda,\ X^* \leq \varepsilon\lambda\} \leq c_1 \exp\left(\frac{-c_2}{\varepsilon}\right) P\{L^* > \lambda\}$

(d) $P\{L^* > 2\lambda,\ S(X) \leq \varepsilon\lambda\} \leq c_1 \exp\left(\frac{-c_2}{\varepsilon^2}\right) P\{L^* > \lambda\}$

A consequence of these is the standard corollary to such good-λ inequalities. (See [18] for a proof).

*Department of Mathematics, Washington University, St. Louis, MO 63130. Current address: Department of Mathematics, Kansas State University, Manhattan, KS 66506. Partially supported by a National Science Foundation Postdoctoral Fellowship.

COROLLARY. *Suppose Φ is an increasing function on $[0,\infty)$ with $\Phi(0)=0$ and suppose that there exists a constant C such $\Phi(2\lambda)\le C\Phi(\lambda)$ for all $\lambda>0$. Then $E(\Phi(X^*))\approx E(\Phi(S(X)))\approx E(\Phi(L^*))$, with all equivalences holding with constants which depend only on c_1, c_2 and C.*

Of course, the most important special case is $\Phi(\lambda)=\lambda^p$ where $0<p<\infty$.

2. Harmonic analysis analogues. Let $u(s,t)$ be a harmonic function on $\mathbf{R}^{n+1}_+$. For $x\in\mathbf{R}$ and $\alpha>0$, set $\Gamma_\alpha(x)=\{(s,t):|x-s|<\alpha t\}$, and set

$$\begin{aligned} N_\alpha u(x) &= \sup\{|u(s,t)|:(s,t)\in\Gamma_\alpha(x)\} \\ A_\alpha u(x) &= \left(\int_{\Gamma_\alpha(x)}|\nabla u(s,t)|^2 t^{1-n}dsdt\right)^{\frac{1}{2}} \end{aligned} \tag{2.1}$$

The nontangential maximal function $N_\alpha u(x)$, and the Lusin area function $A_\alpha u(x)$ are the well known harmonic analysis analogues of X^* and $S(X)$. Gundy [15] proposed the following analogue of L^*. Let $r\in\mathbf{R}$. Then $(u(s,t)-r)^+$ is subharmonic on $\mathbf{R}^{n+1}_+$, so $\Delta(u(s,t)-r)^+$ is a positive measure on $\mathbf{R}^{n+1}_+$. We then set

$$\begin{aligned} D_\alpha u(x;r) &= \int_{\Gamma_\alpha(x)}\Delta(u(s,t)-r)^+ t^{1-n}(dsdt) \\ D_\alpha u(x) &= \sup_{r\in\mathbf{R}} D_\alpha u(x;r) \end{aligned} \tag{2.2}$$

In [17], Gundy and Silverstein show the following change of variables formula:

$$\begin{aligned} &\iint_{\mathbf{R}^{n+1}_+}\Psi(s,t)f(u(s,t))|\nabla u(s,t)|^2dsdt \\ &\quad=\int_{\mathbf{R}}\int_{\mathbf{R}^{n+1}_+}\Psi(s,t)f(r)\Delta(u(s,t)-r)^+(dsdt)dr \end{aligned} \tag{2.3}$$

whenever $\Psi(s,t)$ and $f(r)$ are Borel functions in $\mathbf{R}^{n+1}_+$ and $\mathbf{R}$.
Take $\Psi(s,t)=t^{1-n}\chi_{\Gamma_\alpha(x)}(s,t)$ in (2.3) and we obtain an analogue of (1.1); if we also set $f\equiv 1$ in (2.3), then we obtain an analogue of (1.2). It is for this reason that we call $D_\alpha u(x;r)$ the density of the area integral and $D_\alpha u(x)$ the maximal density. Again, with this choice of Ψ and f in (2.3), and reasoning as in the case of martingales, we obtain the inequality $\|A_\alpha u\|_p\le C_p\|D_\alpha u\|_p$, $0<p<\infty$. As before, the reverse inequality is more difficult. This was done by Gundy [15] when $n=2$, and Gundy and Silverstein [17] (see also Gundy [16]) in all dimensions. Our purpose is to present harmonic analysis analogues of Theorem A. Most of what we state is from Bañuelos and Moore [3].

To describe these results in their greatest generality, we need some more notation. For $(x,y)\in\mathbf{R}^{n+1}$ and $\alpha>0$, set $\Gamma_\alpha(x,y)=\{(s,t):|x-s|<\alpha(t-y)\}$, which is simply a cone with vertex (x,y). If u is harmonic on $\Gamma_\alpha(x,y)$ we define $N_\alpha u(x,y)$ as in (2.1), but with the sup taken over $\Gamma_\alpha(x,y)$. Likewise, we define $A_\alpha u(x,y)$, $D_\alpha u((x,y);r)$, and $D_\alpha u(x,y)$ as in (2.1) and (2.2), but now we integrate over $\Gamma_\alpha(x,y)$ and replace the factor t^{1-n} by $(t-y)^{1-n}$. Then,

THEOREM 1. *Let $\Phi : \mathbf{R}^n \to \mathbf{R}$ be a Lipschitz function with Lipschitz constant M. Let $D = \{(x,y) : x \in \mathbf{R}^n, y > \Phi(x))\}$ be the Lipschitz domain above the graph of Φ. Suppose u is harmonic in D and $0 < \alpha < \beta$. There exists constants k, c_1 and c_2 all depending only on α, β, n and M such that if $\lambda > 0, 0 < \varepsilon < 1$, then*

(a)

$$|\{x \in \mathbf{R}^n : N_\alpha u(x, \Phi(x)) > k\lambda,\ D_\beta u(x, \Phi(x)) \le \varepsilon\lambda\}| \\ \le c_1 \exp\left(\frac{-c_2}{\varepsilon}\right) |\{x \in \mathbf{R}^n : N_\alpha u(x, \Phi(x)) > \lambda\}|$$

(b)

$$|\{x \in \mathbf{R}^n : A_\alpha u(x, \Phi(x)) > k\lambda,\ D_\beta u(x, \Phi(x)) \le \varepsilon\lambda\}| \\ \le c_1 \exp\left(\frac{-c_2}{\varepsilon^2}\right) |\{x \in \mathbf{R}^n : A_\alpha u(x, \Phi(x)) > \lambda\}|$$

For the proof of Theorem 1, we refer the reader to [3]. The proof consists mainly of Green's theorem arguments on subdomains of D constructed over the graphs of cubes in $\mathbf{R}^n$ and is in this regard similar to numerous proofs of similar good-λ inequalities. (See, for example, Burkholder and Gundy [9], Dahlberg [11], Fefferman, Gundy, Silverstein, and Stein [13], Murai and Uchiyama [19] or Bañuelos and Moore [1].) Such Green's theorem arguments give rise to boundary terms which must be controlled; this is usually done by a lemma like the one which follows.

LEMMA. *Suppose $\beta > \alpha$. There exists a constant $C = C(\alpha, \beta, n)$ such that if u is harmonic on $\Gamma_\beta(x)$ then $t|\nabla u(s,t)| \le C D_\beta u(x)$ for all $(s,t) \in \Gamma_\alpha(x)$.*

Once again, we refer the reader to [3] for the proof. We remark that this is an analogue of a lemma in Stein [20], pg 207, which has the same result but with $A_\beta u$ and $N_\beta u$ replacing $D_\beta u$. For $A_\beta u$ and $N_\beta u$ this lemma is an easy consequence of elementary properties of harmonic functions, however, the result for $D_\beta u$ seems to be harder.

To show analogues of Theorem A, (c),(d) in the harmonic analysis setting, we need to consider a different version of the maximal density which is defined in a "smoother" way, but is at any rate comparable to the previous version. Let Ψ be a smooth, positive, radially symmetric function supported on $B(0, \rho)$. Then set

$$\begin{aligned} Du(x;r) &= \int_{\mathbf{R}^{n+1}_+} t\Psi_t(x-s)\Delta(u(s,t)-r)^+(dsdt), \qquad r \in \mathbf{R} \\ Du(x) &= \sup_{r \in \mathbf{R}} Du(x;r) \end{aligned} \tag{2.4}$$

Of course, $Du(x)$ depends on the choice of Ψ and should probably be denoted by something like $D_\Psi u(x)$, but to avoid cumbersome notation we consider Ψ fixed in all that follows, and just indicate the maximal density by $Du(x)$.

THEOREM 2. *Suppose u is harmonic on $\mathbf{R}^{n+1}_+$ and $Du(x)$ is defined as above. Then for all $\lambda > 0, 0 < \varepsilon < 1$,*

(a) If $\rho < \alpha$, then

$$\left|\{x \in \mathbf{R}^n : Du(x) > k\lambda,\ N_\alpha u(x) \le \varepsilon\lambda\}\right| \le c_1 \exp\left(\frac{-c_2}{\varepsilon}\right) \left|\{x \in \mathbf{R}^n : Du(x) > \lambda\}\right|$$

(b) If $\rho < (1+2\sqrt{n})^{-1} \cdot (32 \cdot 64)^{-1}\alpha$, then

$$\left|\{x \in \mathbf{R}^n : Du(x) > k\lambda,\ A_\alpha u(x) \le \varepsilon\lambda\}\right| \le c_1 \exp\left(\frac{-c_2}{\varepsilon^2}\right) \left|\{x \in \mathbf{R}^n : Du(x) > \lambda\}\right|.$$

In both (a) and (b), k, c_1, c_2 depend only on Ψ, α and n.

Here we will prove (b); this is the most difficult. The proof of (a) is shown in [3] and uses techniques similar to those we will use in the proof of (b). The unfortunate aspect of (b) is the restriction $\rho < (1+2\sqrt{n})^{-1}(32 \cdot 64)^{-1}\alpha$; such a result should probably be true under the assumption $\rho < \alpha$, however, we have not been able to show this. If we do only assume that $\rho < \alpha$, then in [3] it is shown that (b) holds with the expression $c_1 \exp\left(-c_2\varepsilon^{-\frac{2}{3}}\right)$ replacing the expression $c_1 \exp\left(\frac{-c_2}{\varepsilon^2}\right)$, but in a sense to be discussed later, this is not as sharp. Another unfortunate aspect of Theorem 2 is that the proof doesn't seem to be adaptable to the case of Lipschitz domains, and although such a result is probably true, we have been unable to prove it. The advantage of both Theorems 1 and 2 is that the exponential expressions on the right hand side allow us to pass immediately to similar theorems where Lebesgue measure is replaced by any measure which satisfies an A_∞ condition with respect to Lebesgue measure. Another advantage of Theorems 1(b) and 2(b), is that such so called "subgaussian" estimates lead to laws of the iterated logarithm involving $A_\alpha u$ and Du. (See [3])

3. The proof of Theorem 2(b). We will prove Theorem 2(b) by reducing it to Theorem 3 which follows, and which is an estimate like that found in Chang, Wilson, and Wolff [10].

For any $\gamma > 0,\ h > 0$ and $x \in \mathbf{R}^n$, we set $\Gamma^h_\gamma(x) = \{(s,t) : |x-s| < \gamma t,\ t < h\}$. We then define $N^h_\gamma u(x)$ and $A^h_\gamma u(x)$ as in (2.1), replacing the $\Gamma_\alpha(x)$ there by $\Gamma^h_\gamma(x)$. Likewise, we define $D^h u(x;r)$, $D^h u(x)$ as in (2.4), but we now restrict the integration to $\mathbf{R}^{n+1}_+ \cap \{(s,t) : t < h\}$.

THEOREM 3. *Suppose u, α, Ψ are as in the statement of Theorem 2. Then there exists constants c_1, c_2 depending only on α, Ψ, and n such that if Q is any cube in $\mathbf{R}^n$ and if $h = \ell(Q)\rho^{-1}$, then $\frac{1}{|Q|}\int_Q \exp[D^h u(x) - c_1(A_\alpha u(x))^2]dx \le c_2$.*

To show how Theorem 2(b) follows from Theorem 3, we borrow a lemma from [3]; the proof of this lemma just involves integration by parts. With h as above, set $D^T u(x;r) = Du(x;r) - D^h u(x;r)$, for $r \in \mathbf{R}$, and set $D^T u(x) = \sup_{r \in \mathbf{R}} D^T u(x;r)$.

LEMMA 1. *If* $t|\nabla u(s,t)| \leq L$ *for all* (s,t) *in a region containing* $(\Gamma_\rho(x) - \Gamma_\rho^h(x)) \cup (\Gamma_\rho(y) - \Gamma_\rho^h(y))$, *then* $|D^T u(x) - D^T u(y)| \leq C\frac{|x-y|}{h}L$, *where* C *depends only on* ρ *and* n.

Proceeding with the proof that Theorem 2(b) follows from Theorem 3, we first note that if $|\{Du > \lambda\}| = \infty$ we are trivially done so we assume the contrary. Then we can choose dyadic cubes $Q \subseteq \mathbf{R}^n$ such that $|\{x \in Q : Du(x) > \lambda\}| > \frac{1}{2}|Q|$ and such that if $\widetilde{Q}$ is the dyadic cube in $\mathbf{R}^n$ with $\ell(\widetilde{Q}) = 2\ell(Q)$ then $|\{x \in \widetilde{Q} : Du(x) > \lambda\}| \leq \frac{1}{2}|Q|$. We will show that for each such Q, $|\{x \in Q : Du(x) > k\lambda,\ A_\alpha u(x) \leq \varepsilon\lambda\}| \leq c_1|Q|\exp(-c_2\varepsilon^{-2})$, then summing over Q gives the result. Fix such a Q; we may assume that there exists an $x_0 \in Q$ such that $A_\alpha u(x_0) \leq \varepsilon\lambda$ and we pick $x_1 \in \widetilde{Q}$ such that $Du(x_1) < \lambda$. Our choice of ρ insures that $\Gamma_\rho(x) - \Gamma_\rho^h(x) \subseteq \Gamma_{\frac{\alpha}{2}}(x_0)$ for all $x \in \widetilde{Q}$. By the Lemma on page 207 of [20], $t|\nabla u(s,t)| \leq c_\alpha \varepsilon\lambda$ for all $(s,t) \in \Gamma_{\frac{\alpha}{2}}(x_0)$, and thus, by Lemma 1, $|D^T u(x) - D^T u(x_1)| \leq C\varepsilon\lambda$ for all $x \in Q$.
Therefore, $|D^T u(x)| \leq C\varepsilon\lambda + \lambda$ for all $x \in Q$. Set $k_1 = k - (C\varepsilon + 1)$. Then $|\{x \in Q : Du(x) > k\lambda,\ A_\alpha u(x) \leq \varepsilon\lambda\}| \leq |\{x \in Q : D^h u(x) > k_1\lambda,\ A_\alpha u(x) \leq \varepsilon\lambda\}|$ and we just estimate the latter quantity. Set $E = \{x \in Q : D^h u(x) > k_1\lambda,\ A_\alpha u(x) \leq \varepsilon\lambda\}$. Then replacing u by $\varepsilon^{-2}\lambda^{-1}u$ in Theorem 3 gives:

$$\begin{aligned} c_2 &\geq \frac{1}{|Q|}\int_Q \exp[\varepsilon^{-2}\lambda^{-1}D^h u(x) - c_1\varepsilon^{-4}\lambda^{-2}(A_\alpha u(x))^2]dx \\ &\geq \frac{1}{|Q|}\int_E \exp[\varepsilon^{-2}k_1 - \varepsilon^{-2}c_1]dx \\ &= \frac{|E|}{|Q|}\exp[-\varepsilon^{-2}(k_1 - c_1)] \end{aligned}$$

and the result follows by taking k_1, and hence k large enough.

4. The proof of Theorem 3. For the proof of Theorem 3, fix a cube Q and let y_0 be the center of Q. We may assume that $u(y_0, h) = 0$, (recall $h = \frac{\ell(Q)}{\rho}$), since both $A_\alpha u(x)$ and $D^h u(x)$ remain unchanged if we add a constant to u. In the interest of brevity, we borrow freely from other sources, and, in some cases will merely give outlines of proofs. In all that follows, the letters, C, c_1 and c_2 represent constants which may change from line to line, but nevertheless, depend only on α, ρ, Ψ and n, and never on u or the choice of Q.

We now consider another version of $D^h u(x)$ which approximates $D^h u(x)$ but will be easier to work with. We borrow the following from [3]; the proof involves numerous integrations by parts.

LEMMA 2. *There exists a vector valued function* $\Phi(x) = (\Phi_1(x), \ldots, \Phi_{n+1}(x))$ *with the following properties:*

(a) *Each* $\Phi_i(x)$ *is supported on* $B(0,\rho)$, *has a mean value 0, and is smooth.*

(b) If we set $\widetilde{D}^h u(x;a) = \int_{\Gamma^h_\rho(x)} \Phi_t(x-s)\cdot\nabla(u(s,t)-a)^+ dsdt,$
then for $|a| \le N^{2h}_{2\rho}u(x)$, *we have* $|\widetilde{D}^h u(x;a) - D^h u(x;a)| \le C\ N^{2h}_{2\rho}u(x)$.

We now set $\widetilde{D}^h u(x) = \sup\{|\widetilde{D}^h u(x;a)| : |a| \le N^{2h}_{2\rho}u(x)\}$. Then (b) implies that $D^h u(x) \le \widetilde{D}^h u(x) + C\ N^{2h}_{2\rho}u(x)$. Then

$$\begin{aligned}\frac{1}{|Q|}\int_Q \exp\left[D^h u(x) - c_1(A_\alpha u(x))^2\right]dx \\ \le \left(\frac{1}{|Q|}\int_Q \exp\left[2\widetilde{D}^h u(x) - c_1(A_\alpha u(x))^2\right]dx\right)^{\frac{1}{2}} \cdot \\ \left(\frac{1}{|Q|}\int_Q \exp\left[2C\ N^{2h}_{2\rho}u(x) - c_1(A_\alpha u(x))^2\right]dx\right)^{\frac{1}{2}} = I\cdot II\ .\end{aligned}$$

We now borrow a result from [2]. This states that if c_1 is large enough, then $II \le c_2$. We will show:

LEMMA 3. *There exists* c_1 *and* c_2 *such that*

$$\frac{1}{|Q|}\int_Q \exp[\widetilde{D}^h u(x) - c_1(A_\alpha u(x))^2]dx \le c_2\ .$$

Lemma 3 will complete the proof of Theorem 3, since then the estimate for I follows by replacing u by $2u$ in Lemma 3 and taking c_1 in I four times as large as in the c_1 in Lemma 3.

To prove Lemma 3, we will break up $\widetilde{D}^h u(x)$ into "smaller" maximal densities. For j an integer, set

$$\widetilde{D}^h_j u(x) = \sup\left\{\left|\widetilde{D}^h u(x;a)\right| : a \in [j, j+1)\right\}\ .$$

Then

$$\widetilde{D}^h u(x) \le \sup\left\{\widetilde{D}^h_j u(x) : |j| \le N^{2h}_{2\rho}u(x) + 1\right\}\ .$$

We claim:

LEMMA 4. *There exists* c_1 *and* c_2 *independent of* j *such that*

$$\frac{1}{|Q|}\int_Q \exp\left[2\widetilde{D}^h_j u(x) - c_1(A_\alpha u(x))^2\right]dx \le c_2\ .$$

Before proving Lemma 4, we show how Lemma 4 implies Lemma 3. For $i = 0, 1, 2, \ldots$ set

$$E_i = \{x \in Q : i \le N^{2h}_{2\rho}u(x) < i+1\}\ .$$

Then

$$\frac{1}{|Q|}\int_Q \exp\left[\widetilde{D}^h u(x) - c_1(A_\alpha u(x))^2\right]dx$$

$$\leq \sum_{i=0}^{\infty}\frac{1}{|Q|}\int_{E_i}\exp\left[\sup_{|j|\leq i+2}\widetilde{D}_j^h u(x) - c_1\,(A_\alpha u(x))^2\right]dx$$

$$\leq \sum_{i=0}^{\infty}\sum_{j=-i-2}^{i+2}\frac{1}{|Q|}\int_{E_i}\exp[\widetilde{D}_j^h u(x) - c_1(A_\alpha u(x))^2]dx$$

$$\leq \sum_{i=0}^{\infty}\sum_{j=-i-2}^{i+2}\left[\left(\frac{1}{|Q|}\int_{E_i}\exp\left[2\widetilde{D}_j^h u(x) - c_1(A_\alpha u(x))^2\right]dx\right)^{\frac{1}{2}}\cdot\right.$$

$$\left.\cdot\left(\frac{1}{|Q|}\int_{E_i}\exp\left[-c_1(A_\alpha u(x))^2\right]dx\right)^{\frac{1}{2}}\right]$$

$$\leq C\sum_{i=0}^{\infty}(2i+4)e^{\frac{-i}{2}}\left(\frac{1}{|Q|}\int_{E_i}\exp[N_{2\rho}^{2h}u(x) - c_1(A_\alpha u(x))^2]dx\right)^{\frac{1}{2}} \leq C$$

where for the last inequality we have again used a result from [2].

So we have reduced matters to Lemma 4. We now state a proposition which will be used several times in the proof of Lemma 4, and in different contexts; consequently, we state it in a general setting. Suppose $q(x) = (q_1(x), \ldots, q_{n+1}(x))$ is a vector valued function on $\mathbb{R}^n$ such that each $q_i(x)$ is smooth, has mean value 0, and is supported on $B(0,\gamma)$. Suppose also that $\varphi(s,t) = (\varphi_1(s,t), \ldots, \varphi_{n+1}(s,t))$ is a vector valued function defined on $\mathbb{R}_+^{n+1}$. Fix a cube $Q \subseteq \mathbb{R}^n$, set $m = \ell(Q)\gamma^{-1}$, and set

$$f(x) = \int_{\Gamma_\gamma^m(x)} q_t(x-s)\cdot\varphi(s,t)dsdt$$

$$S(x) = \left(\int_{\Gamma_{32\gamma}^m(x)} |\varphi(s,t)|^2 t^{1-n}dsdt\right)^{\frac{1}{2}}$$

PROPOSITION. *There exists k_1 and k_2 depending only on γ, q and n such that*

$$\frac{1}{|Q|}\int_Q \exp[|f(x)| - k_1(S(x))^2]dx \leq k_2\ .$$

The proof of this is in [10] or [2]; there $\varphi(s,t) = \nabla u(s,t)$ and $f(x)$ is a Calderón reproducing formula for $u(x,0)$. As an immediate consequence of this, our choice of ρ, and the definition of $\widetilde{D}^h u(x;a)$, we obtain:

LEMMA 5. $\frac{1}{|Q|}\int_Q \exp\left[|\widetilde{D}^h u(x;a)| - c_1(A_\alpha u(x))^2\right]dx \leq c_2$ *for all* $a \in \mathbb{R}$, *with* c_1, c_2 *independent of* a.

We can now proceed with the proof of Lemma 4. Fix j and set

$$B(x) = \left(\int_j^{j+1}\int_j^{j+1}\left|\frac{\widetilde{D}^h u(x;a) - \widetilde{D}^h u(x;b)}{|a-b|^{\frac{1}{2}}}\right|^5 dadb\right)^{\frac{1}{5}}\ .$$

By a lemma of Garsia, Rodemich and Rumsey [14], the Holder continuity of the function $\widetilde{D}^h u(x;a)$, $a \in [j, j+1)$ can be expressed in terms of $B(x)$; more precisely, their lemma implies that $|\widetilde{D}^h u(x;a) - \widetilde{D}^h u(x;b)| \leq \omega |a-b|^{.1} B(x)$ for all $a, b \in [j, j+1)$ where ω is an absolute constant. Thus, $2\widetilde{D}^h_j u(x) \leq 2\omega B(x) + 2|\widetilde{D}^h u(x;j)|$. The Cauchy Schwarz inequality and Lemma 5 now imply that it suffices to show:

$$\text{(4.1)} \qquad \frac{1}{|Q|} \int_Q \exp\left[4\omega B(x) - c_1 (A_\alpha u(x))^2\right] dx \leq c_2.$$

For typographical convenience, we set

$$k(x,a,b) = 4\omega \left| \frac{\widetilde{D}^h u(x;a) - \widetilde{D}^h u(x;b)}{|a-b|^{\frac{1}{2}}} \right| \quad \text{for} \quad x \in Q \ , \ a, b \in [j, j+1).$$

The function $\exp[(y + 4^5)^{\frac{1}{5}}]$ is convex for $y > 0$ and thus:

$$\begin{aligned} &\frac{1}{|Q|} \int_Q \exp\left[4\omega B(x) - c_1 (A_\alpha u(x))^2\right] dx \\ &\quad \leq \frac{1}{|Q|} \int_Q \exp\left[((4\omega)^5 (B(x))^5 + 4^5)^{\frac{1}{5}} - c_1 (A_\alpha(x))^2\right] dx \\ &\quad = \frac{1}{|Q|} \int_Q \exp\left[\left(\int_j^{j+1} \int_j^{j+1} k(x,a,b)^5 da db + 4^5\right)^{\frac{1}{5}} - c_1 (A_\alpha u(x))^2\right] dx \\ &\quad \leq C \sup_{a,b \in [j,j+1)} \frac{1}{|Q|} \int_Q \exp\left[k(x,a,b) - c_1 (A_\alpha u(x))^2\right] dx \end{aligned}$$

Thus, (4.1) will follow from:

$$\text{(4.2)} \quad \frac{1}{|Q|} \int_Q \exp\left[k(x,a,b) - c_1 (A_\alpha u(x))^2\right] dx \leq c_2, \text{ with } c_1, c_2 \text{ independent of } a, b.$$

Let $\eta(x)$ be a positive radially symmetric function supported on $B(0, 64\rho)$ such that $n \equiv 1$ on $B(0, 32\rho)$, and define

$$\widehat{D}^h u(x;a) = \int_{\mathbb{R}^{n+1}_+ \cup \{t<h\}} t\eta_t(x-s) \Delta (u(s,t) - a)^+ (dsdt)$$

for $r \in \mathbb{R}$. Also, note that by definition,

$$k(x,a,b) = \left| 4\omega \int_{\mathbb{R}^{n+1}_+ \cap \{t<h\}} \Phi_t(x-s) \cdot \left(\frac{\nabla(u(s,t)-a)^+ - \nabla(u(s,t)-b)^+}{|a-b|^{\frac{1}{2}}} \right) dsdt \right|$$

Set

$$S(x)^2 = 16\omega^2 \int_{\Gamma^h_{32\rho}(x)} \left| \frac{\nabla(u(s,t)-a)^+ - \nabla(u(s,t)-b)^+}{|a-b|^{\frac{1}{2}}} \right|^2 t^{1-n} dsdt$$

Then

$$S(x)^2 = 16\omega^2 \int_{\Gamma^h_{32\rho}(x)} \chi_{[a,b]}\,(u(s,t))|\nabla u(s,t)|^2 t^{1-n}|a-b|^{-1}dsdt$$
$$\le C \int_{\mathbf{R}^{n+1}_+ \cap\{t<h\}} t\eta_t(x-s)|\nabla u(s,t)|^2 \chi_{[a,b]}\,(u(s,t))|a-b|^{-1}dsdt$$
$$= \frac{C}{|a-b|}\int_a^b \widehat{D}^h u(x;r)dr$$

where the last equality follows from (2.3). So by the proposition:

$$\frac{1}{|Q|}\int_Q \exp\left[k(x,a,b) - c_1\frac{1}{|a-b|}\int_a^b \widehat{D}^h u(x;r)dr\right]dx \le c_2\ .$$

This fact together with the Cauchy Schwarz inequality and Jensen's inequality shows that (4.2) will follow if we can show that

$$\text{(4.3)}\qquad \frac{1}{|Q|}\int_Q \exp[\widehat{D}^h u(x;r) - c_1(A_\alpha u(x))^2]dx \le c_2$$

with c_1 and c_2 independent of r.

To see this, let $\widetilde{Q}$ be a cube in $\mathbf{R}^{n+1}$ with the same center as Q and with $\ell(\widetilde{Q}) = 64\ell(Q)$. Then (4.3) follows if we show a similar estimate with $\widetilde{Q}$ in place of Q. Then $h = \dfrac{l(Q)}{\rho} = \dfrac{\ell(\widetilde{Q})}{64\rho}$. By Lemma 2, there exists a $\widetilde{\widehat{D}}^h u(x;r)$ with properties (a) and (b) (with the ρ there replaced by 64ρ). In particular (b) holds so that $|\widetilde{\widehat{D}}^h u(x;a) - \widehat{D}^h u(x;a)| \le C\ N^{2h}_{128\rho}u(x)$ whenever $|a| \le N^{2h}_{128\rho}u(x)$. But since $\widehat{D}^h u(x;a) = 0$ if $|a| > N^{2h}_{128\rho}u(x)$ then $\widehat{D}^h(x;a) \le |\widetilde{\widehat{D}}^h u(x;a)| + C\ N^{2h}_{128\rho}u(x)$ for all $a \in \mathbf{R}$. Then, reason as before, using Cauchy–Schwarz and Lemma 5 to complete the proof of (4.3).

5. Some remarks. The expressions $c_1 \exp\left(\dfrac{-c_2}{\varepsilon}\right)$ and $c_1\exp(c_2\varepsilon^{-2})$ in the right hand sides of Theorem 2(a) and 2(b) are sharp in terms of decay as $\varepsilon \to 0$; that is, the results would no longer be true with higher powers of ε. For example, integrating Theorem 2(b) gives the estimate: $\|D_\rho u\|_p \le c_p\ \|A_\alpha u\|_p\ ,\ 0<p<\infty$ with $c_p = 0(\sqrt{p})$ as $p\to\infty$. Using this, the estimate $\|A_\rho u\|_p \le \sqrt{2}\ \sqrt{\|D_\rho u\|_p}\ \sqrt{\|N_\rho u\|_p}$, and the known estimate [1]: $\|A_\alpha u\|_p \le c'_p\|N_{2\alpha}u\|_p,\ 0<p<\infty$ with $c'_p = 0(\sqrt{p})$ as $p\to\infty$, then gives, $\|A_\rho u\|p \le \sqrt{2}\ \sqrt{\|D_\rho u\|_p}\ \sqrt{\|N_\rho u\|_p} \le \sqrt{2}\ \sqrt{c_p}\ \sqrt{\|A_\alpha u\|_p}$ $\sqrt{\|N_\rho u\|_p} \le \sqrt{2}\ \sqrt{c_p}\ \sqrt{c'_p}\ \|N_{2\alpha}u\|_p$ with $\sqrt{c_p}\ \sqrt{c'_p} = 0(\sqrt{p})$ as $p\to\infty$, and this known to be the sharpest possible. In a similar way, Theorem 2(a) can also be shown to be sharp. By analogy to the case of martingales, where Theorem A is known to be sharp, it seems that Theorem 1 is probably sharp, but we have been unable to prove this.

Local time has proved to be of fundamental importance in the study of Brownian motion and in the study of martingales in general. Perhaps, when the D function is as well understood as local time, it will play a correspondingly useful role in the study of harmonic functions. Some progress has already been made in this direction. Brossard [7] has shown a Calderón–Stein type theorem relating the finiteness of Du and the existence of nontangential limits of u. Also, Brossard and Chevalier [8] have used D to give another characterization of LlogL.

REFERENCES

[1] R. Bañuelos and C.N. Moore, *Sharp estimates for the nontangential maximal function and the Lusin area function in Lipschitz domains*, Trans. Amer. Math. Soc. 312 (1989), 641–662.

[2] R. Bañuelos and C.N. Moore, *Laws of the iterated logarithm, sharp good-λ inequalities and L^p estimates for caloric and harmonic functions*, Indiana Univ. Math. J. 38 (1989), 315–344.

[3] R. Bañuelos and C.N. Moore, *Distribution function inequalities for the density of the area integral*, Ann. Inst. Fourier (Grenoble), to appear.

[4] M. Barlow and M. Yor, *(Semi) Martingale inequalities and local times*, A. Wahrch. Verw. Gebiete 55 (1981), 237–254.

[5] M. Barlow and M. Yor, *Semi-martingale inequalities via the Garsia–Rodemich–Rumsey lemma, and application to local times*, J. Funct. Anal. 49,2 (1982), 198–229.

[6] R. Bass, *L^p-inequalities for functionals of Brownian motion*, Séminaire de Probabilitiés XXI, Lecture notes in Math. 1247 (1987), Springer–Verlag, New York.

[7] J. Brossard, *Dénsité de l'intégrale d'aire dans $\mathbb{R}^{n+1}_+$ et limites nontangentielles*, Invent. Math. 93 (1988), 297–308.

[8] J. Brossard and L. Chevalier, *Classe LlogL et densité de l'intégrale d'aire dans $\mathbb{R}^{n+1}_+$, Ann. of Math. 128 (1988), 603–618,.*

[9] D.L. Burkholder and R.F. Gundy, *Distribution function inequalities for the area integral*, Studia Math. 44 (1972), 627–544.

[10] S.Y.A. Chang, J.M. Wilson, and T.H. Wolff, *Some weighted norm inequalities involving the Schrödinger operators*, Comment. Math. Helv. 60 (1985), 217–246.

[11] B.E.J. Dahlberg, *Weighted norm inequalities for the Lusin area integral and nontangential maximal functions for harmonic functions in Lipschitz domains*, Studia Math. 47 (1980), 297–314.

[12] B. Davis, *On the Barlow–Yor inequalities for local time*, Séminaire de Probabilitiés XXI, Lecture notes in Math 1247 (1987), Springer–Verlag, New York.

[13] R. Fefferman, R.F. Gundy, M. Silverstein, and E.M. Stein, *Inequalities for ratios of functionals of harmonic functions*, Proc. Nat. Acad. Sci. U.S.A. 79 (1982), 7958–7960.

[14] A.M. Garsia, E. Rodemich and H. Rumsey Jr., *A real variable lemma and the continuity of paths of some Gaussian processes*, Indiana Univ. Math. J. 20 (1970), 565–578.

[15] R.F. Gundy, *The density of the area integral*, Conference on Harmonic Analysis in Honor of A. Zygmund, Beckner, W., Calderón, A.P., Fefferman, R., and Jones, P., editors, Wadsworth, Belmont, California 1983.

[16] R.F. Gundy, *Some topics in probability and analysis*, CBMS #70, (1989).

[17] R.F. Gundy and M.L. Silverstein, *The density of the area integral in $\mathbb{R}^{n+1}_+$*, Ann. Inst. Fourier (Grenoble) 35, 1 (1985), 215–229.

[18] E. Lenglart, D. Lepingle, and M. Pratelli, *Présentation unifiée de certaines inéqualités de la théorie des martingales*, Séminaire de probabilitiés XIV, Lecture notes in Math 784 (1980), Springer–Verlag, New York.

[19] T. Murai and A. Uchiyama, *Good-λ inequalities for the area integral and the nontangential maximal function*, Studia Math. 83 (1986), 251–262.

[20] E.M. Stein, *Singular Integrals and Differentiability Properties of Functions*, Princeton University Press (1970).

RESTRICTION THEOREMS AND THE SCHRÖDINGER MULTIPLIER ON THE TORUS

LUIS VEGA*

Abstract. Restriction theorems for Fourier integrals and series are proven. In the continuous case, a mixed norm inequality for the restriction of the Fourier transform to the sphere is given. On the other hand a discrete version of a result by R.S. Strichartz [10] is found.

1. Introduction and statements. A consequence of the Hausdorff-Young theorem is that the Fourier transform $\hat{f}$ of an L^p-function f is defined almost everywhere in $\mathbf{R}^n$ if $1 < p < 2$. However, it is an interesting fact that for a special range of p's $\hat{f}$ can be defined on submanifolds possessing some curvature. This is also a consequence of *a priori* estimates which, in the case of the unit sphere S^{n-1}, are the following:

$$(1) \qquad \left(\int_{S^{n-1}} |\hat{f}(\omega)|^p d\sigma(\omega) \right)^{1/p} \leq c_{p,q} \left(\int |f(x)|^q dx \right)^{1/p}$$

for some p and q. A standard homogenity argument, implies that $\frac{1}{q} \geq \frac{n+1}{n-1}\frac{1}{p'}$, $\left(\frac{1}{p} + \frac{1}{p'} = 1\right)$. And, by studying the dual extension operator, it can be proved that $q < \frac{2n}{n+1}$ is a necessary condition [6]. If $n = 2$ the problem was solved by C. Fefferman [5] and A. Zygmund [15]. In higher dimensions, the particular case $p = p' = 2, q \leq 2\frac{n+1}{n+3}$ was proved by E. Stein and P. Tomas [11]. Later R. Strichartz extended this L^2 result to quadratic surfaces [10]. In particular, in the case of the paraboloid he proves the following

$$(2) \qquad \left(\int_{\mathbf{R}^n} \left| \int_{\mathbf{R}^{n-1}} e^{i(x_n|\xi|^2 + x\xi)} f(\xi) d\xi \right|^{2\frac{n+1}{n-1}} dx dx_n \right)^{\frac{n-1}{2(n+1)}} \leq C \left(\int_{\mathbf{R}^{n-1}} |f|^2 \right)^{1/2}$$

Observe we have written the extension dual operator instead of the restriction one. On the other hand, there also exist restriction results in Fourier series. Namely,

$$(3) \qquad \left(\int_0^{2\pi} \int_0^{2\pi} \left| \sum_k a_k e^{i(x_2 k^2 + x_1 k)} \right|^4 dx_1 dx_2 \right)^{1/4} \leq C \left(\sum_k |a_k|^2 \right)^{1/2}$$

$$\left(\int_0^{2\pi} \int_0^{2\pi} \left| \sum_{k^2 + j^2 = R} a_{kj} e^{i(x_2 k + x_1 j)} \right|^4 dx_1 dx_2 \right)^{1/4} \leq C \left(\sum_{kj} |a_{kj}|^2 \right)^{1/2}$$

with C independent of R.

*Department of Mathematics, University of Chicago, 5734 South University Avenue, Chicago, IL 60637.

Both theorems were also proved by A. Zygmund in [15]. Let us remark that (2) and (3) can be seen as results for the Schrödinger operator. This one is given by the multiplier $m_t(\xi) = e^{it|\xi|^2}$ which solves the Schrödinger equation

$$\begin{cases} i\frac{\partial}{\partial t}u - \delta u = 0 & x \in \mathbf{R}^n \ t \in \mathbf{R} \\ u(x,o) = 0 = f(x) \end{cases}$$

The purpose of this paper is to prove new results in both $\mathbf{R}^n$ and T^n. In the continuous case we study only the problem in the unit sphere. We are influenced by the works of J.L. Rubio [8] and J.A. Barceló, A. Córdoba [2]. In [8] mixed norm spaces with good spherical symmetry properties are defined. This mixed norm (see Theorem 1 below for the definition) is easy to estimate for the extension operator from the sphere. We prove:

THEOREM 1. *Let f be in $L^2(S^{n-1})$. Then*

$$\left(\int_0^\infty \left(\int_{S^{n-1}} |\widehat{fd\sigma}(r\omega)|^2 d\sigma(\omega)\right)^{p/2} r^{n-1} dr\right)^{1/p} \le C_{p,n}\left(\int_{S^{n-1}} |f|^2 d\sigma\right)^{1/2}$$

where $p > \frac{2n}{n-1}$ and $\widehat{fd\sigma}(r\omega) = \int_{S^{n-1}} f(x)e^{i(r\omega\cdot x)}d\sigma(x)$

Interpolating this theorem with the Stein-Tomas result we get:

THEOREM 2. *Let f be in $L^2(S^{n-1})$. Then,*

$$\left(\int_0^\infty \left(\int_{S^{n-1}} |\widehat{fd\sigma}(r\omega)|^q d\sigma(\omega)\right)^{p/q} r^{n-1} dr\right)^{1/p} \le C_{p,q}\left(\int |f|^2 d\sigma\right)^{1/2}$$

where $p > \frac{2n}{n-1}$ and $\frac{1}{q} \ge \max\left\{\frac{1}{p}, \frac{1}{p}\frac{2n}{n-2} - \frac{1}{2}\right\}$.

Observe that we are working with the extension operator. By duality we get the corresponding restriction result. Recall also that $p > \frac{2n}{n-1}$ is a necessary condition.

In the discrete case we prove the following:

THEOREM 3. *Let N be a natural number and $m \in \mathbf{Z}^{n-1}$, $n \ge 2$. Then*

$$\left(\int_{|t|\le N^{-1}} \int_{T^{n-1}} \left| \sum_{\substack{m_i=1 \\ 1\le i\le n-1}}^{N} a_m e^{it|m|^2} e^{im\cdot x}\right|^p dx\,dt\right)^{1/p} \le C_{p,N}\left(\sum_m |a_m|^2\right)^{1/2} \tag{4}$$

where

$$C_{p,N} = \begin{cases} C_p N^{(n-1)/2-(n+1)/p} & \text{if } p > 2\frac{n+1}{n-1} \\ C & \text{if } p = 2\frac{n+1}{n-1} \\ C_p N^{-\frac{1}{2}+\frac{n+1}{2}\left(\frac{1}{2}-\frac{1}{p}\right)} & \text{if } 2 \le p < 2\frac{n+1}{n-1} \end{cases}$$

and C_p are constants independent of N.

It is clear that the condition $|t| \le N^{-1}$ can be changed to $t \in (\omega, \omega + N^{-1})$ with $\omega \in (-\pi, \pi)$. Also observe that our key estimate is for $p = 2\frac{2+1}{n-1}$ which is bigger than the critical exponent $p = \frac{2n}{n-1}$.

There is a more general version of the above theorem in the case $n = 2$.

THEOREM 4. *Let ψ be a real function defined on an interval $I \subset [1, N]$ such that $|\psi'(s)| \le C_1 s$ and $\psi''(s) \ge C_2$ on I, with C_1, C_2 independent of N. Then:*

$$\left(\int_{|t| \le N^{-1}} \int_0^{2\pi} \left| \sum_{k \in I} a_k e^{i(t\psi(k) + kx)} \right|^p dx\,dt \right)^{1/p} \le C_{p,N} \left(\sum_k |a_k|^2 \right)^{1/2}$$

where

$$C_{p,N} = \begin{cases} C_p N^{\frac{1}{2} - \frac{3}{p}} & \text{if } p > 6 \\ C & \text{if } p = 6 \\ C_p N^{\frac{1}{4} - \frac{3}{2p}} & \text{if } 2 \le p < 6 \end{cases}$$

with C, C_p are independent of N.

As an easy consequence we have:

COROLLARY 5. *Let $R > 1$,*

(i)

$$\left(\int_0^{R^{-1/3}} \int_0^{2\pi} \left| \sum_{R^{1/3} \le |R-k| \le 2R^{1/3}} a_k e^{i\{x_2 (R^2 - k^2)^{1/2} + kx_1\}} \right|^6 dx_1 dx_2 \right)^{1/6} \le C \left(\sum |a_k|^2 \right)^{1/2}$$

(ii)

$$\left(\frac{1}{R} \int_0^1 \int_0^{2\pi} \left| \sum_{|k| \le R/2} a_k e^{i\{x_2 (R^2 - k^2)^{1/2} + kx_1\}} \right| dx_1 dx_2 \right)^{1/6} \le C \left(\sum |a_k|^2 \right)^{1/2}$$

with C independent of R.

A few words about the sharpness of Theorems 3 and 4 are in order. Strichartz' Theorem (2) can be proved using the estimate in Theorem 3 for $p = 2\frac{n+1}{n-1}$. In this sense, Theorem 3 is a discrete version of Strichartz' result. And the same can be said about Corollary 5 and the Stein-Tomas restriction theorem on the sphere S^1. In Proposition 10 below, we sketch the proof of (2), which follows very closely the proof of the classical transference theorem of de Leeuw. Nevertheless, in order to prove (2) it is sufficient to assume a weaker version of Theorem 3 taking the integral in (4) over a region $|t| < N^{-\alpha}$ for any $\alpha < 2$ instead of over $|t| < N^{-1}$ as we do.

On the other hand, Theorem 3 can be substantially improved at least in dimension two. For example, if $a_k = 1$ for $1 \le k \le N$, $n = 2$ and $p = 6$ then (3) is true with $0 \le t \le 2\pi$ and $c_{p,N} \sim (\log N)^{1/6}$. (See [12] p. 104, exercise 7.5.2 and also (7.5) in p. 90). The proof is based on the arithmetic properties of the sequence of the squares and on the fact that the exponent $p = 6$ is even. Therefore the extension of this approach to more general sequences like in Theorem 4 and to bigger dimensions, where the critical exponent $p = 2\frac{n+1}{n-1}$ is not an even integer ($n > 3$), seems more challenging.

We give two other discrete results.

THEOREM 6. *Let $\varepsilon > 0$, then*

$$\left(\int_0^1\int_0^{2\pi}\left|\sum_k a_k e^{i(tk^2+kx)}\right|^4 dx\frac{dt}{t}\right)^{1/4} \le C(1+|\log\varepsilon|)^{1/2}\left(\int\left|\sum_k a_k e^{ikx}\right|^4 dx\right)^{1/4}$$

with C independent of ε.

THEOREM 7. *Let N be a natural number. Then*

$$\left(\int_{-\pi}^{\pi}\left|\sup_{0\le|t|\le N^{-1}}\sum_{k=1}^{N} a_k e^{i(tk^2+kx)}\right|^2\frac{dx}{|x|^{1/2}}\right)^{1/2} \le C\left(\sum_{k=1}^{N}|a_k|^2 k^{1/2}\right)^{1/2}.$$

This last result concerns the discrete version of the Schrödinger maximal function studied by L. Carleson in $\mathbb{R}$ [3]. Using transference methods as in [7], it can be proved that Theorem 7 implies the continuous result obtained by Carleson.

Finally, Theorem 6 is the analogue of:

THEOREM 6′. (See [4]) *Let f be a function in $L^4(-R,R)$ with $R > 1$ and $\varepsilon > 0$. Then,*

$$\left(\int_\varepsilon^1\int_{\mathbb{R}}|T_tf(x)|^4 dx\frac{dt}{t}\right)^{1/4} \le C(\log R+|\log\varepsilon|)^{1/4}\left(\int_{-R}^{R}|f|^4\right)^{1/4}$$

where C is a constant independent of R and ε and

$$T_tf(x) = \int_{\mathbb{R}} e^{-\pi it|\xi|^2}e^{2\pi ix\xi}\hat{f}(\xi)d\xi.$$

Remember that T_t is bounded in $L^p(\mathbb{R}^n)$ if and only if $p = 2$. On the other hand we have:

$$T_tf(x) = t^{-1/2}e^{-\pi i(\frac{1}{4}+tx^2)}\int e^{-2xi(x\xi+|\xi|^2/t)}f(\xi d\xi$$

and the last integral can be seen as the Fourier transform at the point $(x, 1/t)$ of the density f defined on the parabola (ξ,ξ^2). A simple change of variable says that Theorem 6′ is the (L^4, L^4) result of C. Fefferman [5] for the extension operator from the parabola:

$$\left(\int_{2^j}^{2^{j+1}}\int_{-\infty}^{\infty}\left|\int_0^R e^{-i(x_1\xi+x_2\xi^2)}f(\xi)d\xi\right|^4 dx_1dx_2\right)^{1/4} \le Cj^{1/4}(\log R)^{1/4}\left(\int_0^R|f|^4\right)^{1/4}$$

with $j, R > 1$.

This paper is part of my thesis. I want to thank my advisor A. Córdoba for his continuous help and encouragement along these years. I am indebted too to J.L. Rubio de Francia. His kindness and generosity will be very difficult to forget. We also want to thank A. Carbery for calling our attention to the result in [12].

2. Proofs of the results in Fourier series. We shall make use of two lemmas

LEMMA 8. *If φ' is monotone and $|\varphi'| \leq 1/2$, then*

$$\left| \int_a^b r(u)F(u)du - \sum_{a\leq n\leq b} r(n)F(n) \right| \leq A \cdot \max r(u)$$

where r is monotone, $F(u) = e^{2\pi i\varphi(u)}$ and A a universal constant.

See [14, p. 226] for the proof.

LEMMA 9. *If $\varphi'' \geq \rho > 0$ or $\varphi'' \leq -\rho < 0$ then*

$$\left| \sum_{a\leq n\leq b} e^{2\pi i\varphi(n)} \right| \leq (|\varphi'(b) - \varphi'(a)| + 2)(4\rho^{-1/2} + A)$$

where A is a universal constant.

See [14, p. 198] for the proof.

Proof of Theorem 3. The result is trivial if $p = 2$ and $p = \infty$. Therefore we just have to study the case $p = 2\frac{n+1}{n-1}$.

By duality and using P. Tomas' argument it is sufficient to prove

$$(*) \qquad \left(\int_{0\leq|t|\leq N^{-1}} \int_{T^{n-1}} |K * f|^p dx dt \right)^{1/p} < C^2_{p,N} \left(\int |f(x,t)|^{p'} dx dt \right)^{1/p'}$$

where $f \in L^{p'}([-N^{-1}, N^{-1}] \times T^{n-1}), \frac{1}{p} + \frac{1}{p'} = 1$ and

$$K(x,t) = \sum_{\substack{m_i=1 \\ 1\leq i\leq n-1}}^{N} e^{i(m\cdot x + t|m|^2)} \quad (m \in \mathbb{Z}^{n-1}).$$

For $\varepsilon > 0$ let us call $K_\varepsilon(x,t) = \chi_{\{2N^{-1}\geq|t|\geq\varepsilon\}}(t)K(x,t)$ where χ denotes the characteristic function. It is clear that it is enough to prove $(*)$ for K_ε instead of K, with a constant independent of ε.

Finally, for $z \in C, -1 \leq \mathbf{Re}z \leq \frac{n-1}{2}$ we define $K_\varepsilon^z(x,t) = t^z K_\varepsilon(x,t)$ where $t \in \mathbf{R} \setminus (-\varepsilon, \varepsilon)$. Set $f(x,t) = \sum_m a_m(t) e^{im \cdot z}$ with $\operatorname{supp} a_m \subset (-N^{-1}, N^{-1})$, then

$$\int_t \int_{T^{n-1}} g(x,t) \cdot (K_\varepsilon^z * f) dx dt$$

$$= \sum_{m_i=1}^{N} \int_{-N^{-1}}^{N^{-1}} \int_{T^{n-1}} g(x,t) e^{imx} \left(\int_{\substack{0<|t-s|<N^{-1} \\ |s|>\varepsilon}} a_m(t-s) s^z e^{is|\nu|^2} ds \right) dx dt$$

Therefore $\{K_z\}_z$ defines an admissible family of linear operators in the sense of [9].

We will prove $(*)$ using analytic interpolation between the (L^1, L^∞) and (L^2, L^2) estimates.

Let us observe that K_ε^z is a product kernel in the x-variable. Therefore, applying Lemma 8 we conclude that $|K_\varepsilon^z(x,t)| \leq C$ for $x \in T^{n-1}$, $|t| \leq 2N^{-1}$ and if $\mathbf{Re}z = \frac{n-1}{2}$.

Thus

$$\|K_\varepsilon^{\frac{n-1}{2}+i\gamma} * f\|_{L^\infty((-N^{-1},N^{-1})\times T^{n-1})} < C\|f\|_{L^1((-N^{-1},N^{-1})\times T^{n-1})}$$

with C a universal constant.

On the other hand, we need the L^2 estimate:

$$\int_{|t|\leq N^{-1}} \int_{T^{n-1}} |K^{-1+i\gamma} * f|^2$$

$$= \sum_m \int_{|t|\leq N^{-1}} \int_{\substack{|t-s|\leq N^{-1} \\ |s|>\varepsilon}} a_m(t-s) s^{-1+i\gamma} e^{is|m|^2} ds|^2 dt$$

$$= \sum_m \int_{|t|\leq N^{-1}} \left| \int_{|s|>\varepsilon} a_m(t-s) e^{-i(t-s)|m|^2} s^{-1+i\gamma} ds \right|^2 dt.$$

Thus, we have to prove that the operator

$$h \mapsto Th(t) = \int_{M>|s|>\varepsilon} h(t-s) \frac{ds}{s^{1-i\gamma}}$$

is bounded in $L^2(\mathbf{R})$ with an adequate control of the norm. But this is a consequence of the fact that

$$\left| \int_{M>|s|>\varepsilon} e^{isx} \frac{ds}{s^{1-i\gamma}} \right| \leq C e^{-x\gamma}$$

with C a constant independent of x, ε, M and γ. Using the standard interpolation theorem for an analytic family of operators, completes the proof of Theorem 3. □

The proof of Theorem 4 works similarly. The main technical point is the estimate

$$\left| t^{1/2} \sum_{k \in I} e^{i(t\varphi(k)+kx)} \right| \leq C.$$

But this is an immediate consequence of Lemma 9.

PROPOSITION 10. *Strichartz' theorem*

$$\left(\int_{\mathbf{R}} \int_{\mathbf{R}^{n-1}} \left| \int_{\mathbf{R}^{n-1}} e^{i(t|\xi|^2+x\xi)} \hat{f}(\xi) d\xi \right|^{2\frac{n+1}{n-1}} dx dt \right)^{\frac{n-1}{(n+10}} \leq c \left(\int_{\mathbf{R}^{n-1}} |f|^2 \right)^{1/2}$$

follows from Theorem 3.

SKETCH OF THE PROOF. (See [9], Theorem 3.8, p. 260) We can assume $\operatorname{supp} \hat{f} \subset B(0, R)$ and $0 < t < M$ as long as we obtain estimates independent of R and M. Define for $\varepsilon > 0$,

$$\tilde{f}_\varepsilon(x) = \sum_{m \in \mathbf{Z}^{n-1}} \hat{f}(\varepsilon m) e^{im \cdot x},$$

$$\widetilde{T}_{\varepsilon^2 t} \tilde{f}_\varepsilon = \sum_{m \in \mathbf{Z}^{n-1}} e^{i\varepsilon^2 t |m|^2} \hat{f}(\varepsilon m) e^{im \cdot x} \quad [and$$

$$T_t f(x) = \int_{\mathbf{R}^{n-1}} e^{i(t|\xi|^2+x\varepsilon)} \hat{f}(\xi) d\xi.$$

Notice that $\hat{f}(\varepsilon m) = 0$ if $|m| \leq R\varepsilon^{-1}$. Take η to be a positive continuous function with compact support, $\eta(0) = 0$ and $\varepsilon_{m \in \mathbf{Z}^{n-1}} |\eta(x + 2\pi m)|^p = 1$ with $p = 2\frac{n+1}{n-1}$. Then

$$\lim_{\varepsilon \to 0} (2\pi\varepsilon)^{n-1} \varepsilon^{n-1} \widetilde{T}_{\varepsilon^2 t} \tilde{f}_\varepsilon(\varepsilon x) \eta(\varepsilon x) = T_t f(x).$$

On the other hand,

$$\begin{aligned} &\varepsilon^{(n-1)} \int_0^M \int_{\mathbf{R}^{n-1}} |\widetilde{T}_{\varepsilon^2 t} \tilde{f}_\varepsilon(\varepsilon x) \eta(\varepsilon x)|^p dx dt \\ &= \varepsilon^{(n-1)(p-1)} \int_0^M \int_{T^{n-1}} |\widetilde{T}_{\varepsilon^2 t} \tilde{f}_\varepsilon(x)|^p dx dt \\ &= \varepsilon^{(n-1)(p-1)-2} \int_0^{\varepsilon^2 M} \int_{T^{n-1}} |\widetilde{T}_t \tilde{f}_\varepsilon(x)|^p dx dt. \end{aligned}$$

Now taking $\varepsilon < (MR)^{-1}$ and using Theorem 3 with $N = [R\varepsilon^{-1}]$ we obtain by Fatou's lemma

$$\left(\int_0^M \int_{\mathbf{R}^{n-1}} \lim_{\varepsilon \to 0} \varepsilon^{2(n+1)} |\widetilde{T}_{\varepsilon^2 t} \tilde{f}_\varepsilon(\varepsilon x) \eta(\varepsilon x)|^{2\frac{n+1}{n-1}} \right)^{\frac{n-1}{2(n+1)}}$$

$$\leq \lim_{\varepsilon \to o} \varepsilon^{(n-1)/2} \left(\int_{T^{n-1}} |\tilde{f}_\varepsilon|^2 \right)^{1/2} = \left((2\pi)^{1-n} \int_{\mathbf{R}^{n-1}} |\tilde{f}|^2 \right)^{1/2}.$$

Proof of Theorem 6.

$$\int_\varepsilon^1 \int_0^{2\pi} \left| \sum_k a_k e^{i(tk^2 - kx)} \right|^4 dx \frac{dt}{t} = \int_\varepsilon^1 \sum_m \left| \sum_j \bar{a}_j a_{j+m} e^{2itjm} \right|^2 \frac{dt}{t}$$

$$= \left(\sum_j |a_j|^2 \right)^2 + \sum_{1 \leq |m| \leq \varepsilon^{-1}} \int_\varepsilon^1 \left| \sum_j \bar{a}_j a_{j+m} e^{2itjm} \right|^2 \frac{dt}{t}$$

$$+ \sum_{|m| \geq \varepsilon^{-1}} \int_\varepsilon^1 \left| \sum_j \bar{a}_j a_{j+m} \varepsilon^{2itjm} \right|^2 \frac{dt}{t}$$

$$= \left(\sum_j |a_j|^2 \right)^2 + I + II.$$

Then

$$I \leq \int_\varepsilon^{\varepsilon^{-1}} \sum_{1 \leq |m| \leq \varepsilon^{-1}} \left| \sum_j \bar{a}_j a_{j+m} e^{2itj} \right|^2 \frac{dt}{t}$$

$$II = \sum_{|m| \geq \varepsilon^{-1}} \int_{\varepsilon m}^{m} \left| \sum_j \bar{a}_j a_{j+m} e^{2itj} \right|^2 \frac{dt}{t}$$

And by periodicity

$$II \leq C(1 + |\log \varepsilon|) \int_0^\pi \sum_{|m| \geq \varepsilon^{-1}} \left| \sum_j \bar{a}_j a_{j+m} e^{2itj} \right|^2 dt$$

Let us define

$$\Psi(t) = \begin{cases} t^{-1} & \text{if } |t| \geq \varepsilon \\ 1 & \text{if } |t| < \varepsilon \end{cases}$$

Then we have proved:

$$\left(\sum_j |a_j|^2\right)^2 + I + II$$

$$\leq C(1+|\log\varepsilon|)\int_0^{\varepsilon^{-1}}\left(\sum_m\left|\sum_j \bar{a}_j a_{j+m} e^{2itj}\right|^2\right)\Psi(t)dt$$

$$= C(1+|\log\varepsilon|\int_)^{\varepsilon^{-1}}\Psi(t)\int_0^{2\pi}\left|\sum_m\left(\sum_j \bar{a}_j a_{j+m} e^{2ijt}\right)e^{imx}\right|^2 dxdt$$

$$= C(1+|\log\varepsilon|)\int_0^{\varepsilon^{-1}}\int_0^{2\pi}\left|\left(\sum_k a_k e^{ikx}\right)\left(\sum_j \bar{a}_j e^{-ijx}e^{2ijt}\right)\right|^2$$

$$\leq C(1+|\log\varepsilon|^2)\int\left|\sum a_k e^{ikx}|^4 dx.\ \square\right.$$

Proof of Theorem 7. Let $t(x)$ be a real measurable function defined in $[-\pi,\pi]$. We have to prove

$$\int_{-\pi}^{\pi}\left|\sum_{k=1}^{N} a_k e^{i(t(x)k^2+kx)}\right|^2\frac{dx}{|x|^{1/2}} \leq C\left(\sum_{k=1}^{N}|a_k|^2 k^{1/2}\right)$$

with a constant independent of $t(x)$ and N.

Let ω be in $L^2([-\pi,\pi],|x|^{1/2}dx)$:

$$\int_{-\pi}^{\pi}\left(\sum_k a_k e^{i(t(x)k^2+kx)}\right)\omega(x)dx$$

$$\leq \left(\sum|a_k|^2 k^{1/2}\right)^{1/2}\left(\sum_{K=1}^{N} k^{-1/2}\left|\int_{-\pi}^{\pi} e^{i(t(x)k^2+kx)}\omega(x)dx\right|^2\right)^{1/2}$$

$$= \left(\sum|a_k|^2 k^{1/2}\right)^{1/2}\int_{-\pi}^{\pi}\int_{-\pi}^{\pi}\omega(x)\overline{\omega(y)}S_t^N(x,y)dxdy$$

where

$$S_t^N(x,y) = \sum_{k=1}^{N} k^{-1/2} e^{i\{(t(x)-t(y))k^2+(x-y)k\}}$$

We claim that $|S_t^N(x,y)| \le C|x-y|^{-1/2}$ with C a universal constant. Then

$$\int_{-\pi}^{\pi}\int_{-\pi}^{\pi} \omega(x)\overline{\omega(y)}S_t^N(x,y)\,dx\,dy$$

$$\le C\left(\int_{-\pi}^{\pi} |\omega|^2|x|^{1/2}\right)^{1/2}\left(\int \left| \int \overline{\omega(y)}\frac{dy}{|x-y|^{1/2}}\right|^2\right)^{1/2}$$

$$\le C\left(\int_{-\pi}^{\pi} |\omega|^2|x|^{1/2}\right)^{1/2}$$

where the last step is a consequence of a well known weighted inequality for fractional integrals.

Let us prove the claim. Using Lemma 9, it is enough to see

$$|I_\lambda^N(\mu)| = \left|\int_1^N e^{i(\lambda\xi^2+\mu\xi)}\frac{d\xi}{|x|^{1/2}}\right| \frac{C}{|\mu|^{1/2}}.$$

But this last fact is an easy consequence of the stationary phase lemma (see also [3]). □

Remark. There must exist similar results to Theorem 4 and Corollary 5 in dimensions bigger than one. The obstruction is to estimate the L^∞ norm of the kernel

$$\sum_{|m|\le N} e^{i\{t\varphi(|m|)+mx\}}$$

with $\varphi'' \ge \rho > 0$.

3. Proof of Theorem 1. Theorem 1 can be slightly improved:

THEOREM 11. *Let f be in $L^2(S^{n-1})$. then*

$$\left(\int_0^\infty\left(\int_{S^{n-1}}\left|\int_{S^{n-1}} f(\omega)e^{-2\pi i r x\omega}\,d\sigma(\omega)\right|^2 d\sigma(x)\right)^{4/2} r^{n-1}(1+r)^a\,dr\right)^{1/4}$$

$$\le C_a\left(\int_{S^{n-1}} |f|^2\,d\sigma\right)^{1/2}$$

where $a < n-2$.

We get Theorem 1 by interpolating this last result with the following well-known trace theorem.

THEOREM 12. *Let f be in $L^2(S^{n-1})$. Then*

$$\left(\int_0^\infty \int_{S^{n-1}} \left| \int_{S^{n-1}} f(\omega 0 e^{-2\pi i r x \omega} d\sigma(\omega)\right|^2 d\sigma(x) r^{n-1}(1+r)^{-a} dr\right)^{1/2}$$
$$\leq C_a \left(\int_{S^{n-1}} |f|^2 d\sigma\right)^{1/2}$$

with $a > 1$.

We shall make use of the following lemma.

LEMMA 13. *Let J_k be the Bessel function of order k with $k \geq 0$. Then*

$$\int_0^\infty |J_k(r)|^p r dr < C_p < +\infty \qquad \text{if } p > 4$$

where C_p is a constant independent of k.

Proof. Using the integral representation for Bessel functions (see [13] p. 176 formula (4)) and Van der Corput's lemma we get the following estimates:

$$|J_k(r)| \leq \begin{cases} C|k-r|^{-1} & 0 \leq r \leq k - k^{1/3} \\ C2^{-j/4} r^{-1/3} & k + 2^j k^{1/3} \leq r \leq k + 2^{j+1} k^{1/3} \quad j = 0, 1, \ldots, \frac{2}{3} \lg k \\ Cr^{-1/3} & k - k^{-1/3} \leq r \leq k + k^{1/3} \\ Cr^{-1/2} & 2k \leq r \end{cases}$$

where C is a constant independent of k. □

The lemma follows easily from these inequalities.

Proof of Theorem 11. We can write

$$f(x) = \sum_k a_k Y_k(x), \quad x \in S^{n-1}$$

with

$$\int_k |a_k|^2 < +\infty \text{ and } \int_{S^{n-1}} |Y_k(x)|^2 d\sigma = 1,$$

where Y_k denotes a spherical harmonic of degree k.

Then, the Fourier transform of f is (see [9, p. 158]):

$$\widehat{fd\sigma}(r\xi) = 2\pi i^{-k} r^{1-n/2} J_{k-1+n/2}(2\pi r) Y_k(\xi). \tag{5}$$

Let us remark that $\widehat{fd\sigma}(r\xi)| \leq f_{S^{n-1}}|f|$. Therefore, it is trivial to see that

$$\left(\int_{B(0,1)} |\widehat{fd\sigma}|^p\right)^{1/p} \leq \int_{S^{n-1}} |f|^2 p \geq 1$$

On the other hand,

$$\left(\int_1^{\infty}\left(\int_{S^{n-1}} |\widehat{fd\sigma}(r\xi)|^2\right)^{4/2} r^{n-1}(1+r)^a dr\right)^{1/4} \leq C\left(\int_1^{\infty}\left(\sum_k |a_k|^2|J_k(r)|^2\right)^{2} r^{a-(n-3)}dr\right)^{1/4}$$

$$= C\sum_k |a_k|^2 \int_1^{\infty} |J_k(r)|^2\omega(r)r^{a-(n-3)}dr$$

where $\omega(r) \in L^2([1,\infty)r^{a-(n-3)}dr)$. Recall that $a < (n-2)$. Then the above expression is bounded by

$$C_a\left(\sum |a_k|^2 \int_1^{\infty} |J_k(r)|^p r dr\right)^{1/2} \|\omega\|_{L^2([1,\infty)r^{a-(n-3)}dr)}$$

for some $p > 4$ which depends on a. Using Lemma 13 we finish the proof.

For the sake of completeness let us prove Theorem 11.

Proof of Theorem 12. (See [1]) Cutting S^{n-1} into 2^n pieces we can suppose that

$$\widehat{fd\sigma}(\omega) = \int_{B(0,1/2)} f(x)e^{i(x\bar{\omega}+(1-|x|^2)^{1/2}\omega_n)}\frac{|x|}{(1-|x|^2)^{1/2}}dx$$

where $\omega = (\bar{\omega}, \omega_n) \in \mathbf{R}^n$ and $x \in \mathbf{R}^{n-1}$. But

$$\int_{R^n} |\widehat{fd\sigma}(\omega)|^2(1+|\omega|)^{-a} \leq \int_{-\infty}^{\infty} (1+|\omega_n|)^{-a} \int_{\mathbf{R}^{n-1}} |\widehat{fd\sigma}(\bar{\omega},\omega_n)|^2$$

$$\leq C_a \int |f(x)|^2\frac{|x|}{(1-|x|^2)^{1/2}}dx$$

and the last step is a consequence of the Plancherel theorem in $\mathbf{R}^{n-1}$ and the fact that $a > 1$.

4. Further remarks. Some comments on Lemma 13 are in order. It is easy to see that the estimates we get for the Bessel functions in the critical region $(k, 2k)$ are the best possible. This fact says that the sequence $\{J_k(r)\}_k$ studied as multiplier in spherical harmonics is an operator bounded by $r^{-1/3}$ instead of $r^{-1/2}$. However, if we make an average on r things go better and Theorem 1 can be proved. In the case $2 \leq p < 4$ we have the following estimate which can be proved with the same techniques:

THEOREM 14. *Let f be in $L^2(S^{n-1})$:*

$$\left(\int_R^{2R}\left(\int_{S^{n-1}} |\widehat{fd\sigma}(rx)|^2 d\sigma(x)\right)^{p/2} r^{n-1}dr\right)^{1/p} \leq R^{-n(\frac{1}{2}-\frac{1}{p})+\frac{1}{2}}\left(\int_{S^{n-1}} |f|^2 d\sigma\right)^{1/2}$$

with $R > 1$ and $2 \leq p < 4$.

This result has an easy consequence:

COROLLARY 15. *Let f be in $L^2(S^{n-1})$ and $2 \le p < 4$. Then*

$$\lim_{R\to\infty}\left(R^{p\{n(\frac{1}{2}-\frac{1}{p})+\frac{1}{2}\}}\int_R^{2R}\left(\int_{S^{n-1}}|\widehat{fd\sigma}(rx)|^2 d\sigma(x)\right)^{p/2} r^{n-1}dr\right)^{1/p} = C_{p,n}\left(\int_{S^{n-1}}|f|^2 d\sigma\right)^{1/2}$$

where

$$C_{p,n} = 2(2\pi)^{(n-1)/2}\left(\int_1^2 t^{(n-1)(1-p/2)}dt\right)^{1/p}.$$

Proof. If $f = \sum_{k=1}^N a_k Y_k$ the corollary follows from (5) and the asymptotic behavior of the Bessel functions. A standard argument of Functional Analysis and Theorem 14 complete the proof. □

If we take $p = 2$, Corollary 14 is a particular case of a much more general result of S. Agmon and L. Hörmander [1]. However, the curvature allows us to improve the result for $p > 2$. By Lemma 13 it is easy to see that Theorem 14 is false if we take the average on an interval of size R^α with $\alpha < 1$, even in the case $p = 2$. And as a consequence, no convergence result could be proved.

REFERENCES

[1] S. AGMON AND L. HÖRMANDER, *Asymptotic properties of solutions of differential equations with simple characteristics*, Jour. d'Analyse Matématique, 30 (1976).

[2] J.A. BARCELÓ AND A. CÓRDOBA, *Band-limited functions: L^p convergence*, Bulletin A.M.S., 18 no. 2 (1988).

[3] L. CARLESON, *Some analytic problems related to Statistical Mechanics*, Euclidean Harmonic Analysis, Lecture Notes in Mathematics 779, Springer-Verlag (1979), pp. 1–46.

[4] L. CARLESON, P. SJÖLIN, *Oscillatory integrals and a multipler problem for the disc*, Studia Math. 44 (1972), pp. 287–299.

[5] C. FEFFERMAN, *Inequalities for strongly singular convolutions operators*, Acta Mat. 124 (1970), pp. 9–35.

[6] C.S. HERZ, *Fourier transforms related to convex sets*, Ann. of Math. 75, pp. 81–92.

[7] C. KENIG, AND P. TOMAS, *Maximal operators defined by Fourier multipliers*, Studia Mat. LXVII, 1980.

[8] J.L. RUBIO DE FRANCIA, *Transference principles for radial multipliers*, preprint.

[9] E. STEIN AND G. WEISS, *Introduction to Fourier analysis on Euclidean Spaces*, Princeton Univ. Pres, 1971.

[10] R.S. STRICHARTZ, *Restrictions of Fourier transforms to quadratic surfaces and decay of solutions of wave equations*, Duke Math. J. 44 no. 3 (1977).

[11] P. TOMAS, *A restriction theorem for the Fourier transform*, Bull. Am. Math. Soc. 81 (1975).

[12] R.C. VAUGHAN, *The Hardy-Littlewood Method*, Cambridge University Press 1981.

[13] G.N. WATSON, *A Treatise on the Theory of Bessel Functions*, Camb. Univ. Press.

[14] A. ZYGMUND, *Trigonometric Series*, Camb. Univ. Press.

[15] A. ZYGMUND, *On Fourier coefficients and transforms of functions of two variables*, Studia Mat. Vol. L (1974).

NUMERICAL ANALYSIS ON NON-SMOOTH PROBLEMS: SOME EXAMPLES*

LARS B. WAHLBIN†

Abstract. In a partial differential equation roughness in the solution may be caused in various ways, such as by

i) a rough right hand side,

ii) a rough boundary,

iii) rough initial data.

We shall briefly describe two examples from numerical analysis showing that:

Example A) In a linear Poisson problem, roughness introduced by i) or ii) leads to dramatically different behavior in numerical solutions.

Example B) Although for linear and semilinear parabolic problems, in the case of iii) (all other "data" being smooth) there is "no difference" in the sense that the solution is smooth for (some) positive time, in numerical analysis the linear and semilinear problem are miles apart.

While the main point of Example A was "computational folklore" for many years before it was fully proven in 1984, the punchline of Example B was totally unexpected at the time of its discovery in 1987.

More details, and more examples of rough stuff in the numerical analysis of partial differential equations, e.g., in singularly perturbed problems, can be found in WAHLBIN [7].

Example A.

Consider the Poisson problem

$$\begin{cases} -\Delta u & = f \text{ in } \Omega \text{ (bounded domain in } R^d), \\ u & = 0 \text{ on } \partial\Omega. \end{cases} \tag{1}$$

We imagine that we have at our disposal a standard finite element solver, parametrized by $0 < h << 1$, typically the maximum diameter of an element. (There is no sense in making this note a mini-course on finite element theory; hence I shall be quite cavalier in my description.) Upon these implied subdivisions of the domain Ω are erected (continuous) piecewise polynomial function spaces of degree $r-1$ on each element,

$$r \geq 2 \text{ fixed.} \tag{2}$$

These spaces, which incorporate the essential boundary conditions in (1), will be denoted S_h. (Modifications such as "isoparametric" elements may occur at curved boundaries.) The standard solver takes the weak formulation of (1) in H_1^o and

*Supported by the National Science Foundation and by the Army Research Office through the Mathematical Sciences Institute at Cornell

†Department of Mathematics, White Hall, Cornell University, Ithaca, NY 14853

imposes that over S_h in the natural way; integrations are assumed to be exactly performed. In other words, the finite element solution, denoted u_h, is the H_1^o projection of the solution u of (1) into S_h.

We shall further assume that

$$\text{(3)} \qquad \text{“Meshes are unrefined”.}$$

This means that the standard solver does not pay attention to any singularities in the solution of (1), or, all elements throughout the domain Ω have, asymptotically the same diameter, $O(h)$. (In technical lingo, we have a “quasi-uniform” or “quasi-regular” mesh-family.) To reiterate, the parameter h looks the same all over the domain Ω in our present considerations.

Now, a typical property of such finite element spaces S_h is their ability, in principle, to approximate locally. For A a subdomain of Ω,

$$\text{(4)} \qquad min_{\chi \epsilon S_h} \| u - \chi \|_{\infty,A} \leq Ch^k \| u \|_{\infty,k,A_h}, \; 0 \leq k \leq r,$$

where A_h signifies that the domain A may have to be swelled by $O(h)$ (barring boundaries). The top order $r, cf.$(2), in such an estimate will be referred to as “optimal” order. Typically, (4) is accomplished by some “interpolant” or “quasi-interpolant”.

A natural question to ask is now whether the finite element solutions u_h, i.e., the H_1^o projection of u into S_h, duplicates the feat of local approximation described in (4). This is where the answer will differ dramatically according to how singularities are introduced.

In case the roughness in the solution u is introduced by the right hand side f in (1), while $\partial\Omega$ is nice (and “well” approximated by S_h), the finite element solution indeed mimicks the local behavior in (4). E.g., if u is the Green's function centered at a point x_o in Ω, and u_h its discrete analogue, defined by

$$\text{(5)} \qquad (\nabla u_h, \nabla \chi)_{L_2} = \chi(x_o), \text{ for } \chi \epsilon S_h,$$

then

$$\text{(6)} \qquad | (u_h - u)(x) | \leq Cln(1/h)h^r \, | x - x_o |^{-d-r+2}$$

with a constant C independent of h or x. This of course mimicks (4) as well as one may hope for (apart from the mild logarithmic factor which is not necessary for

$r > 2$). We refer to SCHATZ and WAHLBIN [3] for the quite long and technical proof of (6).

The proof of (6) is based on a general principle,

(7)

$$| (u_h - u)(x) | \leq C ln(1/h) min_{\chi \epsilon S_h} \| u - \chi \|_{\infty, A_\delta(x)} + C\delta^{-d/2-s} \| u_h - u \|_{-s, A_\delta(x)}$$

where $A_\delta(x) = \{x+y : \| y \| \leq \delta, x+y \epsilon \Omega\}$ and $\| \cdot \|_{-s,B}$ is the dual norm to $\overset{\circ}{H}_s(B)$ over the pivot space L_2. The first term on the right of (7) is easily seen to give the correct contribution to (6), with $\delta = | x - x_o |$, and the second term is estimated via a (standard in the business) duality argument. This argument is based on smoothness of $\partial\Omega$, to whit, that solutions to (1) inherit two more derivatives than f in L_2.

Let me give a trivial case of such a duality argument. Assume that we know that

$$\| u_h - u \|_{H_1} \leq Ch^{k-1} \| u \|_{H_k}, \; 1 \leq k \leq r. \tag{8}$$

(This is frequently "easy" to show, although I have surpressed certain technicalities involving approximation of curved boundaries.) Then a typical duality argument proceeds as follows:

$$\| u_h - u \|_{-s,\Omega} = \sup_{\substack{v \epsilon \overset{\circ}{H}_s(\Omega) \\ \| v \|_{H_s} = 1}} (u_h - u, v). \tag{9}$$

Let, for each such v, the function w be the solution to

$$\begin{cases} -\Delta w & = v \text{ in } \Omega, \\ w & = 0 \text{ on } \partial\Omega. \end{cases} \tag{10}$$

Then with w_h the H_1^0 projection of w into S_h,

$$\begin{aligned} (u_h - u, v) &= (\nabla(u_h - u), \nabla w) = (\nabla(u_h - u), \nabla(w - w_h)) \\ &\leq \| u_h - u \|_{H_1} \| w - w_h \|_{H_1} \\ &\leq \| u_h - u \|_{H_1} Ch^{s+1} \| w \|_{H_{s+2}} \\ &\leq Ch^{s+1} \| u_h - u \|_{H_1}, \end{aligned} \tag{11}$$

provided $s \leq r - 2$ and

(12) provided the appropriate shift theorem holds in(10).

One obtains a higher order estimate for $\| u_h - u \|_{-s,\Omega}$ than in H_1. (And, e.g., (6) can be proven from (7) by appropriate extensions of the above duality argument.)

This is now where the fun starts in the case that we are on a (say, for simplicity) plane domain, $d = 2$, with polygonal boundary and a maximum reentrant corner of interior angle $\alpha > \pi$. In general, the solution of (1) (now with f as smooth as you like!) will have an

(13) $$a_o r^\beta sin(\beta\theta) \text{ singularity, } \beta = \pi/\alpha < 1,$$

expressed in suitable polar coordinates at the worst vortex. Such a function, under the non-refinement assumption (3) et seg., is approximable in H_1 only to order h^β. Furthermore, the "appropriate" shift theorem in (12) does not gain two derivatives. The upshot of all this is the following after a duality argument, restricted to the situation with a reentrant corner.

Let A be any subdomain of Ω (e.g., where u is C^∞) and s any positive integer. Then, there exists a constant C such that

(14) $$\| u_h - u \|_{-s,A} \leq Ch^{2\beta}.$$

And, (surprise!) conversely, if $a_o \neq 0$ in (13), there is a positive constant $c = c(s, A, a_o)$ such that

(15) $$\| u_h - u \|_{-s,A} \geq ch^{2\beta}.$$

I.e., the estimate (14) based on a "suspect" duality argument is actually sharp.

Thus, even if the solution u is smooth on the subdomain A, the finite element solution will not approximate to optimal order r on that subdomain (since $r \geq 2$ and $2\beta < 2$), however weak an error measure we choose! This is frequently referred to as a pollution effect.

This is the "dramatic" difference in numerical analysis between roughness introduced by right hand sides and rough boundaries, respectively.

Let us conclude this example A by indicating how (15) obtains, of WAHLBIN [6] for more details. It turns out to be sufficient to prove (15) for one function u with $a_o \neq 0$. Due to the maximum principle, the Green's function for (1) centered at a point x_o has, indeed, $a_o \neq 0$. Let G_o denote that Green's function with x_o inside A. Let ω be a smooth cutoff function,

$$\omega(x) = \begin{cases} 1, & \text{for } x \text{ outside } A, \\ 0, & \text{for } x \text{ near } x_o. \end{cases} \tag{16}$$

Take $u_o = \omega G_o$. Indeed than, $a_o \neq 0$. As a consequence of (3) (trivial approximation theory), one has

$$min_{\chi \epsilon S_h} \parallel u_o - \chi \parallel_{H_1} \geq ch^{\beta}, c > 0. \tag{17}$$

Thus,

$$\begin{aligned} c^2 h^{2\beta} &\leq \parallel u_h - u_o \parallel^2_{H_1} = (\nabla(u_h - u_o), \nabla(u_h - u_o)) \\ &= -(\nabla(u_h - u_o), \nabla u_o) = (u_h - u_o, \Delta u_o)_A \end{aligned} \tag{18}$$

where in the last L_2-inner product the integration extends only over A, due to the specific form of $u_o = \omega G_o$. Further, u_o is infinitely differentiable on A. Thus,

$$c^2 h^{2\beta} \leq \parallel u_h - u_o \parallel_{-s,A} \parallel u_o \parallel_{H_{s+2}(A)} \tag{19}$$

and (15) is proven. It may be noted that the singularity in the Green's function, the "good" case here, is worse than that of the "bad" case of a singularity introduced by a reentrant corner.

Pollution effects similar to (15) occur also in numerical analysis of partial differential equations with rough coefficients. The case of roughness introduced by rough boundary data (on smooth $\partial\Omega$) is similar to that of roughness introduced by a right hand side.

Example B.

Consider the parabolic problem

$$\begin{cases} u_t = \Delta u + f(u), \; x \epsilon \Omega, \; t \geq 0, \\ u = 0 \text{ on } \partial\Omega, \\ u(0) = v \end{cases} \tag{20}$$

where f and $\partial\Omega$ are smooth. Indeed, we then have short-time existence of smooth solutions (if, e.g., v is bounded) and since we shall only consider suitably small t, we may assume that f is uniformly Lipschitz. At issue is now when initial data v are nonsmooth (or incompatible) so that the solution $u(x,t)$ is not uniformly nice all the way down to $t = 0$. (It is smooth for $t > 0$.)

The object of this example is to exhibit an "unexpected" difference between the linear and semilinear problem in numerical analysis.

We consider the so-called semi-discrete approximation of (20), i.e., we let $u_h(t) \epsilon S_h$ with S_h as in Example A) for each t to be continuous in time and satisfy

$$(u_{h,t}, x) + (\nabla u_h, \nabla \chi) = (f(u_h), \chi), \text{ for } \chi \epsilon S_h, \tag{21}$$

and

$$u_h(0) = v_h = P_o v \tag{22}$$

where $P_o v$ denotes the L_2-projection into S_h. (The reasons for this choice of approximate initial data can be found in THOMÉE [4] and WAHLBIN [5] and we shall not further elaborate here.)

In the linear case, $f \equiv 0$, the results of HELFRICH [1] say that, in L_2-norms,

$$\| (u_h - u)(t) \| \leq C h^k t^{-k/2} \| v \|, \text{ for } t > 0, k \leq r. \tag{23}$$

The proof of this is too long to enter into here. Basically, in the linear case, the semi-discrete approximation takes full advantage of the smoothness of the solution for positive time to produce an optimal order approximation.

Now the fun starts in the true semi-linear case. With $E(t)$ denoting the linear heat solution operator, a first thought is to use Duhamel's principle; for the solution of (20) then,

$$u(t) = E(t)v + \int_o^t E(t-s) f(u(s)) ds, \tag{24}$$

and similarly for u_h, with $E_h(t)$ the analogous discrete linear evolution operator. With $F_h(t) = E_h(t)P_o - E(t)$ and $e_h(t) = u_h(t) - u(t)$ we have

$$\begin{aligned} e_h(t) &= F_h(t)v + \int_o^t E_h(t-s)[f(u_h(s)) - f(u(s))]ds \\ &+ \int_o^t F_h(t-s) f(u(s)) ds. \end{aligned} \tag{25}$$

Due to (23) we know a lot about F_h. The first term on the right of (25) is easy. The second can be "Gronwalled", due to f uniformly Lipschitz. In the third, we use (23) with k almost, but not quite 2, to keep the integral bounded. The details can be found in JOHNSON, LARSSON, THOMÉE and WAHLBIN [2] and give for $v \epsilon L_2$,

$$(26) \qquad \| (u_h - u)(t) \| \leq Ch^2(t^{-1} + | log(h^2/t) |).$$

Thus, for $r > 2$ we do not have an optimal order result, in contrast to the linear case.

The punchline to this is that (26) is sharp, notwithstanding the suspect argument used in its proof (taking $k \simeq 2$ to keep some integral from blowing up).

The precise sense in which (26) is sharp is as follows: One can exhibit an equation (20), with $f(u) = u^2$ for u small, such that if for all $| u(x,t) | \leq K$, for some t_o,

$$(27) \qquad | (u_h - u)(t_o) | \leq C(K, t_o)h^s,$$

Then of necessity $s \leq 2$. A fullfledged example is given in [2, Section 6], but the essentials can be understood already from the following, [2, Introduction]. Consider the periodic problem over $[-\pi, \pi]$ for $u = (u_1, u_2)$,

$$(28) \qquad \begin{cases} a) \; u_{1,t} = u_{1,xx} + f(u_2), \\ b) \; u_{2,t} = u_{2,xx} \end{cases}$$

where f is smooth with $f(y) = 4y^2$, for $| y | \leq 1$. Let S_h now stand for trigonometric polynomials of degree $(n-1) = 1/h$. The proof of (26) is as before. Now, if (27) holds, we are allowed to use initial data dependent on n. We choose

$$(29) \qquad u_1(0) = 0, \; u_2(0) = cos(nx)$$

so that the approximate solution is $\equiv 0$. The true solution is

$$(30) \qquad \begin{cases} u_2(x,t) = exp(-n^2t)cos(nx) \\ u_1(x,t) = \frac{1-exp(-2n^2t)}{n^2}[1 + exp(-2n^2t)cos(2nx)] \end{cases}$$

and it is clear that the error behaves as $0(n^{-2})$ and not better.

A final remark: In the linear case one may consider a right hand side in (20) of the form $f(x,t)$. One can then show that if f is smooth near the time of interest, optimal order $O(h^r)$ estimates hold near that time. Indeed, to convince one-self that the restriction to $s \leq 2$ in (27) is a genuinely <u>nonlinear</u> effect, it is instructive to consider what happens if u_2 is taken exact in (28b) and then stuffed into (28a), which is then approximately solved as a linear inhomogeneous problem.

REFERENCES

[1] HELFRICH, H.-P., *Fehlerâbschätzungen für das Galerkin Verfahren zur Lősung von Evolutionsgleichungen*, Manuscripta Math. 13, 1974, 219-235.

[2] JOHNSON, C., S. LARSSON, V. THOMÉE AND L.B. WAHLBIN, *Error estimates for spatially discrete approximations of semilinear parabolic equations with nonsmooth initial data*, Math. Comp. 49, 1987, 331-357.

[3] SCHATZ, A.H., AND L.B. WAHLBIN, *Interior maximum norm estimates for finite element methods*, Math. Comp. 31, 1977, 414-442.

[4] THOMÉE, V., Spline approximation and difference schemes for the heat equation in: A.K. Aziz, ed., *The Mathematical Foundations of the Finite Element Method with Applications to Partial Differential Equations*, Academic Press 1972, 711-746.

[5] WAHLBIN, L.B., A brief survey of parabolic smoothing and how it affects a numerical solution: finite differences and finite elements, in: I. Babuška, T.-P. Liu and J. Osborn, eds., *Lectures on the Numerical Solution of Partial Differential Equations*, University of Maryland Lecture Notes 20, 1981.

[6] WAHLBIN, L.B., *On the sharpness of certain local estimates for* $\overset{\circ}{H}{}^1$ *projections into finite element spaces: Influence of a reentrant corner*, Math. Comp. 42, 1984, 1-8.

[7] WAHLBIN, L.B., Local Behavior in Finite Element Methods, in: P.G. Ciarlet and J.-L. Lions, eds., *Handbook of Numerical Analysis*, Vol. II, Finite Element Methods, Elsevier (North Holland), to appear in 1990.